Dedicated to Gabriel Ward Lasker in admiration and friendship and Rachael and Josh Bogin with love.

Contents

ix

Acknowledgments

I am very fortunate to have several colleagues who took time from their own research to read and criticize one or more chapters of this book, offer comments from the first edition, and provide me with results of their own research. For their kindness, and for their forthright comments which have improved the presentation of this work, sincere thanks are extended to Drs. Robert Anemone, Deb Crooks, and her students, Parasmani Dasgupta, Irv Emanuel, Holle Greil, Roland Hauspie, Steve Leigh, Diane Markowitz, Michael McKinney, Larry Schell, Ania Siniarska, Holly Smith, and Napoleon Wolanski. A former student, and now my colleague at the University of Michigan-Dearborn, Matthew Kapell, also read and critiqued several chapters. One of my current students, Veronica Gorden, assisted in the production of this book in several ways, especially reading chapters to make sure they are intelligible to the non-specialist reader.

I also want to thank, again, those colleagues who helped with the first edition of the book, as many of their suggestions for improvement are still found in this second edition: Drs. George Clark, Nancy Howell, Marquisa LaVelle, Michael A. Little, Daniel Moerman, Gerald Moran, Jessica Schwartz, and B. Holly Smith. The professional work and life of Professor Elizabeth Watts and Professor James Gavin inspired much of my interest in the evolution of human growth. Prof. Watts read sections of the first edition and Prof. Gavin engaged me in lively conversation about our disparate ideas about human evolution. Sadly, neither of these colleagues lived to see the drafts of the second edition. Nevertheless, their writings and conversations with me leave a mark on this book.

Parts of the first edition of this book were written during a research leave from the University of Michigan-Dearborn. The author appreciates the assistance of Drs. Eugene Arden, Victor Wong, and Donald Levin in helping to arrange for the leave. A generous grant from the American Philosophical Society helped to defray research expenses at that time. The second edition was written without the benefits of a leave from teaching, nor special financial assistance. Of course Cambridge University Press, in the person of my editor Dr. Tracey Sanderson, provided much support, even some money, but especially kind understanding of the inevitable delays in writing.

Sandra Bogin contributed several of the line drawings that illustrate the text and, more importantly, provided physical and emotional support and encouragement for the writing. Professor Gabriel W. Lasker read every word of the draft of the text, and engaged the author in many discussions about the intellectual and technical content of the book. His involvement in all aspects of the production of the book are appreciated deeply.

Introduction

It is the purpose of this book to describe and interpret some of the evolutionary, physiological, cultural, and mathematical patterns of human growth. Given this purpose, the title of this book requires some explanation. A cell biologist might think of the phrase 'patterns of growth' in terms of a series of genetically controlled cell duplication and division events. An embryologist might think of patterns of cell differentiation and integration leading to the development of a functionally complete human. The clinician interprets patterns of growth, especially deviations from expected or 'normal' growth, as evidence of disease or other pathology in the patient. All of these concepts of 'pattern' may be biologically valid and useful in their own areas of specialization, but this book is about none of them. The goal of this account is to consider the growth of the human body in a unified and holistic manner. The result, it is hoped, will be a synthesis of the forces that shaped the evolution of the human growth pattern, the biocultural factors that direct its expression in populations of living peoples, the intrinsic and extrinsic factors that regulate individual development, and the biomathematical approaches needed to analyze and interpret human growth.

The study of human growth in relation to evolutionary biology, biocultural factors, intrinsic and extrinsic factors, such as genes, hormones, the physical and social environment, and mathematics may seem like a strange brew of topics. In fact, it is a common mix for biological anthropologists. The rest of this book is designed to show the reader that the anthropological blend of scholarship and research is, in fact, a practical and rewarding combination.

Introductory students of human growth often assume that the field is primarily a part of pediatric medicine. Indeed, until the publication of the first edition of *Patterns of Human Growth* in 1988, all but one of the leading introductory texts were written by physicians, and were written with the medical student in mind, or as a practical guide for parents. The one exception is the book *Child Growth* (Krogman, 1972), written by a biological anthropologist, but focused primarily on pediatric topics. While it may seems logical for human growth to be a subfield of medicine, it is more accurate, however, to view pediatric medicine and 'parenting' as subfields of the study of human growth. In turn, human growth is a part of a much broader discipline, namely anthropology. A little bit of history, and an applied example, are provided here to justify this statement. Chapter 1 includes a more detailed history of the study of human growth.

Anthropology and growth

The study of human growth has been a part of anthropology since the founding of the discipline. European anthropology of the early to mid-nineteenth century was basically anatomy and anthropometry, the science of human body measurements (Malinowski & Wolanski, 1985). Early practitioners of American anthropology, especially Franz Boas, are known as much for their studies of human growth as for work in cultural studies, archaeology, or linguistics. Boas was especially interested in the changes in body size and shape following migration from Europe to the United States. At the time of those studies, around 1910, most anthropologists and anatomists believed that stature, and other measurable dimensions of the body such as head shape, could be used as 'racial' markers. The word 'race' is set in inverted commas here because it refers to the scientifically discredited notion that human beings can be organized into biologically distinct groups based on **phenotypes**[1] (the physical appearance and behavior of a person). According to this fallacious idea, northern European 'races' were tall and had relatively long and narrow heads, while southern European races were shorter and had relatively round skulls. Boas found that, generally, the children of Italian and Jewish European migrants to the United States were significantly taller and heavier than their parents. The children of the migrants even changed the shape of their heads; they grew up to have long narrow heads.

In the new environment of the United States, the children of recent southern European migrants grew up to look more like northern Europeans than their own parents. Boas used the changes in body size and shape to argue that environment and culture are more important than genes in determining the physical appearance of people. In terms of environment, life in the United States afforded better nutrition, both in terms of the quantity and the variety of food. There were also greater opportunities for education and wage-paying labor. These nutritional and socioeconomic gains are now known to correlate with large body size. In terms of culture, in particular child-rearing practices, there were other

---

[1] Formal definitions for all words in bold type are found in the Glossary.

changes. In much of Europe infants were usually wrapped up tightly and placed on their backs to sleep, but the American practice at the turn of the century was to place infants in the prone position. In order to be 'modern' the European immigrant parents often adopted the American practice. One effect on the infant was a change in skull shape, since pressure applied to the back of the infant's skull produces a rounder head, while pressure applied to the side of the skull produces a longer and narrower head. The sleeping position effect on skull shape was demonstrated first in Europe by Walcher (1905).

The work of Boas and his colleagues, shows that an interest in human growth is natural for anthropologists. This is because the way in which a human being grows is the product of an interaction between the biology of our species, the physical environment in which we live, and the social/economic/political environment that every human culture creates. Moreover, the basic pattern of human growth is shared by all living people. That pattern is the outcome of the four million year evolutionary history of the **hominids**, living human beings and our fossil ancestors. Thus, human growth and development reflect the biocultural nature and evolutionary history of our species.

Maya in Disneyland

The biocultural nature of human growth may be appreciated by the following example based on my own research in Guatemala and the United States on the impact of the economic and political environment on the growth and development of Maya children (Bogin & Loucky, 1997). Two samples of Maya are compared; one a group living in their homeland of Guatemala, and the other a group of migrants living in the United States. Both groups include individuals between the ages of 5 and 14 years old. The Guatemala sample live in a village with an irregular supply of water, no safe drinking water, and unsanitary means for waste disposal. The parents of these Maya children are employed, predominately, as tailors or seamstresses by local clothing manufacturers and are paid minimal wages. There is one public health clinic in the village, which administers treatment to infants and preschool children with clinical undernutrition – an omnipresent problem. The incidence of infant and childhood morbidity and mortality from infectious disease is relatively high. Deaths due to political repression, especially the civil war of late 1970s to early 1980s, is common for the Maya of Guatemala. The residents of this particular Maya village were caught up in the military hostilities of that time, but escaped the worst of the civil war. They also suffered from reduced food availability due to the collapse of the Guatemalan economy during the 1980s (Bogin, 1998).

The United States sample reside in two places, Indiantown, a rural agricultural community in central Florida, and Los Angeles, California (hence the title for this section 'Maya in Disneyland'). The political status of the Maya in the USA is heterogeneous. Some have applied for, and a few have won, political asylum. Others have temporary legal rights to reside and to work, but many remain undocumented. Adults in the Florida community work as day laborers in agriculture, landscaping, construction, child care and other informal sector jobs. Many of the Los Angeles Maya work in the 'sweatshops' of the garment industry, although a few have jobs in the service sector or technical professions.

The growth in height, weight and body composition of Maya children and youths living in Guatemala is significantly retarded compared with United States National Center for Health Statistics (NCHS) reference data.[2] Figure I.1 illustrates the mean height and weight of the Guatemalan Maya from the village of San Pedro during two time periods. Some researchers argue that the small size and delayed maturation of mal-nourished populations such as the Maya is a genetic adaptation to their poor environmental conditions. If this argument were true, then a change in the economic, social, or political environment would not influence growth. The notion that the small size of the Maya is primarily genetic is clearly wrong, for as also shown in Figure I.1, the United States-living Maya are significantly taller and heavier than Maya children living in Guatemala. The Maya in the United States attain virtually the same weight as the NCHS sample. The average increase in height is 5.5 cm between Maya in the United States versus Maya in Guatemala. This increase occurred within a single generation, that is, as children moved from Guatemala to the United States. Moreover, the change in average stature is, perhaps, the largest such increase ever recorded. By contrast, the immi-grant children measured by Boas averaged about 2.0 cm taller than their European-born parents.

The reason for the increase in body size of the Maya children is the same as for the European immigrant children measured by Boas. In the United States there is both more food and a greater variety of food than in rural

[2] NCHS reference data represent the growth status of a healthy, well-nourished population from the United States in the year 1977. These reference data are recommended by the World Health Organization for the evaluation of human growth for all populations so as to provide a common baseline for international comparison. The NCHS reference data are used throughout this book for all such comparisons.

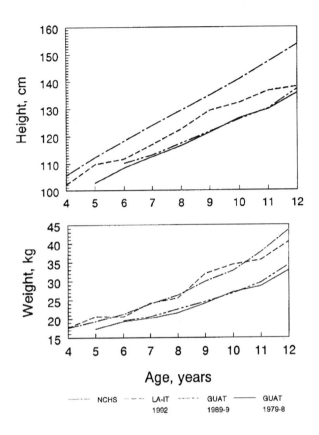

Figure I.1. The mean height, or weight, by age of the Los Angeles and
Indiantown Maya sample (LA-IT), the Maya samples from Guatemala measured
in 1979–80 and 1989–90 (GUATE), and the NCHS reference data. The NCHS
references were developed by the United States National Center for Health
Statistics, and represent the growth of healthy, well-nourished children. Data for
boys and girls within all samples are combined. Note that there is virtually no
change in the mean size of the Guatemala-living Maya. This indicates that the
generally poor environmental conditions for growth remained unchanged for
that decade. The Maya in the USA are significantly larger than Maya in
Guatemala.

Guatemala. In the USA there are also social services that are unavailable in
rural Guatemala, including health care, food supplementation programs,
schools, job training programs. All of these differences improve the biological and social environment for human growth. The most important differences, however, are safe drinking water and the conditions that go with
less political repression. The public supply of safe drinking water in the

USA eliminates the constant exposure to bacteria, parasites, agricultural pesticides, and fertilizers that contaminate drinking water in rural Guatemala. The relative political freedom for Maya in the USA allows parents to pursue their goals for the healthy growth and development of their children. In 1977, Robert LeVine, an anthropologist of the family and of children, proposed a universal evolutionary hierarchy of human parental goals. The primary goal is to encourage the survival and the health of a child. Secondary goals relate to developing the child into a self-supporting adult and instilling cultural beliefs and behavioral norms. Economic and political conditions in Guatemala make it difficult for parents to achieve these goals for their children. The political economy of the United States offers real possibilities for success, and Maya parents seize upon these, just as other immigrants have done before them.

As Boas argued for nearly 50 years, the study of human growth provides a mirror of the human condition. Reflected in the patterns of growth of human populations are the 'material and moral conditions of that society' (Tanner, 1986, p. 3). The forces holding back growth in Guatemala are severe indeed, and the growth differences between Maya of Guatemala and the United States may be used as a measure to assess the magnitude of change in political and socioeconomic conditions.

Growth and evolution

The pattern of human growth serves as another type of mirror; one that reflects the biocultural evolution of our species. **Biological evolution** is the continuous process of genetic adaptation of organisms to their environments. **Natural selection** determines the direction of evolutionary change and operates by **differential mortality** between individual organisms prior to reproductive maturation and by **differential fertility** of mature organisms. Thus, genetic adaptations that enhance the survival of individuals to reproductive age, and that increase the production of similarly successful offspring, will increase in frequency in the population over time.

Human biocultural evolution produced the pattern of growth and development that converts a single fertilized cell, with its complement of **deoxyribonucleic acid (DNA)** into a multicellular organism composed of hundreds of different tissues, organs, behavioral capabilities, and emotions. That process is no less wondrous when it occurs in an earthworm than in a whale or a human being. Indeed, many growth processes that occur in people are identical to those in other species and attest to a common evolutionary origin. The discovery of PAX-6, a 'master-control gene' for growth and development of the eye (Halder *et al.*, 1995) common to species as diverse as marine worms, squid, fruit flies, mice, and people, is powerful evidence for the common evolutionary origin of the eye. Other organs, and the genetic mechanisms that control their growth and development, also are shared among diverse species (Chapter 7 discusses the genetics of growth). Nevertheless, some events in the human life cycle may be unique, such as a distinct childhood growth stage and menopause (discussed in Chapters 2, 3, and 4), and these attest to ongoing evolution of our species.

Dobzhansky (1973) said that, 'nothing in biology makes sense except in the light of evolution'. Human growth, which follows a unique pattern among the mammals and, even, the primates, is no exception to Dobzhansky's admonition. A consideration of the chimpanzee and the human, two closely related (genetically) extant primates, shows the value of taking an evolutionary perspective on growth. Huxley (1863) demonstrated many anatomical similarities between chimpanzees and humans. King & Wilson (1975) showed that such anatomical similarities are due to a near identity of the structural DNA of the two species. One interpretation of King & Wilson's findings is that the differences in size and shape between chimpanzee and human are due to the regulation of gene expression, rather than the possession of unique genotypes. Of course, humans are not descended from the chimpanzee, but both species did have a common ancestor some five, or more, million years ago. During evolutionary time, mutations and selective forces were at work on the descendants of this ancestor shaping their genetic constitution and its expression in their phenotypes.

The anatomical differences between human and chimpanzee that result from alterations in gene regulation are achieved, in part, through alterations in growth rates. D'Arcy Thompson showed in 1917 that the differences in form between the adults of various species may be accounted for by differences in growth rates from an initially identical – one might better say 'similar' – form. Thompson's transformational grids (Figure I.2) of the growth of the chimpanzee and human skull from birth to maturity, show how both may be derived from a common neonatal form. Different patterns of growth of the cranial bones, maxilla, and mandible are all that are required to produce the adult differences in skull shape. Of course, the differences in skull growth are related to size and shape of the brain, and size of the dentition (both species have the same number and types of teeth). In a similar manner, the differences in the post-cranial anatomy between chimpanzee and human being result from unequal rates of growth for common skeletal and muscular elements.

Despite the anatomical and biochemical evidence for the evolutionary

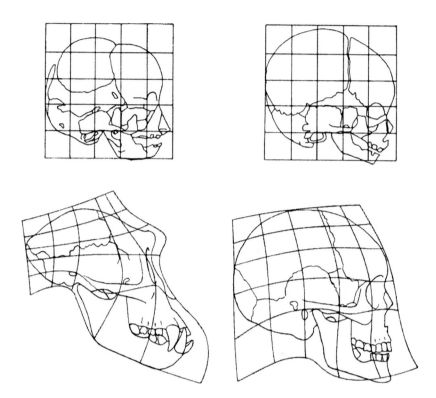

Figure I.2. Transformation grids for the chimpanzee (left) and human (right) skull during growth. Fetal skull proportions are shown above for each species. The relative amount of distortion of the grid lines overlying the adult skull proportions indicate the amount of growth of different parts of the skull (inspired by the transformational grid method of D'Arcy Thompson, 1942, and redrawn from Lewin, 1993).

origins of the human growth pattern, most works on human growth give little consideration to this topic. A paragraph or two is all that may be found in the current physiological and medical texts devoted to human growth. Recent textbooks in biological anthropology, however, do give space and emphasis to the evolution of the human pattern of growth (e.g., Relethford, 1997; Boaz & Almquist, 1997). In this book, two chapters are devoted to an account of the evolution of the human pattern of growth. Moreover, the theme of evolution runs throughout this book because it is the evolutionary perspective which makes sense of the rest.

Growth theory

This is also a book about a theoretical approach to the study of human growth. The literature in the general area of animal growth is rich in both hypothesis testing and theory (e.g., Thompson, 1917; Huxley, 1932; Brody, 1945; Weiss & Kavanau, 1957; Bertalanffy, 1960; Goss, 1964, 1978; Bonner, 1965; Snow, 1986). By contrast, prior to 1980 only a few workers had published hypotheses about the course and regulation of human growth (e.g., Bolk, 1926; Tanner, 1963; Frisch & Revelle, 1970; Grumbach *et al.*, 1974; Gould, 1977; Bogin, 1980). An example of some important contributions to growth theory, and some of its impact on later research, is given in Box I.1.

Box I.1.   Models of the regulation of human growth

A conceptual model of growth

In 1963, James Tanner proposed a conceptual model for the biological regulation of human growth. Basic elements of the model are represented in Figure BI.1. A major feature of this model is that growth is target seeking and self-stabilizing. The curve labeled 'Time Tally 1' represents a hypothetical mechanism that provides a 'target size' for body growth, and also, keeps track of biological time during infancy and childhood. Biological time is measured in units of maturation, with the clock started at conception and stopped when some functionally mature state is reached. The curve labeled 'Inhibitor' represents the concentration in the body of a hypothetical substance, perhaps a by-product of cell division or protein synthesis, that acts upon the time tally to regulate growth rate. The amount of mismatch, 'M', between the two curves determines the rate of growth at each chronological age.

Tanner's model accounts for the deceleration of growth velocity during infancy and childhood and explains the phenomenon of catch-up growth (Prader *et al.*, 1963) following serious illness or starvation. During the normal postnatal growth of an infant, the amount of mismatch between the size the infant might attain and its actual size is large, and growth rate is rapid. As the child grows in size the amount of mismatch decreases and the concentration of inhibitor increases. As a result, the rate of growth slows. Under non-normal conditions, for instance starvation, the

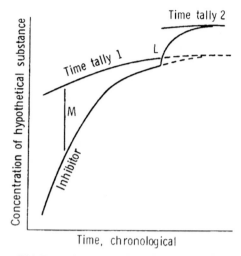

Figure BI.1. Tanner's conceptual model for the regulation of human growth. Rate of growth is determined by the amount of mismatch (M) between the concentration of a hypothetical inhibitor substance and a time tally. Time Tally 1 controls growth during childhood. Time Tally 2 controls growth during adolescence. At point L, the switch between time tallies occurs, and the adolescent growth spurt is initiated (Tanner, 1963).

rate of growth slows or stops during the period of insult. The concentration of inhibitor remains constant during this time as well. The time tally continues to register the mismatch between actual size and target size. When the insult to the child is removed, in this case upon refeeding, there is a rapid increase in growth rate to restore the balance between the time tally, the expected amount of mismatch, and the concentration of inhibitor. When this balance is restored the rate of growth assumes the normal velocity for that child, as if the insult had not occurred.

To account for the abrupt change in the velocity of growth that occurs at adolescence, Tanner suggests the existence of a second time tally. 'Time Tally 2' (Figure BI.1) operates in the same manner as time tally 1, but both the mechanism controlling the new tally and its inhibiting substance are assumed to be distinct from the old one. Tanner believes that the switch to the new tally occurs when a minimum velocity of growth, or minimum mismatch, on the old tally is reached. This is labeled point 'L' in Figure BI.1. After the switch occurs, a new larger mismatch is established and a rapid increase in growth rate results. This is the adolescent growth spurt. As the mismatch is reduced, and the concentration of inhibitor is increased, the rate of growth slows once more. Variation in the timing of the adolescent spurt between individuals is explained by changing the point at which the switch between tallies takes place.

This model is as conceptually stimulating today as when it was first proposed. In 1980, I published a paper that built on Tanner's model by incorporating the, then popular, mathematical concepts of catastrophe theory. Since then, much more has been learned about the biological nature of the time tallies and the inhibitor substances that regulate growth both before and after puberty. This research is reviewed in Chapter 7 – Genetic and endocrine regulation of human growth.

Where is the growth regulator located?
A complete and exact knowledge of the biological mechanisms involved in the regulation of amount of growth and rate of maturation are unknown. Tanner's model proposes that the control of amount of growth is related to the concentration of a hypothetical inhibiting substance that acts on a time tally located in the brain, perhaps in the hypothalamus. Goss (1978) suggests that overall size of the body is regulated by a genetically programmed amount of growth for certain visceral organs. For instance, the heart or the kidneys may grow to a predetermined size and the functional limits of these organs to support the operation of other organs and systems of the body may determine the size to which these tissues, and the body as a whole, may grow. Snow (1986) interprets the results of experimental studies of growth-controlling mechanisms to support Goss' hypothesis. However, Snow finds that growth control in a developing organism is evident before organogenesis occurs during embryonic development. This fact, Snow suggests, indicates that control may lie in a tissue that differentiates relatively early in development and is distributed widely throughout the body, rather than in the visceral organs. A tissue composed of cells derived from the neural crest of the embryo, such as the central nervous system, is Snow's choice of a likely tissue for this function.

Some recent research with insects adds a further complication to the question of how body size is regulated. Nijhout & Emlen (1998) experimented with the growth of body parts in buckeye butterflies and dung beetles. In one experiment they removed embryonic cells that would have developed into the hind wing of butterflies. The adult butterflies had fore wings, but these were significantly larger than expected. Experiments with beetles manipulated the size of their horns by exposing them to hormones that were known to shrink or enlarge horn size. As larger horn size was induced, the size of the eyes decreased, and vice versa. Their findings indicate that there is an interaction between body parts in their growth. A greater amount of investment in the size of one part may limit the size of another part. If this is true, then there is no simple genetic determination of the final size and shape of an organism. Nor is there a single, central time- or size-regulating mechanism. Instead, animals may be able to adjust their final size and shape during development in response to many external stimuli.

Until about 1970, most research and writing in the area of human growth and development was descriptive and atheoretical. A typical anthropological or public health study would be based on a sample of children, youths, or adults measured for some variable(s) (height, weight, etc.). The data would be presented, described and compared to similar data for another group of people. Or, in the fields of medicine and psychology, an unusual case study of growth, perhaps resulting from physical or psychosocial pathology, was described. In another realm, growth data derived from statistically representative samples of human populations were used to construct reference standards of height, weight, and other physical dimensions. These standards have value in public health work to assess the growth, development, and nutritional status of populations that are 'at risk' for growth failure or malnutrition, and to monitor the effectiveness of intervention programs designed to improve the health and growth status of such populations.

Historically, most disciplines begin with this sort of descriptive phase. Human growth research was no exception, as documented in two books that detail the historical development of the field (Boyd, 1980; Tanner, 1981). One sign of a maturing discipline is when it begins to develop hypotheses to examine the nature of the processes that account for the descriptive data. An early example of the use of hypotheses in growth research is the classic series of studies by Boas (see Boas, 1912, 1922) on the growth of the children of migrants to the United States, as discussed above. The scholarly value of the research lies in Boas' use of the scientific method. Boas challenged the dogmatic belief that physical types, or 'races', were fixed and biologically distinct. His hypothesis was that a change in environment brought about by migration from rural areas in Europe to urban centers of the United States would bring about alterations in the amount and rate of growth. The value of his hypothesis-testing method lasted, and is still being used in the area of migration research to test the relationship between migration history and human biology (Boyce, 1984; Mascie-Taylor & Lasker, 1988).

Mature sciences are noted for their ability to synthesize several hypotheses that have been verified independently into one or more comprehensive theories that can explain the known data and, in turn, indicate the kind of observations that should be made in further research. Despite the early work of Boas, hypothesis testing in human growth research was only sporadic until the late 1960s. Since that time, the relationships between nutrition and growth, physical development and chronic disease, and environmental stress and growth, among other topics, began to be actively pursued. Unfortunately, a dreadful number of purely descriptive studies were still produced in the last third of the twentieth century.

One very important reason for the paucity of hypothesis testing and experimentation in human growth research is the ethical dilemma. In non-human animal research, the genetic and environmental determinants of growth may be studied in the laboratory. The growth of the entire organism, anatomical regions, or even isolated populations of cells may be studied with great precision. Using human beings in experimental research is almost always unacceptable ethically. It is even illegal to use human tissues in research in some countries. Moreover, laboratory controlled experimentation is conceptually unrealistic for many questions relating to human growth. Normal human growth and development require the complexities of a normal social and cultural milieu.

Even so, students of human growth and development are not restricted to descriptive studies. Human growth may be studied with the same intellectual excitement, experimental approach, and theoretical research as is done by laboratory biologists. Creative and rigorous field research that makes use of 'natural experiments', for example, the migration of southern Europeans or Maya to the United States, can be employed. These natural experiments, in combination with powerful new statistical and computational methods, the study of pathological growth disorders, and controlled experimentation within allowable and ethically justifiable limits, can be used to achieve a sophisticated understanding of human growth.

Advances in technology are also forging a new conceptualization of growth. Scientific advances in fields from cell biology to computerized imaging allow for human growth research that can proceed without some of the ethical and moral limitations that restricted experimentation on people. Molecular biology has reached the stage where the biochemical substances that control growth and development can be identified, their gene sequences decoded, and synthesized. The invention of the polymerase chain reaction by Kary Mullis in 1983 (Mullis *et al.*, 1986) allowed for the nearly unlimited duplication of segments of DNA (deoxyribonucleic acid). The identification of regulatory growth genes, master genes for development, and genes that cause specific growth pathologies, all of which were proposed as possible prior to 1983, have actually been identified since then.

Another major advance is the synthesis of data from many branches of biology, including ecology, demography, genetics, physiology, **phylogeny** (evolutionary history of a species) and **ontogeny** (growth and development of individuals), into a type of grand unification called **life history theory**. The life history approach attempts to unravel the reasons why different species of animals follow different sequences of development. Why, for example, do closely related species have different durations of each life stage, such as infancy or adulthood? Why do new life stages evolve, such as the juvenile stage of the social mammals? Life history theory also tries to understand the differences between species in the timing of reproduction, the spacing between births, the number of total offspring produced, and the amount of investment in time and energy parents provide for their off-spring. As will be shown later in this book, some 'mysteries' of human growth, such as the nature of childhood, adolescence, and menopause in women, can be understood in terms of life history strategies for efficient reproduction.

Human auxology

This book is a synthesis of methods and knowledge about human growth gleaned from evolutionary biology, from reports of the growth of human populations living under various ecological regimes, from statistical and mathematical applications, from medical pathology, and from experimental biology. Theoretical perspectives from anthropology, economic history, political economy, and life history are used to order the data derived from observation and experiment. Drawing upon these areas of research and theory, this book strives to include human growth within the field of auxology. The term auxology refers to the study of biological growth. It could be the study of any type of growth, and some botanists and most veterinary and farm animal zoologists use this term to refer to growth research in their fields. Botanists use the term 'auxins' to refer to the hormones of plants that promote an increase in size of stems, leaves, and other structures. During the last three decades, European human biologists began to use the phrase 'human auxology' to refer to human growth research (Borms *et al.*, 1984; Bogin, 1986). An ideal human auxology will combine the results of descriptive studies, experimental research, and hypothesis testing into a comprehensive theory of the structural and functional elements of growth.

The organization of this book

The aim of this book is to show that the tempo and mode of human growth are basic to the understanding of our species' place in nature. This is done by dividing the patterns of human growth into five areas. The first area is an overview of the history of the study of growth and the basic biological principles of growth and development during the human life cycle, presented in Chapters 1 and 2. The second area is the evolutionary foundation of the human growth pattern, treated in Chapter 3 and Chapter 4. Chapter 3 describes the evolution of the human growth curve. The mammalian foundations of the human growth curve, and the non-human primate embellishments upon that foundation, are presented conceptually and mathematically. In Chapter 4 the pattern of human growth is considered from an ecological and evolutionary perspective. The relation of growth rates to feeding and reproductive adaptations is examined using data from paleontology, paleoecology, demography, ethology and ethnology – in other words from a life history perspective. The unique features of the human pattern of growth, including the evolution of the childhood growth period and the human adolescent growth spurt in height, are detailed. The result is a comprehensive exploration, and, occasionally, explanation for the functional and adaptive significance of human growth patterns.

The third area is variation among human populations in growth patterns. In Chapter 5, several cases of such variation are described and the adaptive value of population differences in growth is discussed from an evolutionary perspective. In Chapter 6, some of the physiological, environmental and cultural reasons for population variation in growth are explored using the literature of field and experimental research on human growth.

The fourth theoretical area is covered in Chapter 7, which describes and analyzes the genetic and endocrine factors that regulate the growth of individual human beings. The facts, explanations, and hypotheses presented in the previous chapters give evolutionary and functional meaning to the pattern of human growth, but they do not explain how the amount and rate of growth are controlled. Genes determine the structural elements (e.g., proteins) for growth and form. Genetic information also interacts with factors from the environment, and these factors help to provide guidelines for genetic expression. The gene–environment interactions may not, however, directly regulate growth and development. Rather, their influence is often mediated by the endocrine system. In Chapter 7, several examples of hormonal mediation are given, including the control of small body size in African pygmies, and the effects of nutritional and psychological stress on growth.

Interspersed throughout the book is discussion devoted to a fifth theoretical area, mathematical and biological models of the process of human growth and development. A scientific understanding of the pattern of human growth requires detailed information of the biological factors that determine development. However, it also requires the precision and economy of analysis that is provided by mathematics. These are not separate

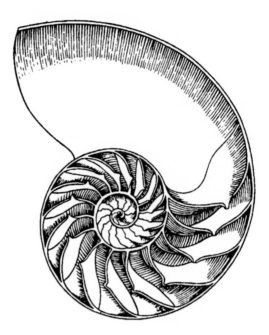

Figure I.3 Shell of *Nautilus pompilius*, sagittal section (from D'Arcy Thompson, 1942).

realms of knowledge, for the growth and form of an organism often display a clear relationship between biological structure and mathematical regularity. D'Arcy Thompson (1942) described the biological form of the growth of the *Nautilus* shell, seen in Figure I.3, and several other spiral shapes in nature, with a mathematical function called an equiangular spiral. In a similar fashion, the form of the curve of growth of the human being can also be reduced to one or more mathematical functions. The mathematical treatment of human growth, or the growth of any other organism, is made possible by the predictability of biological development. Growth must produce a biological form that meets the ecological requirements of life for the species. Thus, in terms of growth and form, including the morphology and physiology of organisms, new individuals resemble other members of the same species more than they resemble members of other species. Due to this predictability, growth and form are amenable to the precision of mathematical description. Where appropriate and when needed there are sections of boxed text that describe some of the classic and recent innovative quantitative and qualitative approaches to the study of growth regulation.

A recapitulation and synthesis of the major themes of the book is given in Chapter 8. As a challenge for the future, Chapter 8 ends with an invitation for readers of this book to consider growth and development research as a professional career. Younger researchers will be able to build on the information now available to develop accurate, dynamic, and imaginative models of growth regulation that combine quantitative and qualitative elements from mathematics, molecular biology, neuroendocrine physiology, and the sociocultural environments in which human beings live.

1 Background to the study of human growth

People, like all animals, begin life as a single cell, the fertilized ovum. Guided by the interaction of the genetic information provided by each parent and the environmental milieu, this cell divides and grows, differentiates and develops into the embryo, fetus, child, and adult. Though growth and development may occur simultaneously, they are distinct biological processes. **Growth** may be defined as a quantitative increase in size or mass. Measurements of height in centimeters or weight in kilograms indicate how much growth has taken place in a child. Additionally, the growth of a body organ, such as the liver or the brain, may also be described by measuring the number, weight, or size of cells present. **Development** is defined as a progression of changes, either quantitative or qualitative, that lead from an undifferentiated or immature state to a highly organized, specialized, and mature state. **Maturation**, in this definition, is measured by functional capacity, for example, the development of motor skills of a child that result in mature state of bipedal walking. Though broad, this definition allows one to consider the development of organs (e.g., the kidney), systems (e.g., the reproductive system), and the person.

Why grow and develop?

A large mammal, such as a human being, is composed of about 10^{12} cells, which result from approximately 2^{38} mitoses (cell divisions) since the moment of fertilization. During mitosis, cells differentiate into dozens of types of tissues and organs. Why must multicellular organisms of the five kingdoms of life – animals, plants, fungi, protists, and monera – undergo this process of cell multiplication and specialization? The answer is, because all such living things are mortal. Reproduction is necessary to replace those organisms that die, and reproduction requires both growth and development. Some forms of life reproduce asexually, in which one or a few cells are contributed by a parent and those cells eventually grow and develop into a new mature individual. Most species reproduce sexually, requiring a single cell with some biological material from each of two parents. In either case, the initial contribution of cells cannot look or behave in any way like the parent. To be like their parents, '. . . the new organisms will have to suffer changes before they become something approaching replicas of the old' (Newth, 1970 p.1). Newth, an embryologist, adds that the process of growth and development is arduous, often prolonged, and generally hazardous. Sex, growth, development, and death are some of the prices to be paid for multicellularity.

Historical background for the study of human growth

Like the fertilized ovum, the study of human growth and development has undergone an arduous, prolonged, and hazardous history of intellectual development. Arduous and prolonged because it has taken more than three thousand years of study to arrive at our present state of knowledge. Some of the highlights of these three millennia are described in the next paragraphs of this section. The history of growth research is hazardous because misinformation and gaps in our knowledge lead to tragic consequences for human beings. The thalidomide drug disaster of the 1960s and the decline in breast-feeding in 'modern' societies are just two recent examples of these hazards. Thalidomide is a sedative and hypnotic drug that was used to treat some of the discomforts of pregnancy. It was withdrawn from sale after it was found to cause severe birth defects, especially of the limbs. The thalidomide case promoted tougher standards for testing the safety of new drugs and more research on environmental influences of embryonic development.

A decline in breast-feeding during the twentieth century is a typical occurrence in both developed and developing nations. The development of milk-based formulas to feed infants and aggressive promotion of these formulas as 'modern', and sometimes superior to breast milk, led partly to the decline. In the past decade, a body of research has accumulated showing that there are harmful effects for infants who are never breast-fed. These include increased incidence of respiratory and inner ear infections, accumulation of more body fat and later risk of heart disease (Cunningham, 1995; Scariati *et al.*, 1997), and reduced mental development (Dettwyler, 1995) in formula-fed infants. After reaching an all-time low in the early 1970s, the incidence of breast-feeding in the United States has increased as both the medical profession and the public have become aware of the benefits of nursing (Ryan, 1997a).

These examples of hazards in the history of the study of human growth and development provide reason for caution in the future. Another reason to study the history of any discipline is that one learns what topics have been studied, when such inquiry first occurred, and which problems are in need of further study. Historical study is important from a conceptual perspective as well, as it may help explain why scholars and practitioners have been interested in human development. In particular, the history of study of human growth makes clear the relevance of growth research to medicine, epidemiology, and public health. More generally, an understanding of the history of the study of human growth reveals connections between economics, art, law, politics, philosophy, and other fields of knowledge that influence the course of human events.

The following is a review of some of the major historical events in the study of human growth, with special emphasis on those which still have an influence on growth research today. Boyd (1980) and Tanner (1981) provide book-length histories of the study of human growth. Lowery (1986) offers a brief review focused on pediatric medicine. Bogin & Kapell (1997) present a concise history of the study of normal human growth emphasizing topics of anthropological interest. The *History of Physical Anthropology: An Encyclopedia* (Spencer, 1997) and the *Cambridge Encyclopedia of Human Growth and Development* (Ulijaszek *et al.*, 1998) provide many entries relating to the history of human growth research.

Prehistory and early historic period

Small stone sculptures, often called 'Venus figurines', and cave paintings from Europe depict people and animals that may be pregnant. The earliest of these artistic renderings date from about 25 000 years BP (before present). There are also rock paintings from southern Africa and Australia, older than the European cave art, that depict people and other animals of various ages and sexes. The fossil record includes Neanderthal and early modern human skeletons of children, including some with developmental pathology. One Upper Paleolithic skeleton from Italy seems to have been from an adolescent with a type of dwarfism caused by acromesomelic dysplasia (Frayer *et al.*, 1987). This type of dwarfism results in severe deformity and physical impairment, but normal intelligence. The affected individual would have been unable to contribute much labor to a hunting and gathering group. Frayer and colleagues believe that this individual's survival to the teenage years indicates both tolerance and care for impaired infants and children. This may be so, but a detailed understanding of how Paleolithic peoples may have interpreted the meaning of pregnancy, normal and pathological growth, and human development in general are matters for speculation.

The earliest written records about human growth date from Mesopotamia, about 3500 BP. Inscribed myths recount the act of fertilization, the nine months of pregnancy, and both full term and premature birth. Concerns about low birth weight or prematurity, birth defects, and twinning also are recorded. The Sumerians divided postnatal life into several stages that correspond to modern ideas of infancy, childhood, youth, adulthood and old age. There is no direct evidence that Sumerians measured the dimensions of the body. Some of the art works seem to depict accurately size differences between children and adults. Other works of art depict high status people, such as male elders, as disproportionately taller than lower status people, such as women. Several texts also make mention of a positive relationship between health, social status, and stature. Thus, both in Sumerian art and in life there was a relationship between growth and biosocial conditions. That this relationship appears in the earliest writing on human growth is fascinating, for the study of this association is still a very active area of research today (see Chapters 5 and 6).

The ancient Egyptian, Chinese, Hindu, Greek and Mesoamerican civilizations follow many of these Sumerian traditions. Written records and art work show that the earliest interest in the biology of children was primarily a concern with the preservation of life. Greek, Roman and Arab physicians prescribed regimes of physical activity, education, and diet to help assure the health of children. Their advice was more often guided by the needs of their societies for military personnel and by religious dogma about children rather than by empirical observations of the effect of child-rearing practices or child growth, development, and health. Of course there were marked differences between these cultures in specific cultural values, but the universal biological nature of pregnancy, birth, and infancy (this nature is reviewed later in this chapter) meant that all human societies must converge on some basic strategies for the care and feeding of their young.

Some of the early civilizations, such as the Egyptians and the Hindu Indians, showed careful concern for measurement of the body, including children and youths. Egyptians used a grid system to carefully render body proportions correctly. Other cultures, such as the Chinese and Early Jewish Tradition, emphasized more spiritual aspects of human development in their concern for the young. One matter of repeated concern for these ancient societies is the number of stages of life. Numbers vary, but by the time of the Romans 'seven ages of life' becomes a frequent blueprint for human development. Today, research into life history theory, which is concerned in part with the number of stages in the life cycle of organisms, is very popular and productive. Human life history theory is a central concern of Chapters 3 and 4 of this book.

The Latin West and the Renaissance

Egyptian, Greek, and Roman artwork, especially three-dimensional sculptures, depict infants and children fairly accurately in terms of body size and proportion (infants have relatively larger heads and shorter arms and legs than children and adults). In some of this art the children are depicted at play. Viewing this art can give a sense that infancy, childhood, youth, and the other 'seven stages of life' were each accorded its own special biology and behavior. Scholarly concern with the stages of life continued following the collapse of the Roman Empire, but there seems to have been a shift in the status accorded to children and youths. Medieval physicians, clerics, and artists began to follow a tradition of treating the child as a miniature adult. In this tradition, the growth and development of infant to adulthood involved only an increase in size and maturity during the growing years.

There is some debate as to how, and why, children of the Medieval Age were perceived and treated. Nevertheless, Kaplan (1984, p. 46) writes that, 'Plague, pestilence, ignorance, extraordinary poverty, drudgery, starvation, perpetual warfare were . . .' some of these reasons why children lost special status. Ariès (1965) reconstructs the social world of that time and concludes that after the age of seven years, children were forced to enter the social world of adults. Ariès believes that this accorded children great freedom, but Kaplan views this as a kind of abandonment. In either case, the social reality was that young people were expected to become adult-like at a fairly young age. In the art works of the time, especially paintings and mosaics, young people are depicted with the same body proportions as adults. The 'Rucellai Madonna', attributed to the Italian artist Duccio (1285?), the 'Madonna of the Trees' by Bellini (1487), and 'Peasant Dance' by Pieter Bruegel the elder (1568?) are all in this stylistic tradition.

A few words of caution need to be interjected at this point. Our current interpretation of the history growth research, and of 'children' and 'childhood' (in fact human development from fetus to young adult), is distorted by our own ethnocentric ideals and beliefs. Ethnocentric notions, of course, clouded the perceptions and behavior of people in the past toward children as well. Sommerville (1982) mentions that the Greeks made note of the *absence* of infanticide among the Egyptians. After invading the Americas, the Spanish were impressed that all Aztec, Maya, and Inca infants were breast-fed, and that nursing, along with other foods, continued until four years of age. Even women of royal status nursed their own babies (Shein, 1992). It seems that in sixteenth century Spain, and elsewhere in Europe, breast-feeding was viewed as 'too natural' and thus not becoming to people. This was especially so for those of high social status who often contracted lower social status women as wet nurses. So, when historians of the past 40 years debate the alleged brutalities against children of past ages, readers should be careful not to commit the ethnocentric fallacy of judging other cultures by one's own standards. What late twentieth century writers note with amusement about the past may be telling us more about our own unconscious assumptions, or wanting behaviors, toward our own children.

It is also important to interpret the art of the past with some caution. A critic of the first edition of this book stated, 'Art is a figurative medium and cannot often be used to make objective assumptions about culture' (Kathryn Stark, personal communication). The same critic wonders if it is correct to state that Picasso (b1881–d1973) paints children 'better' than Bruegel (b1525?–d1569)! Despite the fact that it may not be possible to know exactly how infants, children, and youth were treated, and perceived, during the Medieval period, it is clear that human growth and development were not studied scientifically. During the Renaissance period there was a revival of the classical Greek concept of the dynamics of growth, which is amusing as this concept was never accepted popularly by the Greeks or Romans. The scholar Giordano Bruno (ca. b1550–d1600) wrote, '. . . We have not in youth the same flesh as in childhood, nor in old age the same as in youth; for we suffer transmutation, whereby we receive a perpetual flow of fresh atoms and those we have received are ever leaving us' (Boyd, 1980, p. 176). This is a remarkably modern statement of the constant turnover of cells and the constituents of cells in the human body. Leonardo da Vinci (b1452–d1519) proposed that new studies of human growth and development, from conception onwards, needed to be undertaken. Leonardo initiated his own human dissections, including his study of a seventh-month fetus, the placenta, and stillborn full-term infants (Figure 1.1). In the year 1502, the physician Gabrielo de Zerbis (ca. b1460–d1505) published a description of the anatomical differences between child and adult. Many other medical and scientific publications quickly followed.

Leonardo used his scientific studies of human growth to produce drawings that correctly rendered adult and child body proportions. Building on the work of Vitruvius, a first-century BC Roman architect and writer, Leonardo developed canons, or rules, for drawing human proportions (Tanner, 1994). Albrecht Dürer (b1471–d1528), a German artist, devised a method of geometric transformations that he used to accurately render proportions of the human head and face. With his geometric methods, Dürer could draw not only the canonical types, but any manner of human variation in size or proportion. He applied his method to drawings of men, women, children and infants (Figure 1.2). Including women and children in

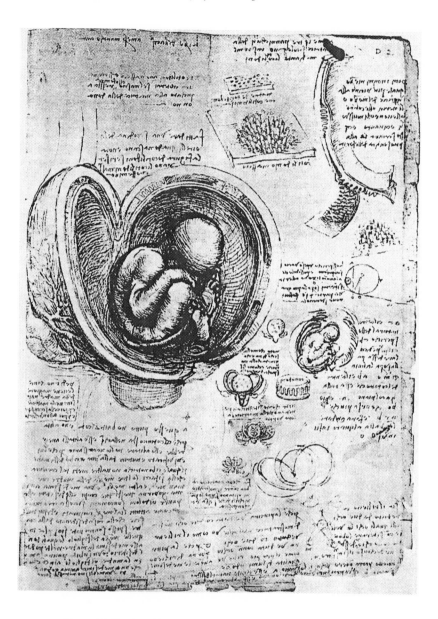

Figure 1.1. Fetal positions and structures of the placenta shown in sketches by Leonardo da Vinci. Windsor collection, folio 8r. From *Quaderni d' anatomia*, volume III.

PROPORTION DE L'HOMME. LIVRE I. 30

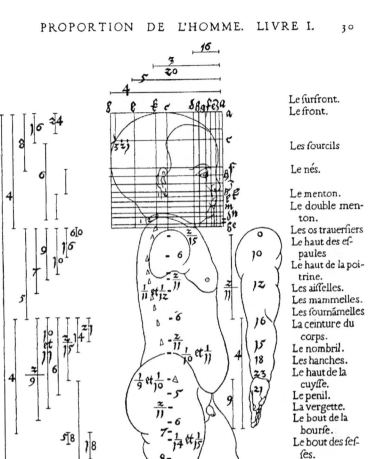

Le surfront.
Le front.

Les sourcils

Le nés.

Le menton.
Le double men-
ton.
Les os trauersiers
Le haut des es-
paules
Le haut de la poi-
trine.
Les aisselles.
Les mammelles.
Les soumámelles
La ceinture du
corps.
Le nombril.
Les hanches.
Le haut de la
cuysse.
Le penil.
La vergette.
Le bout de la
bourse.
Le bout des ses-
ses.
Le surgenouil.
Le mygenouil.

Le sougenouil.
Le bas du gras.

Le coup du pied.

La plante.

Figure 1.2. Proportions of the infant, diagram by Albrecht Dürer. From the 1557 edition of *Les quarte livres d'Albert Dürer*.

this type of methodological work was an innovation, as most artists followed the teachings of Cennino Cennini (*ca.* b1400) who wrote that women do '. . . not have any set proportion' (Boyd, 1980, p. 202). Children, it seems, were too inconsequential for Cennini to even mention! As a reflection of the art and scholarship of the time, the work of Leonardo and Dürer portended a major change in the concept of children and research into human growth.

After the year 1600, the post-Renaissance painters begin to depict children with normal proportions and also with growth pathologies. The Flemish artist Van Dyck depicts three normal children in the painting 'The Children of Charles I' (1635). The painting 'The Maids of Honor' by Diego Velazquez (1656) depicts a normal child, a woman with achondroplastic dwarfism (normal-sized head and trunk with short arms and legs) and a man with growth-hormone deficiency dwarfism (proportionate reduction in size of all body parts). At the time of these paintings, of course, the biological control of normal and pathological growth was not known. Nor is it clear today how physicians and scientists of that time regarded different types of dwarfism.

Embryonic and fetal development

Ancient, Classical, and Medieval scholars had written a great deal on human growth and development prior to birth. Some of this was undoubtedly based on observation of human and non-human fetuses, but much was also the product of imagination and myth. The actual process of growth and development from fertilized ovum to the birth of a human child is so counterintuitive to our expectations, based on our experience with child growth after birth, that through much of human history scholars and physicians did not know or believe that it occurred. It was not until the year 1651 that the physician William Harvey helped establish that the embryo is not a preformed adult. Harvey showed that during prenatal development there are a series of embryological stages that are distinct in appearance from the form visible just before and after birth. The Greek physician Galen (*ca.* b130–d200 AD) wrote about the appearance of the fetus in the later stages of pregnancy. However, the first accurate drawings of the fetus were made by Leonardo da Vinci, as mentioned above. Other descriptions of fetal anatomy and physiology followed Leonardo's work, notably the studies published by Vesalius in 1555 and Volcher Coiter in 1572. Coiter studied a fetus less than 3 cm in length, indicating that the fetus had developed for about 10 weeks since conception.

During the seventeenth and eighteenth centuries descriptive anatomical

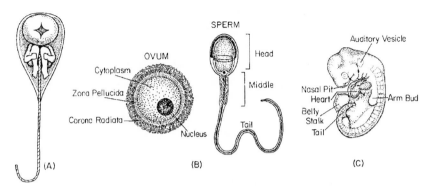

Figure 1.3. (A) Preformationist rendering of a human spermatozoon (Hartsoeker, 1694; from Singer, 1959). (B) Diagram of human ovum and spermatozoon. (C) Diagram of human embryo 32 days after fertilization.

studies continued, with most of the work being done on fetuses in the last trimester of pregnancy (last three months). The fetus of this age is of unmistakable human appearance, so these studies failed to appreciate the physical changes that take place earlier in prenatal life. Some biologists continued to believe in preformation, and a few extended that concept beyond pregnancy to the formation of spermatozoa (Figure 1.3). In 1799 S. T. Sommerring published drawings of the human embryo and fetus from the fourth week post-fertilization to the fifth month, clearly showing that the embryo is not a preformed, or miniature, human being.

The scientific study of the cellular mechanisms of fertilization and embryonic development has its roots in the work of Karl Ernst von Baer (b1792–d1876), published in 1829 (see Baer, 1886). He described the 'germ layers' of the embryo, properly called the endoderm, the mesoderm, and the ectoderm. The endoderm cells give rise to the internal organs, cells from the mesoderm form the skeleton and the muscles, and the skin and the teeth develop from ectoderm cells. The work of von Baer removed the need to invoke mystical 'vital forces' to explain embryological transformations, replacing these with more mechanistic processes. However, it was not until the twentieth century that an understanding was achieved of the highly complex nature of the physical, chemical, and biological processes that occur during prenatal growth.

Longitudinal studies of the eighteenth century

Another post-Renaissance advance was a growing interest in how early life events could influence later development. For instance, by the 1700s physicians pursued the study of birth weight and its relation to child health. Prenatal and neonatal influences on later development remain topics of research interest today. A new research strategy had to be developed to study the relationship between early influences and later growth outcomes. This is the longitudinal method of research. A longitudinal approach requires that the same individuals be examined on at least two occasions, separated by some period of time. Prior to the use of longitudinal methods, the predominate strategy was the cross-sectional approach in which each individual is examined only once. The cross-sectional approach has the advantage that many people can be measured in a short period of time. The disadvantage is that the dynamics of growth, such as changes in rate of growth, cannot be properly studied.

The Count Philibert Guéneau du Montbeillard (b1720–d1785) of France, measured the stature of his son every six months from the boy's birth in 1759 to his eighteenth birthday. George-Louie Leclerc de Buffon (b1707–d1788) included the measurements, and his commentary on them, in a Supplement to his *Histoire Naturelle* in 1777. These data are usually considered to constitute the first longitudinal study of human growth and, due to Buffon's commentary, the most famous study. The growth in height of this boy, both in terms of achieved stature by age and rate of growth at any age, are illustrated in Figure 1.4. The original data were reported in antiquated French units of measurement, but Scammon (1927) converted these to modern metric units. The metric data are drawn here as mathematically smoothed curves (the cubic spline technique was applied to the data given by Scammon by the present author). The smoothing makes the important features of the curve more easily seen.

The curve in the figure labeled A is the boy's total height at each measurement. If growth is viewed as a motion through time, then this graph may be called the **distance curve** of growth. The boy's rate of growth between successive measurements is graphed in part B of the figure, commonly called the **velocity curve** of growth. Buffon had earlier written on the adolescent spurt in growth (the rapid acceleration in growth velocity around the time of sexual maturation) and on the general advancement of maturation of girls compared with boys. With the data on Montbeillard's son, Buffon noted the seasonal variation in rate of growth; the boy grew faster in the summer than in the winter. Buffon also wrote of the daily variation in stature; the boy was taller in the morning after lying at rest during the night than he was in the evening after working and playing during the day. Since Buffon's time, it has become necessary to take these variations in seasonal growth and daily stature into account when designing or analyzing longitudinal growth studies.

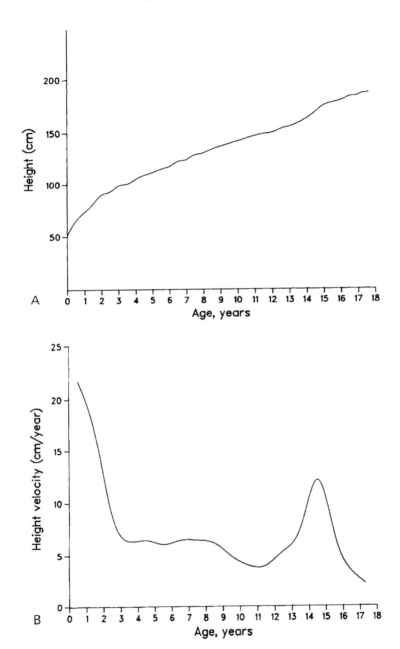

Figure 1.4. Growth in height of the son of the Count Montbeillard during the years 1759–77. (A) Amount of growth achieved during each six months from birth to age 18 years. (B) Rate of growth in height during each six months from birth to age 18 years.

Except for the adolescent growth spurt, Buffon made little mention of changes in growth velocity that are clearly seen in Figure 1.4. Montbeillard's son shows four distinct phases of growth velocity. The approximate duration of each phase, its name, and its tempo are: from birth to three years of age, the **infant phase**, growth decelerates rapidly from its maximum value of 22 cm/year to 6 or 7 cm/year; from three to about seven years of age, the **childhood phase**, growth rate remains fairly constant; after age 7 years until about 11 years, the **juvenile phase**, growth rate decelerates again, at first rather slowly but then much faster; from about 11 to 18 years of age, the **adolescence phase**, there is a classic adolescent growth spurt, with an acceleration period from about 11 to 14.5 years followed by a deceleration period that continues until 18 years and beyond. Note that the acceleration period is less steep, that is the tempo of growth changes less rapidly, than the deceleration period. A more detailed discussion of these phases of postnatal growth is presented in the next chapter.

Another eighteenth century longitudinal study of growth is that of the students of the Carlschule, conducted between the years 1772 and 1794. The pupils of this high school, founded by the Duke of Wurttenburg, included sons of the nobility and of the bourgeoisie. The growth data showed that the former were, on average, taller than the latter during the growing years but both groups achieved approximately equal height at 21 years of age. Thus, the sons of the nobility experienced an advancement of the rate of growth. This study, and the work of Buffon, clarify the important difference between amount of growth achieved at a given time and the rate of growth over time.

Statistical approaches of the nineteenth century

In 1835 Lambert Adolphe Quetelet (b1796–d1874) published the first statistically complete study of the growth in height and weight of children. Quetelet was the first researcher to make use of the concept of the 'normal curve' (commonly called today the normal distribution or 'bell-shaped' curve) to describe the distribution of his growth measurements, and he also emphasized the importance of measuring samples of children, rather than individuals, to assess normal variation in growth. Quetelet's statistical approach was followed in Europe by Luigi Pagliani (b1847–d1932). Pagliani began his studies on the size and fitness of Italian military personnel. He later applied his methods to children, and in 1876 demonstrated that the growth status and vital capacity (the maximum volume of air that can be inspired in one breath) of orphaned and abandoned boys, ages 10 to 19, improved after they were given care at a state-run agricultural colony.

Pagliani also noted that children from the higher social classes were taller, heavier, and had larger vital capacities than poverty-stricken children. Finally, Pagliani followed Buffon in taking longitudinal measurements of the same children. From these Pagliani noted that menarche (the first menstruation of girls) almost always followed the peak of the rapid increase in growth that takes place during puberty. He concluded that reproduction was delayed in young women until growth in size was nearly finished. This, he considered, was a proper relationship, for the nutritional and physiological demands of growth would interfere with similar demands imposed by pregnancy.

Politics, heredity, environment and growth

During the nineteenth century growth research was used for the first time to inform political and legal decisions regarding the treatment of children. The growth of European cities during the eighteenth century led to a flow of rural-to-urban migrants. Urban life dislocated many people from traditional rural family social organization. One result was an increase in the number of infants and children who became wards of the parish or were placed in foundling hospitals. The growth and health of these abandoned infants was extremely poor and many died. The physician William Cadogan (b1711–d1797) published *An Essay Upon Nursing and the Management of Children from Their Birth to Three Years of Age* (1750) that instructed the women working in foundling hospitals and parish orphanages. In a sense, this book was the first practical pediatric guide to baby care. The need for such books has not diminished over time. Indeed, the current best-seller, Benjamin Spock's (b1903–d1998) *Baby and Child Care* (first published in 1946), sold over 30 million copies in its first 30 years in the Untied States (only the Bible sold more copies). By 1998 the book had sold more than 49 million copies worldwide, in 39 languages.

The publication of Cadogan's *Essay* reflected a broad concern for infant health in England, where by 1767 laws were passed regulating the operation of foundling homes. On the Continent new concerns for infant care was sparked by the publication of Jean Jacques Rousseau's (b1712–d1778) book *Emile* in 1762. Rousseau advocated a 'return to nature', including the breast-feeding of infants by their own mother. Artificial feeding devices ranging from cow's horns to clay vessels had been used for centuries, but were becoming more common in cities. Such devices were difficult to clean, and the animal milk or other liquids fed to infants was surely not maintained under hygienic conditions. The inevitable result was intestinal infection for the infant. The higher social classes, especially in large cities such as

Paris, often 'farmed out' their infants to wet-nurses. Since these nurses might have several infants to feed, including her own infant, it is likely that some or all of her charges received too little breast milk, or were fed artificially. Rousseau's book was highly critical of these practices and, '. . . profoundly influenced thinking on child care and education throughout Europe' (Boyd, 1980, p. 270).

Another social force working against the welfare of children was the Industrial Revolution. Between the years 1765 to 1782 James Watt (b1736–d1819) developed a commercially viable steam engine, which forever changed the nature of human labor. The factory system ushered in under steam power reduced the need for human muscle power, and allowed children to be employed for many tasks. One survivor of childhood labor described his life to a British Parliamentary investigation in 1832 (quoted from Sommerville, 1982):

> Have you ever been employed in a factory? – Yes.
> At what age did you first go to work? – Eight.
> Will you state the hours of labour at the period when you first went to the factory, in ordinary time? – From 6 in the morning to 8 at night.
> When trade was brisk what were your hours? – From 5 in the morning to 9 in the evening.
> With what intervals at dinner? – An hour [once per day].
> During those long hours of labour could you be punctual; how did you awake? – I seldom did awake spontaneously; I was more generally awoke, or lifted from bed, sometimes asleep, by my parents.
> Were you always on time? – No.
> What was the consequence if you had been too late? – I was most commonly beaten.
> Severely? – Very severely, I thought.

The testimony continues at some length and describes a life in the factories of fatigue and hunger, punctuated by beatings.

Medical hygienists, or what we call today Public Health professionals, established the decline of health associated with urbanization and industrialization, by measuring the height and weight of people. In France, Louie-René Villermé (b1782–d1863) found in 1829 that military conscripts from mill or factory areas were too short, and suffered too many disabilities to make them fit for military service. Edwin Chadwick (b1800–d1890) published data on growth and health of factory children in his *Report of the Commissioners on the Employment of Children in Factories* (1833). Some of these data are reproduced in Figure 1.5, which compares the average height deficit of the English factory children (as reported by Tanner, 1981) against the international reference data for stature published by the United States

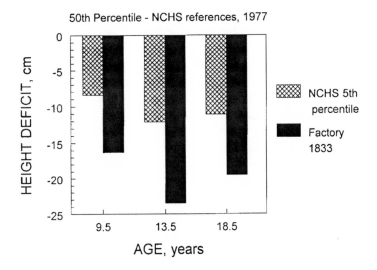

50th Percentile - NCHS references, 1977

Figure 1.5. Height of English factory children in 1833 compared with the National Center for Health Statistics (NCHS) references. The heights of the factory children are shown as deficits, in cm, to both the 50th percentile and the 5th percentile of the NCHS references.

National Center for Health Statistics (NCHS) (Hamill *et al.*, 1977). In this figure, the '0' line represents the fiftieth percentile height of the reference population. The factory children are 16.3 to 23.5 cm shorter than the NCHS fiftieth percentile, and 7.9 to 11.4 cm below the NCHS fifth percentile. In a group of children an average height below the fifth percentile is an indication of major growth delay and stunting. This magnitude of stunting is usually seen only in children with serious pathology. Even children growing up under conditions of poverty in the least developed nations of the world today have average heights above the fifth percentile of the NCHS references. At 18.5 years of age only the Pygmy populations of central Africa have smaller average heights – about 150 cm for Pygmies versus 158 cm for the factory children.

In response to these findings Friedrich Engels (b1820–d1895) campaigned extensively against the employment of children in English factories. Engels cited evidence of stunted growth, spine and bone deformities, and the physical and sexual abuse of child workers. Some physicians, Engels proclaimed, said that the factory districts would produce 'a race of pigmies' (Sommerville, 1982, p. 146) by a kind of evolutionary decline. Engels may have misunderstood the workings of evolutionary biology, but his social approach to growth stunting was correct. Chadwick's report included

Engels' concerns and that report, along with personal testimonies, led to the passing of the Factories Regulation Act (1833) in England. The Act prohibited the employment of children under the age of nine and stipulated that periods for eating and rest must be provided for older children during the work day. Age was determined by the state of dental maturation as assessed by the eruption of permanent teeth.

The relationship between tooth eruption and chronological age was verified in an extensive survey by Sir Edwin Saunders published in 1837. Saunders assessed the state of permanent molar eruption in English school children at ages 9 and 13 years, and showed that dental eruption was a better indicator of chronological age than height. By this work Saunders helped establish that development of the dentition is less influenced by the environment, while height, and therefore skeletal development, is more affected. In later years an appreciation of these differences would lead to the crucial concept of 'biological versus chronological age' and research leading to the production of atlases of dental and skeletal development (topics discussed in more detail later in this chapter).

Race and growth

In the United States a highly contentious political debate involved the use of growth data. Starting in 1875, Henry Pickering Bowditch (b1840–d1911) gathered measurements of height and weight, taken by school teachers, of 24 500 school children from the Boston, Massachusetts area. In a series of reports published in 1877, 1879 and 1891, Bowditch applied modern statistical methods (following Quetelet) to describe differences in growth associated with sex, nationality, and socioeconomic level between different samples of children. Bowditch was the first person to construct percentile growth charts, which he published in 1885. The NCHS growth charts discussed above are the modern descendants of Bowditch's work. With the first use of these charts Bowditch found that children of the laboring classes were smaller than children from the non-laboring classes. To account for this fact Bowditch preferred an environmental, rather than a genetic, explanation. He said the non-laboring classes were taller because of the '. . . greater average comfort in which [they] live and grow up . . .' (Boyd, 1980, p. 469).

This conclusion ran counter to that of Francis Galton (b1822–d1911). In his book *Natural Inheritance* (1889) Galton demonstrated the heritability of stature and other physical traits. Galton's work led some to believe that heredity was the all-powerful determinant of human form and functional capabilities. Galton's work was used to support the eugenics movement, a pseudo-scientific political movement that claimed to be able to improve the human species by controlled breeding. Eugenicists held that the laboring classes were genetically inferior to the non-laboring classes. One proof of this inferiority was their short stature. Eugenicists also believed that the 'race', or ethnic origin, of American-born children could easily be determined on the basis of physical measurements, and that 'racial' admixture, especially between 'Anglo-Saxons' and people from southern and eastern Europe, would bring about a physical degeneration of Americans.

Previously mentioned in the Introduction to this book was the research of Franz Boas (b1858–d1942), a German-born anthropologist working in the United States. Boas demolished the position of the eugenicists with his studies of migrants to the United States and their children. Boas found that children of recent immigrants grew up to look much like the 'good old Americans' (older generations of immigrants) due to modifications in the process of growth and development as a response to environmental change. Accordingly, Boas concluded that, 'we must speak of plasticity (as opposed to permanence) of types' (1910, p. 53). The term 'types' is a synonym for 'race'. Eugenicists believed that each 'race' could be defined by genetically fixed sizes and shapes of the human body or parts of the body. The plastic changes in growth discovered by Boas applied to both the laboring and non-laboring classes. Boas ascribed this plasticity to the better health care and nutrition received by the children in the United States. Bowditch had earlier found the same effect of migration on growth but in smaller samples of children, for example, that German-Americans were taller on average than Germans in Germany. What both men were able to show was that the American-born offspring of immigrant parents grew up to look more like each other in body size than they looked like their parents.

Thus, Bowditch and Boas proved statistically that the eugenicists claim that ethnicity could be determined by physical measurements was not true. Bowditch concluded that his research disputed the '. . . theory of the gradual physical degeneration of the Anglo-Saxon race in America' (Boyd, 1981, p. 469) as a result of admixture (intermarriage) taking place there. The results of Boas' research agreed with those of Bowditch, but despite their work many eugenicists and politicians called for quotas on the immigration of so-called inferior peoples into the United States. In 1911 Boas presented to the US Congress a report titled, 'Changes in the bodily form of descendants of immigrants', which explained his research and may have helped delay the imposition of limitations on immigration. Nevertheless, the American Congress eventually passed the 'Immigration Restriction Acts' in 1921 and 1924, which specifically targeted southern and eastern Europeans and Asians for migration quotas (Gould, 1981).

Boas and the environmentalists may have lost that political battle, but their work influenced three generations of anthropologists, public health workers, epidemiologists and others. Undernutrition, poor health, illiteracy, and poverty are still rampant today. Documentation of the pernicious effects of these conditions on the physical and mental growth of children continues to be carried out, and, in the tradition of Villermé, Chadwick, Bowditch, and Boas, recommendations for action to alleviate this suffering are made by researchers who have, '. . . a feeling of responsibility for the children's welfare' (Borms, 1984).

Twentieth-century research

Boas' scientific discoveries also include his research into the methodology of growth studies (1892, 1930). One of his enduring contributions is the importance of calculating growth velocities from the measurements of individuals rather than from sample means. As shown in Figure 1.6, the former method gives an accurate estimate of average growth rate, while the latter method mixes data from early, average, and late maturing children and results in a mean velocity curve that underestimates the actual velocity of growth of all children during the adolescent growth spurt. Boas also provided the concept of **tempo of growth** to understand the difference between early and late maturing individuals. The effects of maturational timing are most evident at the time of the adolescent growth spurt, but they are present at all stages of life – even during the prenatal period. Early maturers are always ahead of late maturers in skeletal development and other indicators of biological maturation. Tanner (1990, p. 75) explains that the concept of tempo of growth is a metaphor from classical music, '. . . some children play out their growth *andante*, others *allegro*, a few *lentissimo*.' With both methodological and conceptual advances such as these, and the descriptive knowledge gained since the time of Buffon, the modern era of growth measurement and analysis began.

In the first half of the twentieth century several large-scale longitudinal studies of growth were started in the United States and Europe. In addition to better quantification of amounts and rates of growth of healthy children, these studies made use of new technologies in radiology, physiology, and psychology to characterize the biological maturation of body systems. Equally important, these longitudinal studies represented a new philosophy about human growth and development. Research workers and politicians became interested in the causes of individual differences between people. Perhaps this interest was a consequence of the work of Boas on the environmental determinants of growth and physical development. It may

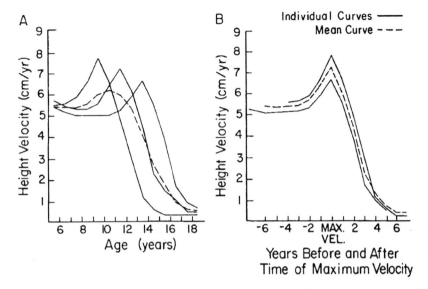

Figure 1.6. (A) Individual velocity curves of growth (solid lines) and the mean velocity curve during the adolescent growth spurt. The mean velocity curve does not represent the true velocity of growth of any individual. (B) The same curves plotted against time before and after peak height velocity (maximum velocity) of each individual. The mean curve accurately represents the average velocity of the group.

also be a reflection of an American cultural ideal, the 'rugged individual' and how to ensure continued production of the same. Lester W. Sontag (b1901–d1991), who directed the Fels Longitudinal Study in Yellow Springs, Ohio from 1929 to 1970, writes, '. . . that modern understanding of the growth, health, behavior, and effectiveness of human beings could only be understood if the nature and significance of individual characteristics of each child's physiological, biochemical, nutritional, educational, and environmental characteristics could be assessed and integrated into a total picture' (1971, p. 988). Sontag summarizes this approach as the study of 'the whole child'.

The American studies

The American longitudinal studies were supported in their early years by private donors. In particular the Rockefeller Foundation and the Laura Spellman Rockefeller Memorial Fund were major sources of financial support. Lawrence K. Frank (b1890–d1968) was an administrator at both funds and helped start and maintain virtually all the major longitudinal studies. Eventually there was also public support for longitudinal research on human growth. The United States National Research Council created the Committee on Child Development in 1923 leading to several White House Conferences on Child Health. Specialized journals such as *Child Development*, started in 1929, and *Growth*, first published in 1937, also appeared. By this time the study of normal growth was of national importance, both for its scientific and political value.

Several large, expensive long-term studies were initiated. These include the Fels Study mentioned above, the University of Iowa Child Welfare Station Study, the Harvard Growth Study, the University of Colorado Child Research Council Study, the Brush Foundation Study of Western Reserve University (Cleveland, Ohio), and several studies at the University of California, Berkeley. Sontag (1971) states that in general these research programs shared several features. First, they were interdisciplinary with physicians, psychologists, anthropologists, and others taking a global approach – they studied 'the whole child'. Second, they were fastidious in terms of the methodology of data collection. Third, they collected data as an end in itself, as they posed no research questions or scientific hypotheses about human growth to be addressed. Fourth, they planned to continue data collection for 15 years or more.

With one exception, all of the American longitudinal studies ended when either funding disappeared, the justification for data collection without purpose could not be sustained, or, most importantly, the philosophy of the 'whole child' research approach was abandoned. The only one of these studies still active today is the Fels Longitudinal Study, which began in 1929. The sample of the Fels study are healthy, well-nourished boys and girls, living in small urban communities and rural areas of southwestern Ohio. Participants in the study are measured longitudinally, ideally once a year or more often, from birth to maturity for height, weight, and a variety of other physical and psychological characteristics. Alex F. Roche (b1921–), became director of the Fels study after Sontag (Roche, 1992). Roche states that the Fels study remained viable because members of its staff were willing to use the data to answer important questions about human growth, development and health. As an example both Sontag and Roche cite the work of Stanley M. Garn (b1922–), a member of the Fels team from 1952 to 1968, on the development of fatness from birth to old age, and the relation of growth in early life to health and disease in adulthood (see Garn, 1958, 1970). These were new directions in research in the 1950s and 1960s, and have become 'normal science' today.

Other notable consequences of the American longitudinal research program is the work of Frank K. Shuttleworth (b1899–d1958) who used the data of the Harvard Growth Study to design new statistical methods to analyze longitudinal data. His first major report on this was published in 1937, and many of his methods are still used standardly today. Howard Meredith (b1903–d1985), long associated with the University of Iowa Growth study, focused much attention on population, geographic, and sex-related differences in growth. He was instrumental in applying many new mathematical techniques to the description of growth curves and did much to sort out some of the genetic and environmental determinants of human growth.

The Brush Foundation and Spelman Fund supported a longitudinal study at Western Reserve University in Cleveland, Ohio. Katherine Simmons (no dates found) and T. Wingate Todd (b1885–d1938) produced reference standards from ages three months to 13 years for height and weight from these data. They also analyzed the correlation between growth in height, weight, and sexual maturation over time. Simmons and Todd came to the important conclusion that weight was not as reliable an indicator of healthy growth as height. This is because weight is the sum of many body tissues, and also includes the amount of fat and water in the body. All of these tissues and components of the body can vary with age, sex, nutritional status, state of health, physical activity, etc., which make weight too imprecise a measure of health.

Todd also used the hand–wrist radiographs collected during the study, and an extensive series of human skeletons he collected, to publish the first major *Atlas of Skeletal Maturation* (Todd, 1937). Todd's method to assess skeletal maturation was based on first selecting a 'representative' radiograph for each age and sex. Todd and others had already found that the rate of skeletal development of girls is advanced over that of boys. For example, from all of the radiographs of eight-year-old girls, Todd found the one that had both an equal degree of development of all the bones in the hand and wrist (and there are 28 such bones, or ossification centers to consider), and also had a degree of development that was average among all the eight-year-old girls (see Figure 2.4 for an illustration of hand-wrist radiographs). This process was repeated separately for eight-year-old boys, and then repeated again at every age. These 'average' radiographs then became the standard against which all other radiographs were judged. Todd's *Atlas* was used for many years and was revised in the 1950s into the form still used today (Greulich & Pyle, 1959).

Interest in the study of body composition may have been stimulated by Simmons and Todd's work. Major studies of body weight and its division, '. . . into "fat" and "lean" components . . .' (Baumgartner, 1997) began in earnest in the 1940s with the work of Albert R. Behnke. In the 1950s Ansel

Keys and Joseph Brozek expanded these studies to separate body composition into components, '. . . of water, protein, minerals, carbohydrates, or glycogen . . .' (ibid).

In 1928 the first of several longitudinal studies, The Berkeley Growth Study, was started at the University of California, Berkeley. Nancy Bayley (b1899–) was director until 1954, and she fully subscribed to the 'whole child' philosophy of these studies. Bayley was trained as a psychologist and one of her most notable contributions is the Bayley Infant Scales of Motor and Mental Development, which she created in 1940. She also received training in anthropometry from Sarah Idel Pyle (b1895–d1987) at Iowa. Bayley not only collected physical growth data (22 different measurements) but also analyzed it in novel ways. There were 61 infants recruited for the study, 31 boys and 30 girls, and 47 were followed to maturity. After three years of age, and then again after ten years and at maturity, Bayley calculated correlation coefficients between the measurements. These correlations indicate the amount of relationship between the measurements. Stated another way, the correlations can be used to estimate the degree of predictability of one measurement from another measurement, say the predictability of weight if stature is known.

Bayley was the first to find that correlations are much lower from birth to six months of age than from six months to one year of age. Correlation between measurements continues to rise until the age of two years (Bayley & Davis, 1935). Thus, it seemed that an infant's individual pattern of growth is a bit disorganized just after birth, but becomes better organized during the first two years of life. This discovery had important practical implications for the prediction of adult size at early ages. In 1952 Bayley and S. R. Pinneau published the first tables for predicting adult height using height and skeletal age measured at earlier ages. These prediction tables are still used by physicians and researchers.

In addition to work on longitudinal patterns of growth in external body size, several studies of the growth of internal organs and chemical composition by Richard E. Scammon (b1883–d1952) and Edith Boyd (b1895–d1977) are of importance. Figure 1.7 reproduces one of Scammon's classic illustrations of growth of the internal organs, and the body as a whole, during postnatal life. By necessity, this work was based on cross-sectional samples of cadavers, but Scammon and Boyd took a longitudinal approach when presenting their data. Their studies of the internal milieu of the human body started with the embryo and continued until old age. With this research the scientific and medical world gained its first accurate understanding that in terms of chemical composition, 'The organic body is not a closed system. It receives and liberates materials constantly' (from

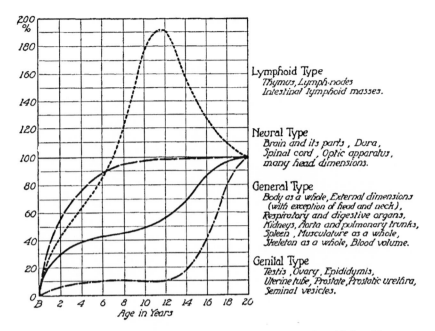

Figure 1.7. Four major types of postnatal growth curves from birth to 20 years, expressed as the percent of the total increment of growth (Scammon, 1930).

Scammon's 'Introduction' to Boyd, 1980). These studies of internal organs showed that the parts of the body do not develop or grow according to one grand pattern, rather there are several different curves of growth for different types of tissues.

The European studies

European longitudinal growth research had a slower start than in the United States and was probably delayed by the World Wars. In the 1920s there were two notable studies with a longitudinal component. The first was the Aberdeen, Scotland study conducted by Alexander Low (b1868–d1950). Low personally measured 21 dimensions of 900 newborn babies in 1923. He then remeasured 65 of the boys and 59 of the girls annually until they were 5 years old. The data were never properly analyzed until Tanner *et al.* (1956) produced a study of correlations between measurements taken at different ages. As part of their analysis, the authors also searched for the original participants of the Aberdeen study and found 42 of the men and 38 of the women. The adults were measured and correlations between the birth to age five year data and the adult data were calculated. Tanner and colleagues generally confirmed similar work by Nancy Bayley (discussed above). However, the Tanner group was the first to try to explain the pattern of increasing correlations from birth onwards. They emphasized that late in pregnancy, the fetus may be 'deflected very considerably' from its growth trajectory. After birth the infant slowly recovers toward that trajectory, 'somewhat after the manner of a growing animal who has passed through a period of not too severe malnutrition' (quoted material from Tanner, 1981, p. 347). A discussion of this perinatal 'malnutrition' effect is presented in the section on birth, and birth size, in Chapter 2.

The second longitudinal study of the 1920s was conducted by Rachel Mary Fleming (no dates known). She recorded the annual measurements of stature and head dimensions on children and youth three to 18 years old in England and Wales. She published the data in 1933 and her analysis included a longitudinal curve of growth in stature, head length and head breadth for each participant in the study – there were more than 100 participants. With this individual approach Fleming noted that some children grew throughout the teenage years, while others came to an early stop. Some children grew steadily during puberty and others experienced a rapid burst of growth followed by an end to growth. Some children were tall until puberty but then ceased growing and remained shorter than average as young adults. Today these patterns of growth are well know and reflect normal individual differences in the 'mode and tempo' (as Boas, 1930 called them) of growth.

Fleming also noted 40 individuals with unusual stature graphs. These children, of different ages, showed a longitudinal pattern of normal growth, a period of growth arrest, and then a resumption of normal growth. None of these children had been identified as growth-delayed during the cross-sectional medical examinations by the school physician. All attended a small village school and lived so far away from the school that they had to bring lunches to school, whereas other children attending the school, but who lived in the village, went home for dinner. The bag lunches were usually bread and jam, while the home dinners, the main meal of the day, were usually a hot lunch of meat and vegetables. The typical evening meals for all the children were bread, jam, and tea. Thus, only on weekends did the bag lunch children receive a full meal at dinner. When the school instituted a lunch program of meat and vegetable stew the stunted children returned to normal growth. The cause of the growth stunting seemed to be a combination of inadequate food intake, plus the energy expended in the commute to school.

Fleming's serendipitous finding of alterations in the pattern of growth associated with a change in school policy was to be repeated by another British researcher, Elsie M. Widdowson (b1906–), in the aftermath of World War II. Widdowson's research, discussed in detail in Chapter 6, found that a negative emotional environment during school mealtimes can suppress physical growth. The point to make here is that much scientific discovery is unexpected and found only when the individual researcher is willing to take the risk of exploring new research territory, trying new strategies and methods, and thinking in new ways about how things work. Rachel Fleming is not well known today, but she exemplified these characteristics of noteworthy research. She played a part in changing social policy, especially the introduction of free school lunch programs for undernourished children and the need to identify such children.

Fleming's work formed a basis for the discovery that children's nutritional requirements are divided between maintenance and repair of the body, work, and growth. When food intake is inadequate to meet all of the requirements, it is usually growth that suffers first. Later in life, growth-stunted children become adults with reduced performance in both physical and mental work capacity. In retrospect this seems obvious, but it took decades of research, tens of millions of dollars, and much intellectual debate to reach this conclusion (see Chapters 5 and 6). With the advantage of her longitudinal data, Fleming also performed statistical analyses to reveal several features of the human adolescent growth spurt; '. . . The results show that the fast-growing period for girls starts sooner, finishes earlier, and is less intensive than for boys. The sexes are equal up to 11 years of age, between 11–14 girls are taller than boys, but from 14 onwards the boys become steadily taller than the girls' (Boyd, 1980, p. 374). These well-known features of human growth at adolescence were new discoveries to late nineteenth and early twentieth century growth researchers. Many were puzzled by the growth spurt, some denied it really existed, and others had their Victorian sensibilities offended by the transitory 'ascendancy' in stature (as Tobias, 1970 refers to it) of girls over boys during adolescence.

During World War II the Oxford Child Health Survey was started by John Ryle (b1899–d1950). A total of 470 infants and children were recruited and measured between the ages of one month to five years of age. The data include anthropometric measurements, illness histories, social changes in the families of the participants, and radiographs of the hand and wrist, knee, and chest. From an historical perspective, the X-ray data are most important. Roy Acheson (b1921–) was a member of the team analyzing the data. To make use of the radiographs Acheson developed a new system to rate skeletal maturation. He was dissatisfied with the Todd method of assessing a bone age from the entire hand–wrist or a group of bones. Acheson wanted a simpler system and found that individual bones of the hand–wrist could be used to determine maturation. Acheson determined the number of identifiable stages of appearance of each bone as it matured. Each stage was given a number, 1, 2, 3, etc. A maturity score for any hand-wrist radiograph was easily calculated by giving each bone its stage number and adding up all of the numbers. A higher numerical total meant a higher skeletal maturity score. The score could be treated just like any other measurement, such as height in centimeters, for statistical analysis. This simplified procedure for skeletal maturity was refined and standardized by Tanner and colleagues (1983b) into a widely used clinical method.

Soon after World War II, longitudinal research in Europe began in earnest. The British Harpenden study started in 1949 and continued until 1971 (Tanner, 1981). Other longitudinal studies were started in Paris, Zurich, Stockholm, London, Brussels. These studies were coordinated by the International Children's Center (ICC) in France. There was an ICC sponsored study in the United States as well, the Louisville (Kentucky) Study. This was, and is, a longitudinal study of twins (results are discussed in Chapter 7). All of the ICC studies followed a standardized procedure for anthropometric measurement and data recording. This was a major advance as earlier studies often used methods of measurement and recording that made comparisons between data sets difficult. The focus of these ICC studies, and the Harpenden Study as well, was medical, and the research personnel were mostly pediatricians and endocrinologists. James M. Tanner (b1920–), a member of the ICC board, states that in contrast to the 'whole child' approach of the American studies, the European longitudinal research was concerned more with growth pathologies. Important medical discoveries were made by these studies, but there were also advances in understanding the basic biology of human growth. It is commonplace in biology that we come to understand how a normal system operates by studying what happens when things go wrong. Tanner advanced the cause of growth research in Europe by promoting many of these studies and by publishing several major works. His book *Growth at Adolescence* (1955 and 1962) summarized the state of knowledge of normal and pathological growth derived from all of the research of the first half of the twentieth century. Tanner and colleagues, especially his long-time research partner Reginald Whitehouse (b1911–d1987), also published some of the most widely used modern references for growth in height and weight, the stages of sexual maturation, and for skeletal maturation (Tanner & Whitehouse, 1975; Tanner *et al.*, 1983b).

Longitudinal studies in the developing world

The ICC also sponsored longitudinal studies in Africa; the Dakar and Kampala Studies. By the early 1960s some results of these studies were published, and the findings were puzzling. In some respects African infants and children were advanced over Europeans, such as in skeletal development. In other measures, such as height, weight, and fatness, Africans were smaller than Europeans. Moreover, the differences between populations increased with age. Some researchers were content to ascribe these growth differences to genetics, but other scholars looked more to the ecology for human growth of Africa and its history. Poverty is the word often chosen to best describe the human ecology of Africa. The history of Colonialism was cited by some workers to explain the origins of much of this poverty. There was little hope for extensive longitudinal research in Africa to settle the question of 'genes versus environment'. By the 1960s, Europe was losing its colonies in Africa and there was little popular sentiment in Europe, or North America, to invest the resources needed to conduct research.

In Central America research was underway to examine the environmental and genetic determinants of human growth. Robert MacVean (b1917–), an American who settled in Guatemala after World War II, helped to establish the American School of Guatemala as a laboratory school (one that conducts research and training in addition to teaching) in 1948. A longitudinal study of child, juvenile, and adolescent development was chosen as the basic research project of the laboratory school program. Data collection started in 1953 and until 1963 was limited to all of the students at the American School, an expensive private school. Beginning in 1963 other schools, representing middle and low socioeconomic status (SES) families in Guatemala City, were added to the study. All of the students attending each school were measured once each year for height, weight, and hand grip strength, and eruption of the permanent teeth; a hand–wrist X-ray was taken. Several tests of cognitive development and school performance (e.g., IQ and reading tests) were also administered. As the measurements were taken from the entire school population of each year, this was not a pure longitudinal study. Children might have left or entered the school at any time, although many of the students did continue through the elementary grades and on to complete secondary school. This study, now called 'The Longitudinal Study of the Growth and Development of the Guatemalan School Child', became one of the first large-scale mixed-longitudinal investigations of human growth.

Actually, the mixed-longitudinal study is considered a very powerful statistical design for growth research. In this design, subjects of different starting ages are measured for several years. The overlap in age means that data covering much, or all, of the growing years may be collected in just a few years rather than two decades. In addition, each age group serves as a check against other age groups to ascertain if the data collected are representative, that is, that the particular individuals of any age group are typical in terms of the measurements taken. In a pure longitudinal study, peculiarities of the sample of children, such as a disease or social stigma, may invalidate application of any findings to the population at large.

Masses of data were collected, but little analysis was conducted until Francis E. Johnston (b1931–), an American anthropologist, became a consultant to the study in the late 1960s. From 1974 to 1976 the present author (Barry Bogin, b1950–), one of Johnston's students, did his doctoral dissertation research in Guatemala and became a consultant to the longitudinal study as well. I suggested adding measures of body composition (triceps and subscapular skinfolds and arm circumference) to better estimate nutritional status, and these have been taken on all participants since 1976. I also suggested adding a school with a **Maya** population. Maya are the majority ethnic group of Guatemala. Maya ethnicity is characterized by language (there are more than 20 Maya languages in Guatemala), traditional clothing styles, religious practices, and rules for cultural behavior (such as kinship and family organization). In many ways the present-day Maya of Guatemala are the cultural descendants of the Classic Period Maya who constructed the ceremonial centers of pre-contact Mesoamerica (places such as Tikal and Palenque). Prior to 1976, the schools participating in the longitudinal study were comprised primarily of **Ladinos** (attending all of the schools) or Europeans/North Americans (attending the American School). Ladinos are the second largest ethnic group in Guatemala. In a sociocultural sense, Ladinos are descendants of the Spanish conquistadors who ruled Guatemala for the past 400 years. In 1979 a Maya school, comprised mostly of children from very low SES families, was added to the study.

The important results of this study are not discussed here, but are incorporated into many discussions of human growth and development throughout this book. From an historical perspective it is important to state now that the American School study, and its enlarged successor, was the first major research program of its type in any developing nation. Moreover, the study is still operating today, making it one of the longest-lived projects of its type, a resource of unparalleled data which remains to be fully exploited.

Guatemala was also the site of the second major longitudinal study of human development in the developing world. This was the research program of the Institute of Nutrition of Central America and Panama (IN-CAP), and is sometimes called the 'Four Village Study' because participants were recruited from two small (about 500 person) and two larger villages (about 900 persons) in rural Guatemala (Martorell *et al.*, 1975). Except for population size, the large and small villages were fairly homogeneous in terms of ethnicity (all people were Ladinos) and other social and economic factors. This was a study of growth and development from birth to age seven years. Participants were recruited when their mothers were pregnant, at their birth, or soon thereafter. This was also an experimental study. The participants in one large and one small village received a dietary supplement called *atole*, the local term for a gruel usually made with corn meal. The participants in the other villages received a low calorie *fresco*, the name for a cool refreshing drink. When the study started in 1969, the *atole* supplement contained protein, carbohydrates, vitamins and minerals. The *fresco* contained only carbohydrates, water, and artificial flavor. From 1971 to 1977, when the study ended, it was decided to increase the vitamin and mineral content of the *atole* and add equal amounts of these vitamins and minerals to the *fresco*, so that the analysis could be confined to the energy containing nutrients, the protein and carbohydrates.

The major research question of the INCAP study was to what extent a nutritional supplement, especially of protein, could enhance the physical and mental development of children. The infants and children living in the four villages suffered from both kwashiorkor, an acute type of undernutrition that can kill, and chronic mild-to-moderate undernutrition, not severe enough to kill a person directly. In the 1960s it was hypothesized that chronic mild-to-moderate malnutrition could retard growth and development and reduce the body's ability to fight off diseases, which could kill. There were also hypotheses that poor, rural people of less-developed countries like Guatemala had 'adapted' in some way to a limited food supply, and would not benefit from additional food. However, the INCAP study proved that lack of adequate nutrition is a major factor retarding the physical growth and mental development of Guatemalan children. The infants and children supplemented with *atole* grew significantly taller and heavier than the children receiving the *fresco*. The *atole*-supplemented children also performed better on cognitive tests and were more likely to enter school. By extension, these results could also apply to children living in poverty in Africa, Asia, and the developed nations of Europe, North America, Australia, and Japan. To test this, several studies were started, especially in Asian countries, that followed in the steps of the INCAP project (specific findings of the INCAP study and the others will be presented in later chapters).

Other basic research related to growth

During the twentieth century, research into genetics of growth became possible. The rediscovery of Mendelian principles in the year 1900 and the characterization of the DNA molecule (deoxyribonucleic acid) in 1952, are two of the major historical events influencing growth research. An early emphasis on 'racial genetics', *à la* Galton, gave way to modern population genetics by the early 1950s. Methods to study the influence of genes at the individual and family level were also developed, including studies of monozygotic and dizygotic twins and studies of family pedigrees (see Chapter 7 for details). By the 1960s research on the effect of chromosomal variations and abnormalities had appeared. To date, the only evidence that specific genes exist for growth in size, body proportions, body composition, and rate of maturation is derived from studies of sex chromosomes. Garn & Rohmann (1962) proposed that genes on the X chromosome control some aspects of the development of skeletal and dental tissue. J. German and colleagues (1973) proposed that genes on the Y chromosome stimulate the growth of skeletal tissue to produce the greater average stature, arm length, and biacromial breadth (shoulder width) of men versus women. The specific sequences of DNA responsible for these effects are still not known.

An important corollary of these genetic studies was the discovery of new ways in which the environment can produce effects on growth that seem to be hereditary. In 1964, both in Poland (Malinowski & Wolanski, 1985) and the United States (Bloom, 1964) researchers found that the heritability of some phenotypic characteristics (such as stature) is higher in groups of parents and children living under low socioeconomic status (SES) conditions, but that the heritability is lower when children live under more favorable SES conditions than did their parents. Benjamin S. Bloom (b1913–) formalized this observation into what he called the 'powerful environment hypothesis'. By this Bloom meant that when succeeding generations of people grow up under the same, or similar, environment of extreme deprivation or privilege, each generation will develop similar physical and cognitive characteristics. A high correlation between parent and offspring in any measurable trait is often taken as evidence of a genetic effect. When living under powerful environments, however, the traits of both generations may be altered in similar ways without any genetic contribution. Polish researchers were able to show that when the negative influence of the environment is ameliorated the correlation between generations declines, sometimes effectively to the point where there is no correlation (reviewed by Wolanski, 1967). In many ways the powerful environment hypothesis complimented and extended the work of Boas and other anti-eugenics researchers.

Technological developments

The technical basis for all of the research on growth lies in the precision of the instruments used to measure lengths, weights, circumferences, and other dimensions and the accuracy and reliability of the methods of measurement. Growth research technology and methods, beginning with the invention of the anthropometer (a device to measure stature) by Johann S. Elsholtz (b1623–d1688) in 1654, are reviewed by Noel Cameron (1984). Cameron also reviews the development of skinfold calipers, radiography, photogammetry, and data analysis. These devices are mentioned here briefly, and then discussed in greater detail later in this chapter and other chapters where relevant.

The anthropometer was developed to measure lengths of the body, especially stature. Eventually, it was modified to measure lengths of body segments (arms, legs, etc.) and then was used to measure body breadths, such as the **biacromial** ('shoulder') and **bicristal** ('hip') breadths. Skinfold caliper measurements, first used by Ludwig W. Kotelman (b1839–d1908) in Germany in 1879, have become the most widely used method to evaluate subcutaneous fat, and its relation to growth, body composition, health, and behavior. Modern anthropometers and calipers are designed to produce both accurate and reliable measurements, which are needed especially in longitudinal research where the same individual is measured repeatedly.

The discovery of X-rays in 1895 was soon followed by applications to document skeletal and dental development. Several works depicting normal skeletal development appeared starting in 1904 and culminating in Todd's 1937 *Atlas of Skeletal Maturation* (discussed above). Todd's atlas was revised by William Walter Greulich (b1901–d1987) and Sarah Idel Pyle (b1895–d1987) and published in 1950 as the *Radiographic Atlas of Skeletal Development of the Hand and Wrist*, which is still widely used for basic and clinical research (Greulich & Pyle, 1959). Longitudinally collected radiographs from the Oxford Child Health Survey (see above, Acheson, 1954), the Fels Study, and the Harpenden Study were used to create newer methods for the assessment of skeletal maturation (Roche *et al.*, 1975a; Tanner *et al.*, 1983b). These newer methods are more precise in the prediction of adult stature, which are a concern of parents, pediatricians, the military, and the ballet – ballerinas should not be too short or too tall.

Photogammetry is a method of growth evaluation based upon '. . . photographs of a subject posed in a particular position to facilitate either

visual appraisal of the body . . . or detailed measurement of body parts . . .'
(Cameron, 1984, p.142). One of the primary areas of use of photogammetry
is **somatotyping**, which as developed by William H. Sheldon (b1898–d1977)
relates human morphology to physical and psychological behavior (Sheldon, 1940). Progress in imaging, such as PET scans (positron emission
tomography) and MRI scanning (magnetic resonance imaging) allow for
the non-invasive examination of internal soft tissue and for the examination of internal structures of fossil human ancestors without damaging the
specimen. There are also devices under development that will scan the
human body and automatically derive many standard anthropometric
measurements (Jones, 1995). These devices, along with advances in data
collection strategies, statistical processing, and computerization have, and
will continue to, revolutionize growth research.

Endocrines and growth control

In the early 1960s, Tanner stated, 'There exists at present no entirely
convincing and coherent theory of endocrinology of adolescence. . .' (1962,
p. 176). It may be added that an understanding of the endocrine regulation
of growth at all other stages of life, prenatal and postnatal, was equally
poor. In 1974 Melvin Grumbach and colleagues published *Control of the
Onset of Puberty*, which contains several 'coherent and convincing' theories
of the endocrinology of growth. Today, models and theories of hormonal
control exist for all other stages of growth (Box 1.1 describes the meaning
and use of models in the study of growth). The rapid pace of research in this
field is due both to technological advances in the assay of hormonal factors
and advances in understanding how hormones exert their influences on
human growth and development. The history of human growth hormone
(hGH) and the insulin-like growth factors (IGFs) are examples. The existence of hGH was demonstrated in 1944, and the first IGF in 1957. The
process by which these hormones work was largely speculative until several lines of research were combined into the 'duel-effector' model (Green *et
al.*, 1985 – Chapter 7 details this model). Anthropological interest has been
stimulated by the discovery of human population differences in the presence of these hormones. For example, African pygmy populations seem to
be deficient in one of the IGFs (Merimee *et al.*, 1981), or its receptor
(Baumann *et al.*, 1989).

There are at least six hypothalamic hormones, eight pituitary hormones,
a dozen or more hormones secreted from other endocrine glands (e.g.,
thyroid, adrenal, ovary and testis) and a host of growth factors produced
throughout the body that regulate human growth. Each of these has a

Box 1.1.   Models in biology: description and explanation

Models are representations that display the pattern, mode of structure, or formation of an object, a process, or an organism. A model may also serve as a standard for comparison, between hypotheses that test human understanding of how some physical or biological phenomenon operates and the actual nature of that phenomenon. For example, models are often employed to study aspects of human growth, such as the adolescent growth spurt, and aspects of maturation, such as the onset of puberty. No model can represent a complete understanding of the regulation of human growth and maturation, for that would require inclusion of all of the many genetic, endocrine, and environmental factors, and their interactions, that influence the developmental process. However, models are easier to construct, to understand, and to test than total reality, since the models usually represent only a small portion of total amount of this detail.

Two types of models are used by growth researchers. The first type describes a result. For instance, a series of longitudinal measurements of the height of a child may be compared to a mathematical formula that fits a curve to the growth data. The mathematical parameters of the curve are chosen to model the actual increases in height of the child over time. The fit of the curve may be quite precise, that is, the model describes growth very well. However, this type of model does not explain why increases in height occur or when changes in the rate of growth in height are likely to take place.

A second type of model attempt to describe a result as well as explain some of the determinants of the observations. An example is D'arcy Thompson's model for the growth of the Nautilus, presented in Figure I.3. The equiangular spiral, the mathematical function that describes the growth of the *Nautilus* shell, implies that the proliferation of shell material occurs at a constant rate; proportional to the amount of tissue already produced. Thus, volumetric growth rate constantly accelerates, producing ever-larger chambers in the shell. Unlike a purely descriptive representation, D'arcy Thompson's model predicts, with great precision, the size and volume of the next chamber to develop.

A predictive model of growth has both mathematical and biological meaning, and is preferred to descriptive models. Applied to human growth, such a model might attempt to describe and predict growth in height from birth to maturity. For instance, if it is assumed that a genetic program aims the growth of a child toward an 'ideal' size, as might be the case with strong genetic selection for size, then the rate of growth should be proportional to the difference between present size and ideal size. Departures from the predicted rate of growth, such as the mid-growth spurt during childhood and the adolescent growth spurt, may be explained by additional hypotheses about the regulation of the growth. In this way a model building process is encouraged, which leads to the testing of hypotheses against observations of growth and maturation and the eventual formulation of a theory of human development.

history of discovery and understanding similar to hGH and the IGFs (reviewed in Chapter 7). Initially, most of the information on hormone regulation was derived from studies of children with endocrine pathologies. Newer, non-invasive assay methods (e.g., detecting hormones in saliva rather than blood or tissue) now permit the study of large samples of normal children. This should lead to a more comprehensive understanding of endocrinology and growth in the next decade or so.

Growth theory

The public health work of the nineteenth century and the large-scale longitudinal studies of the twentieth century provided a wealth of growth data. Advances in fields such as molecular biology, endocrinology, nutrition, and the social sciences allowed scientists and physicians alike to turn the study of human growth into a research and medical specialty. However, all of these data and technical advances were primarily descriptive in nature. They told us how children grew and how their growth was affected by heredity and the environment, but they could not tell us why. To understand the 'why' of growth and development a theoretical approach was needed. For example, the cell theory of Matthias Jakob Schleiden (b1804–d1881) and Theodor Schwann (b1810–d1882), proposed in 1838, made it possible to understand the earlier work of von Baer, who had described the different germ layers of the developing embryo. These layers were distinct types of cells that gave rise to the different tissues and organs of the body. With the publication of Charles Darwin's (b1809–d1882) *Origin of Species* in 1859 biological research became a modern theoretical science. The scientific method of experimentation and hypothesis testing was increasingly applied to biological questions, including the control of growth and development.

One example is the work of Ernst Haeckel (b1834–d1919), who proposed the theory of recapitulation during embryological development – that is, the development of the individual organism follows the evolutionary history of life (see Haeckel, 1874). In contrast, von Baer in his 1827 study *De Ovi Mammalium at Hominis*, emphasized that during development embryos from different classes of animals move away from common forms, thus there is no recapitulation. This dispute stimulated experimentation and refinements to both hypotheses. Eventually, recapitulation was discredited, but Haeckel's work gave rise to interest in growth theory.

Many other scientists contributed to growth theory during the late nineteenth and early twentieth centuries, but the work of one person stands out more than any other. D'Arcy Wentworth Thompson's (b1860–d1948) book, *On Growth and Form* (1917, 1942, 1992), is a *tour de force* combining the classical approaches of natural philosophy and geometry with modern biology and mathematics to understand the growth, form, and evolution of plants and animals. Thompson visualized growth as a movement through time. Scientists from Buffon to Boas had studied the velocity of growth; Thompson made it clear that growth velocities in stature or weight were only special cases of a more general biological process. The development of flower parts in plants or the evolution of antler size in mammals were also examples of growth as a movement through time. Thompson developed the concept and methodology of using **transformational grids** to quantify the process of growth during the lifetime of an individual or during the evolutionary history of a species (see Figure I.2). Until the advent of high-speed computers, which are needed to carry out the mathematical procedure of the method, the transformational method was difficult and slow to apply and, hence, little used by other biologists (Bookstein, 1978). Even so, *On Growth and Form* provided an intellectual validity to growth and development research and stimulated succeeding generations of growth researchers to think about growth in new ways (e.g., see Huxley, 1932; Tanner, 1963; Thom, 1983; Bogin, 1980, 1991).

Conclusion

This highlights some advances in the study of human growth. By the 1940s many of the basic principles of physical growth and development were known. Since then, researchers have been making progress, often slowly, in unraveling the underlying biology of physical growth. One fact is clear – all normal, healthy, and well nourished children follow the same basic pattern of growth from birth to maturity. Research also shows that a common pattern of human development and growth occurs during the prenatal period as well. The next chapter describes these major features of human growth and development from conception to death.

2 *Basic principles of human growth*

Stages in the life cycle

Many of the basic principles of human growth and development are best presented in terms of the events that take place during the life cycle. One of the many possible ordering of events is given in Table 2.1, in which growth periods are divided into developmentally functional stages. For convenience, the life cycle may be said to begin with fertilization and then proceed through prenatal growth and development, birth, postnatal growth and development, maturity, senescence, and death. In truth, however, the course of life is cyclical – birth, the onset of sexual maturation in the adolescent boy or girl, and even death are each fundamental attributes of the cycle of life. In the fetus, child, and adult, old cells die and degrade so that their molecular constituents may be recycled into new cells formed by mitosis. At the population level, people grow, mature, age, and die even as new individuals are conceived and born. Declaring one moment, such as fertilization, to be a beginning to life is arbitrary in a continuous cycle that passes through many stages, both in the individual person and in generation after generation.

In the following sections the stages of prenatal and postnatal life are described. The timing of growth events is presented, usually in the form of mean, median, or modal ages. The reader should bear in mind that these ages indicate the central tendency and not the normal range of variation that occurs naturally in the timing of many growth events. Research may find, for example, that the median age at menarche (first menstruation) is 12.6 years in a sample of girls. The actual age at menarche for individual girls in the sample may range from 8.0 to 15.0 years, and both the earliest and latest ages represent perfectly normal variation. When the range of variation in the timing of growth events is important in terms of the basic principles of human growth it will be given in this chapter. Otherwise, individual and population variation in growth are discussed in more detail in Chapter 5.

Table 2.1. *Stages in the human life cycle*

Stage	Growth events/Duration (approximate or average)
Prenatal Life	
Fertilization	
First trimester	Fertilization to twelfth week: Embryogenesis
Second trimester	Fourth through sixth lunar month: Rapid growth in length
Third trimester	Seventh lunar month to birth: Rapid growth in weight and organ maturation
Birth	
Postnatal Life	
Neonatal period	Birth to 28 days: Extrauterine adaptation, most rapid rate of postnatal growth and maturation
Infancy	Second month to end of lactation, usually by age 36 months: Rapid growth velocity with steep deceleration in velocity with time, feeding by lactation, deciduous tooth eruption, many developmental milestones in physiology, behavior, and cognition
Childhood	Third to seventh year: Moderate growth rate, dependency for feeding, mid-growth spurt, eruption of first permanent molar and incisor, cessation of brain growth by end of stage
Juvenile	Ages seven to ten for girls, or 12 for boys: Slower growth rate, capable of self-feeding, cognitive transition leading to learning of economic and social skills
Puberty	Occurs at end of juvenile stage and is an event of short duration (days or a few weeks): Reactivation of central nervous system mechanism for sexual development, dramatic increase in secretion of sex hormones
Adolescence	Five to eight years after the onset of puberty: Adolescent growth spurt in height and weight, permanent tooth eruption virtually complete, development of secondary sexual characteristics, socio-sexual maturation, intensification of interest and practice in adult social, economic, and sexual activities
Adulthood	
Prime and transition	From 20 years of age to end of child-bearing years: homeostasis in physiology, behavior, and cognition, menopause for women by age 50 years
Old age and senescence	From end of child-bearing years to death: decline in the function of many body tissues or systems

Prenatal stages

The course of pregnancy may be divided into three-month periods, or **trimesters**. During the first trimester, one of the major events is the multiplication of a single cell, the fertilized ovum, into tens of thousands of new cells. At first, cell division may produce exact copies of the original parent cell, but within hours of the first division, distinct groups of cells begin to form. Variations in the rate of cell division may be seen in the separate groups. Eventually these groups of cells form different kinds of tissue (the 'germ layers' of endoderm, mesoderm, and ectoderm) that will constitute the growing embryo. Thus growth (an increase in cell number or cell size) and development (cellular differentiation) begin soon after conception. After the initial embryonic tissues are formed, the first trimester is taken up with **organogenesis**, the formation of organs and physiological systems of the body. During the first few weeks after conception the embryo has an external appearance that is 'mammalian', that is, many mammal embryos share these same external features. By the eighth week the embryo has many phenotypic characteristics that may be recognized as human.

Though the human body is composed of dozens of different kinds of tissues and organs, their generation and growth during prenatal life, and postnatal life as well, takes place through a few ubiquitous processes. Goss (1964) described two types of cellular growth, **hyperplasia** and **hypertrophy**. Hyperplasia involves cell division by mitosis. For example, epidermal cells of the skin form by the mitotic division of **germinative cells** (undifferentiated cells) in the deep layers of the skin. Hypertrophic growth involves the enlargement of already existing cells, as in the case of adipose cells growing by incorporating more lipid (fat) within their cell membranes. Goss (1964) also described three strategies of growth employed by different tissues: renewal, expansion, and stasis. **Renewing tissues** include blood cells, gametes (sperm and egg cells), and the epidermis. Mature cells of renewable tissue are incapable of mitosis and have relatively short lives; for example, red blood cells (erythrocytes) survive in circulation for about six months. The supply of red blood cells is constantly renewed by a two-step process: first, the mitotic division of pre-erythrocyte cells of the bone marrow, and second by the differentiation of some of these into mature red blood cells. Goss pointed out that the undifferentiated cells are sequestered in a growth zone that is 'spatially distinct from the differentiated compartment' (1986, p. 5). This two-step process and the growth zone for undifferentiated cells are common physiological features of many types of renewing tissues.

Expanding tissues include the liver, kidney and the endocrine glands, the cells of which retain their mitotic potential even in the differentiated state. In the liver, for example, there is no special germinative layer or compartment, and most liver cells are capable of hyperplasia to replace other cells lost by damage. Relatively large portions of a diseased liver may be excised surgically, as a proliferation of new cells will eventually occur to meet physiological demands. Perhaps it is no coincidence that in Greek mythology the gods punished Prometheus, who gave the secret of fire to the mortals, by binding him to a rock and sending an eagle to peck out his liver each day for all eternity.

Mitotically **static tissues**, such as nerve cells and striated muscle, are incapable of growth by hyperplasia once they have differentiated from precursor germinative cells. Because the reserve of the germ cells is limited and usually depleted early in life, the pool of static tissues cannot be renewed if damaged or destroyed. However, unlike renewable tissues, which have short lives, static tissues usually live as long as the person survives. Static tissues can grow by hypertrophy, for example, individual nerve cells may grow to relatively great lengths during normal development, and, if not fatally damaged by accidents or surgery, regrow new interconnections. The physique of body-building enthusiasts results from hypertrophy of existing muscle cells and not from the formation of new muscle cells. These tissues are static in the sense that they cannot undergo mitosis, but they are reconstituted by a turnover of material at the subcellular and molecular levels. Studies of dietary intake and excretion of nitrogen provide an estimate of protein metabolism in the body. This is because most of the body's store of nitrogen is in protein molecules. As muscle tissue forms the largest mass of protein in the body, measures of nitrogen balance may be used as indicators of muscle turnover and renewal. Data published by Cheek (1968) and Young *et al.*, (1975) indicate that in young adult men, about 2–3 percent of the muscle mass is renewed each day. In infancy, when new muscle tissue is forming by hyperplasia, the rate of protein renewal is about 6–9 percent per day. The magnitude of this metabolic renewal may be appreciated by the fact that much of the basal metabolic rate of the body (which may be measured by the heat that the body produces when at complete rest) is due to protein turnover (Waterlow & Payne, 1975). A similar turnover of cellular material occurs in other 'static' tissues, such as nerve cells in the body and in the brain and in expanding tissues. The renewing tissues may also undergo a turnover of protein molecules during their relatively short lives, but the major metabolic dynamic of these tissues is their mitotic proliferation and eventual death.

Thus, the biological substrate of the individual is not permanent, and from embryonic life through adulthood the human body is in a constant state of decomposition and reorganization. Tanner (1978, p. 26) observed that, 'This dynamic state enables us to adapt to a continuously changing environment, which presents now an excess of one type of food, now an excess of another; which demands different levels of activity at different times; and which is apt to damage the organism. But we pay in terms of the energy we must take in to keep the turnover running . . . Enough food must be taken in to provide this energy, or the organism begins to break up'. To be sure, different tissues turnover at different rates, so that in muscle cells nitrogen is replaced in a few days to a few weeks, while the calcium in bone cells is replaced over a period of months. During the years and decades of life, sufficient turnover and renewal of the molecular constituents of the body's cells must take place to rejuvenate the entire human being.

The metabolic dynamic of the human organism is most active during the first trimester of prenatal life. The multiplication of millions of cells from the fertilized ovum, and the differentiation of these cells into hundreds of different body parts, makes this earliest period of life highly susceptible to growth pathology caused by either the inheritance of genetic mutations or exposure to harmful environmental agents that disrupt the normal course of development (e.g., certain drugs, malnutrition, disease, etc. that the mother may experience). Because of this, it is estimated that about ten percent of human fertilizations fail to implant in the wall of the uterus, and of those that do so about 50 percent are spontaneously aborted (Bierman *et al.*, 1965; Werner *et al.*, 1971). Most of these spontaneous abortions occur so early in pregnancy that the mother, and father, are usually unaware that they have happened.

By the start of the second trimester of pregnancy the differentiation of cells into tissues and organs is complete and the embryo is now a fetus. During the first trimester, the embryo grows slowly in length, often measured as crown–rump length (CRL). At 18 days post-conception the embryo has a CRL of about 1.0–1.5 mm and by 12 weeks post-conception the CRL is about 53 mm (Meire, 1986). The rate of growth in length increases during the second trimester. By the fourth month crown–rump length is about 205 mm, by the fifth month 254 mm, and by the sixth month between 356 and 381 mm, which is about 70 percent of average birth length (Timiras, 1972). Increases in weight during this same period are also rapid. At eight weeks the embryo weighs 2.0 to 2.7 g (O'Rahilly & Muller, 1986) and at six months the fetus weighs 700 g. This is about 20 percent of birth weight (Timiras, 1972), so relative to size at birth the growth in length during the second trimester exceeds the growth in weight. During the third trimester of pregnancy growth in weight takes place at a relatively faster rate. During the last trimester, the development and maturation of several physiological systems, such as the circulatory, respiratory, and digestive systems, also occurs, preparing the fetus for the transition to extra-uterine life following birth.

Birth

Birth is a critical transition between life *in utero* and life independent of the support systems provided by the uterine environment. The **neonate** moves

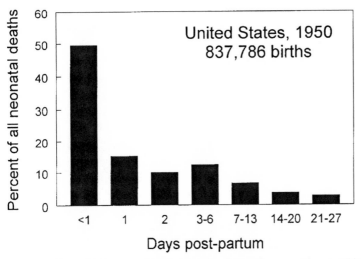

Figure 2.1. Percent of deaths occurring during the neonatal period (birth to day 28). Data from the United States, all registered births for 1950 (Shapiro & Unger, 1965). These data are presented in lieu of more recent data as the technology for extraordinary neonatal medical care in existence today reduces neonatal deaths.

from a fluid to a gaseous environment, from a nearly constant external temperature to one with potentially great volatility. The newborn is also removed from the supply of oxygen and nutrients that has been provided by the mother's blood passed through the placenta, and which also handles the elimination of fetal waste products, and must now rely on his or her own systems for digestion, respiration, and elimination. The difficulty of the birth transition may be seen in relation to the percentage of deaths by age during the **neonatal period** (the first 28 days postpartum), as shown in Figure 2.1. Of course, most of these deaths were not due to the birth process itself; rather the leading factor associated with neonatal death is inadequate growth and development during the prenatal period. Excessive prenatal growth also carries higher risk for neonatal death, but the discussion here is confined to inadequate growth as this is correlated with leading cause of death in all human populations. The most common indicator of inadequate prenatal growth is **low birth weight** (defined as a weight less than 2500 grams at birth). During the 1980s, about four out of five neonatal deaths in the United States were of low birth weight babies (Emanuel *et al.*, 1989). An index of relative mortality during the neonatal period by birth weight is given in Figure 2.2. Relative mortality is defined as the percentage of deaths in excess of the number that occur for infants within the normal birth weight range of 3.0 to 4.5 kilograms. These data are for infants at all gestational ages. **Prematurity**, defined as birth prior to 37 weeks gestation, may cause additional complications that increase the chances of neonatal death. Some infants are small for gestational age (SGA), which is often defined as a birth weight less than the tenth percentile for completed week of gestation. Newborns who are both SGA and premature are usually at a greater risk of death than a premature child of the expected weight for gestational age (Gould, 1986).

The causes of low birth weight, prematurity, and SGA are many and not all are well understood. It is known that one set of causes may be **congenital** (hereditary or inborn) problems with the fetus. Another set of causes may be placental insufficiency, or maternal conditions including undernutrion, disease, smoking, or alcohol consumption. Whatever the cause, low birth weight, with or without prematurity, is associated with **socioeconomic status** (SES) of the mother. SES is a concept devised by the social sciences to measure some aspects of education, occupation, and social prestige of a person or a social group. A higher or lower SES does not, by itself, influence birth weight; rather the attitudes and behaviors of people are associated with SES and it is these attitudes and behaviors that influence diet, health, smoking, alcohol consumption, and other determinants of birth weight. Any measure of SES will only partially account for all of the social and cultural factors that affect growth. Nevertheless, SES serves as a proxy for these more specific determinants and at a gross level the effect of SES is obvious; for example, the incidence of low birth weight in the wealthier, higher SES nations averages 5.9 percent of all live births; in the poorer, lower SES nations the incidence averages 23.6 percent (Villar & Bellizan, 1982). The Netherlands has one of the lowest rates at 4.0 percent; Bangladesh has the highest rate at 50 percent (World Resources Institute, 1994).

At a finer level, within the developed nations, the socioeconomic relationship with birthweight is still strong. When educational attainment is used to estimate general SES, researchers in the United States found that 10.1 percent of births to women with less than 12 years of schooling are low birth weight compared with 6.8 percent of births to women with 12 years, and 5.5 to those with 13 or more years of formal education (Taffel, 1980). Blacks and other non-white minority groups show consistently lower average birth weights compared with whites, and part of this difference is accounted for by the lower socioeconomic status of the non-white groups. However, when white (European-American) and black (African-American) women are matched for socioeconomic status, black women still give birth to a higher percentage of low birth weight infants. To some researchers, the

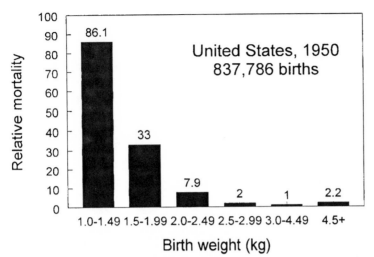

Figure 2.2. Index of relative mortality by birth weight (kg) during the neonatal period. Data from the United States, all registered births for 1950 (Shapiro & Unger, 1950). Relative mortality is calculated as the risk of death for an infant born at a given weight. Infants of normal birth weight, 3.0 to 4.5 kg, have a relative mortality of 1.0. Infants born at 2.5 to 3.0 kg have twice the risk of death as normal birth weight infants, and so on.

white–black differences suggest that genetic factors are major determinants of birth weight (Taffel, 1980; Collins & Butler, 1997; Collins *et al.*, 1997), but others point to non-genetic factors (David & Collins, 1991; Emanuel *et al.*, 1989; Emanuel, 1993).

To put this debate in perspective it is first necessary to state more clearly what is known about the factors that produce variation in birthweight. Based on a variety of evidence and many research studies, Robson (1978) estimates the variance in birthweight due to fetal genotype to be 10 percent and the variance due to parental genotype to be 24 percent. Most of the remaining 66 percent of the total variance is due to non-genetic maternal and environmental factors. There are many environmental factors that act directly on the developing embryo and fetus. Some obvious factors, briefly mentioned above, are smoking (Garn, 1985; Schell & Hodges, 1985), alcohol consumption (Able, 1982), and famine (Stein *et al.*, 1975). There is also evidence that exposure to urban pollution in the form of noise, lead, and polychlorinated biphenals can reduce gestation length leading to premature births and cause low birth weight in full-term infants (Schell, 1991b).

Emanuel and colleagues find that these relationships may hold across so-called 'racial' lines, influencing births to both white and black women, but the relative amounts of influence on birth weight are not equal (the term 'race' and similar terms are set in quote marks to indicate that for human beings 'racial' categories denote socially defined groups and not biologically or genetically justifiable classifications). It is well-established that in any given year more black infants are low birth weight than white infants, even when black and white mothers are matched for SES variables (income, education, housing, occupation, etc.). Even more curious are the findings for so-called 'biracial' infants. One study by Migone *et al.* (1991) examined all 'biracial' single births reported in the United States in the year 1983. Such births were to either white mothers and black fathers, or black mothers and white fathers. The percentage of low weight births for these 'biracial' couples was compared to samples of single births to same 'race' parents. The results, shown in Table 2.2, indicate a statistically significant trend. The 'race' of either parent contributes to this trend, however the mother's 'race' is a stronger predictor of birth weight than the 'race' of the father. The data in Table 2.2 are unadjusted for known confounding variables, but the significant trend remained even when the authors of the study adjusted for mother's age, education, marital status, prenatal care, live birth order, previous fetal deaths, baby's sex, and gestational age. Another study reports that in a sample of low SES women (defined as likely to be unmarried and live on incomes of less than $10 000 per year) 14 percent of infants born to black mothers with white fathers were low birth weight, compared with 9 percent of infants born to white mothers with black fathers (Collins & David, 1993).

What does the 'race' effect seen in these studies mean? The social conditions of life for same 'race' or different 'race' parents are likely not to be equivalent. This is especially true for the two groups of women studied by Collins & David. Moreover, the differences or similarities in living conditions between the various combinations of parents are likely to be quite complex and not easily summarized by a statistic such as current SES. Recognizing these complexities, Emanuel and colleagues try to explain the reason for the persistence of lower birth weights for infants born to black mothers by the 'intergenerational effect hypothesis'. By this, the researchers mean that the SES matching is valid only for the current generation of adult women. The mothers and grandmothers of these black and white women were less likely to be equally matched for SES. Given the social history of the United States and Britain, previous generations of black women were likely to be of lower SES than their white counterparts. The intergenerational effect hypothesis predicts that the poor growth and development of women from older generations will have a lasting effect on the current generation.

Table 2.2. *Total number of births, number of low weight births (< 2500 g), and proportion of low birth weight births for each parental 'racial' group*

Parental groups (mother–father)	Total births	Low weight births	%
White–white	24 059	1027	4.3
White–black	18 004	1108	6.2
Black–white	5 617	459	8.2
Black–black	15 220	1629	10.7

There is considerable support for the intergenerational effect hypothesis. In several studies, Emanuel and his colleagues working with both British (Emanuel *et al.*, 1992) and United States data (Sanderson *et al.*, 1995) find the mother's own birth weight, her health history during infancy and childhood, and her adult stature (which reflects the total history of her growth and development) are strong predictors of the birth weight of her offspring. These results were confirmed in a similar study of all births in Norway since 1967 (Skjaervern *et al.*, 1997). The authors of the study linked the birth weight records of women with the birth weight of their own infants and produced a sample of 101 264 mother–infant birth weight pairs. Mother's birth weight was strongly associated with the weight of infant.

Other measures of growth at birth

Weight at birth is just one measurement that is commonly taken to indicate the amount of growth that took place during prenatal life. Recumbent length, the circumference of the head, arm, and chest, and skinfolds are others. Recumbent length is similar to stature, however the person measured is lying down and is stretched out fully by having the examiner apply pressure to the abdomen and knees. The maximum distance between the vertex of the head and the soles of the feet constitutes the measurement. This can be measured at a very young age, or under other circumstances when stature (standing height) is impossible to determine. Circumferences measure the contribution made by a variety of tissues to the size of different body parts. For example, head circumference measures the maximum girth of the skull and hence, indirectly, the size of the brain. This is because of the intimate conformity between the brain and the tissues which surround and protect it, and the dominant role of the brain in determining head size. Similarly, arm circumference includes the measurement of bone, muscle, subcutaneous fat, and skin. For infants of the same weight and length, variations in arm circumference are chiefly due to variations in the amounts of muscle and, especially, subcutaneous fat.

Some representative data for several measures of size at birth and at 18 years of age are given in Table 2.3. These average figures show that at birth, boys are a bit longer, heavier and larger-headed than girls, but the girls have slightly more subcutaneous fat at birth than the boys. In reality there is such a wide range of variation in actual birth dimensions that the small average sexual dimorphism in size is biologically insignificant. At 18 years of age, on average, men and women display well-marked sexual dimorphism in all of these growth variables, except head circumference. Another difference between the infant and the adult is in body proportions. For children born in the United States, head circumference at birth averages about 70 percent of length at birth. By age seven years, head circumference averages 42 percent of length and at maturity the average value falls to about 30 percent. The reason for this change in percentage over time is that during the fetal, infant, and childhood stages of life the growth of the brain proceeds at a faster rate than the growth of the body (Scammon, 1930). For the average child in the United States, head circumferences reaches 80 percent of mature size by about 7 years of age, though length of the body is only 68 percent complete at the same age (Nellhaus, 1968; Hamill *et al.*, 1977). There are also proportional changes in the length of the limbs, which become longer relative to total body length during growth. The proportional changes are illustrated in Figure 2.3.

The composition of the newborn's body in terms of adipose tissue and muscle tissue has been determined. The newborn's total body weight is about 12 percent body fat and 20 percent muscle mass. By adulthood, men average 15 to 17 percent body fat and 40 percent muscle mass; women average 24 to 26 percent fat and 35 to 37 percent muscle mass (Katch *et al.*, 1980; Holliday, 1986). The composition of the newborn, the adolescent, and the adult in terms of other soft tissues, and a variety of chemicals, is also available (Widdowson & Dickerson, 1964; Fomon *et al.*, 1966; Forbes, 1986), and show marked differences between birth and adulthood. The study of the formation and maturation of hard tissues, such as the skeleton, during prenatal and postnatal life is another means towards describing different stages of development. Most bone forms from cartilage, which becomes calcified (calcium and phosphorous minerals are added) and, then, ossified (hardened) into mature skeletal tissue. Bone formation takes place throughout the growing years. A record of the process can be captured on radiographs, since at certain X-ray exposure levels cartilage is 'invisible', but calcified and ossified bone is radio-opaque (Figure 2.4). Radiographs of skeletal development of normally growing children from

Table 2.3. *Mean size at birth and at age 18 years for children born in the United States. Data from various sources of nationally representative statistics*

	Birth		18 years	
	Boys	Girls	Boys	Girls
Recumbent length/stature (cm)[a]	49.9	49.3	176.6	163.1
Weight (kg)[a]	3.4	3.3	71.4	58.3
Head circumference (cm)[b]	34.8	34.1	55.9	54.9
Triceps skinfold (mm)	3.8[c]	4.1[c]	8.5[d]	17.5[d]
Subscapular skinfold (mm)	3.5[c]	3.8[c]	10.0[d]	12.0[d]
Arm muscle area (mm²) at age one year[e]	20.4	19.6	75.0	57.2

[a]Hamill *et al.*, 1977; [b]Nellhaus, 1968; [c]Johnston & Beller, 1976; [d]Johnson *et al.*, 1981; [e]Frisancho, 1990.

the United States and England have been compiled into atlases, which may be used to assess the stage of bone maturation of other children (Todd, 1937; Greulich & Pyle, 1959; Roche *et al.*, 1975a; Tanner *et al.*, 1983b).

The importance of these contrasts between early and later life is twofold. First, they allow clinicians and researchers to assess a child's stage of biological maturation for different organs, tissues, or chemicals independent of chronological age. Biological maturation is used to help determine if a child is developing too slowly or quickly, either of which may indicate the presence of some disorder. Second, the contrasts between early and later life are also conceptually important. They show that the infant may take any one of several paths for growth, maturation, and functional development. Adult human morphology, physiology, and behavior as well, are **plastic** (Lasker, 1969) and in no way rigidly predetermined. **Plasticity** refers to the ability of an organism to modify its biology or behavior to respond to changes in the environment, particularly when these changes are stressful. People, of course, cannot sprout wings or breath water, but the sizes, shapes, colors, emotions, and intellectual abilities of people can be significantly altered by environmental stress, training, and experience.

When the biology and behavior of people are considered together (a biocultural perspective), it seems that human beings are, perhaps, the most plastic of all species. In part this is because the human lifespan is long, relative to most other animals, and the adult human phenotype is achieved after many years during which a variety of factors may influence their final outcome. This makes people highly variable and adaptable.

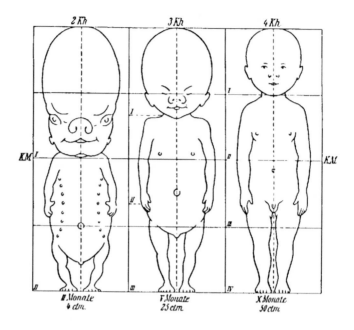

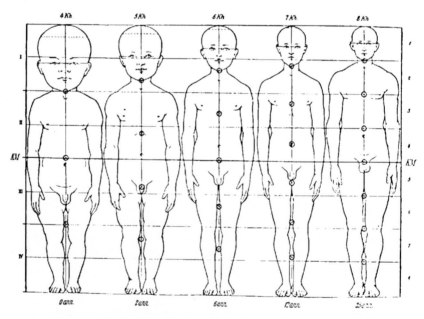

Figure 2.3. Diagrams illustrating the changes in body proportions of human beings that occur during prenatal and postnatal growth. Human body segments (head, trunk, arms, legs) grow at different rates and mature, generally, from a head to foot direction (from Stratz, 1909).

Postnatal life

In contrast to the widely used trimester system for prenatal growth and development, many ways have been proposed to divide life after birth into distinct periods. An historical discussion of some of these was given in the previous chapter (such as the Roman 'seven stages of life'). A five-stage model of human postnatal growth and development was proposed in the first edition of this book (Bogin, 1988b). These five stages are infant, child, juvenile, adolescent, and adult. Although any stage model is somewhat arbitrary (life is a continuous process) the use of stages aids description and analysis of the life cycle. The rationale for the five-stage model begins with an analysis of the amount and rate of growth from birth to adulthood. To visualize the amount and rate of growth that takes place during each of these stages, the growth in height, or length, for normal boys and girls is depicted in Figure 2.5 (growth in weight follows very similar curves). The stages of growth are also outlined in Table 2-1. In Figure 2.5 the **distance curve** of growth, that is the amount of growth achieved from year to year, is labeled on the right-side *y*-axis. The **velocity curve**, representing the rate of growth during any one year, is labeled on the left side. Below the velocity curve are symbols indicating the average duration of each stage of development. It is readily apparent that changes in growth rate and tempo are associated with each stage of development. Such changes in rate and tempo were shown in the previous chapter in relation to the growth of Montbeillard's son (Figure 1.4). Each stage may also be defined by characteristics of the dentition, by changes related to methods of feeding, by physical and mental competencies, and by maturation of the reproductive system and sexual behavior.

Infancy stage

The **infancy** stage occupies the first three years of human post-natal life. Infancy is characterized by the most rapid velocity of growth of any of the post-natal stages. During the first year of postnatal life infants may add 28 centimeters in length and seven kilograms in weight, which represents more than 50 percent of birth length (about 50 cm) and 200 percent of birth weight (about 3.4 kg). The rate of decrease in velocity, or deceleration, is also very steep, which makes infancy the life stage of most rapidly changing rate of growth. The infant's curve of growth, rapid velocity and deceleration, is a continuation of the fetal pattern, in which the rate of growth in length actually reaches a peak in the second trimester and then begins a deceleration that lasts until childhood (Figure 2.6).

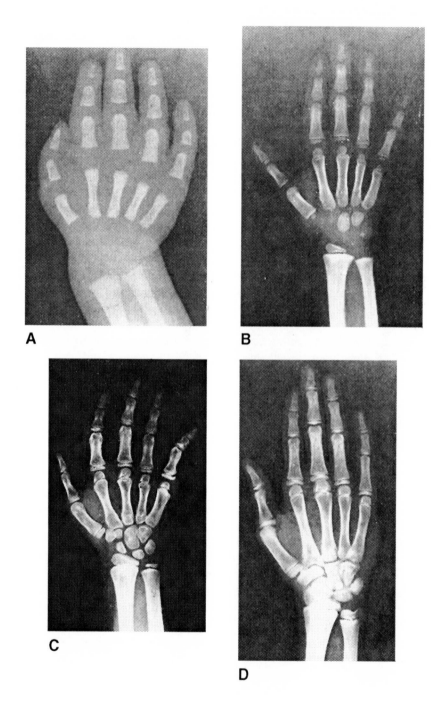

Figure 2.4. (Opposite) Radiographs of the hand and wrist at different skeletal ages illustrating the sequence of bone maturation events. (A) Newborn; at this age most wrist bones, and the growing ends of the finger bones (called epiphyses), are formed of cartilage. At certain X-ray exposures this cartilage is 'invisible'. The newborn has no ossification centers (places where bone is present) in the wrist and no visible epiphyses. (B) Three years old; some wrist ossification centers present, epiphysis of radius present, most epiphyses of hand calcified (i.e., forming bone). (C) Eight years old; all ossification centers calcified, epiphyses of the radius and ulna are not as wide as their diaphyses. (D) Thirteen years old; all bones have assumed final shape, epiphyses of the radius and ulna are almost as wide as their diaphyses, the growth in size and closure of epiphyses remains to be completed (from Lowery, G. H. *Growth and Development of Children*, 8th edn. © 1986 Yearbook Medical Publishers, Inc., Chicago).

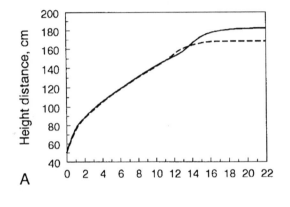

A

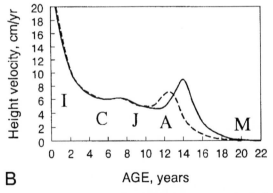

B   AGE, years

Figure 2.5. Idealized mean distance (A) and velocity (B) curves of growth in height for healthy girls (dashed lines) and boys (solid lines) showing the postnatal stages of the pattern of human growth. In graph B note the spurts in growth rate at mid-childhood and adolescence for both girls and boys. The stages of postnatal growth are abbreviated as follows: I, infancy; C, childhood; J, juvenile; A, adolescence; M, mature adult. Data use to construct the curves come from Prader (1984) and Bock & Thissen (1980).

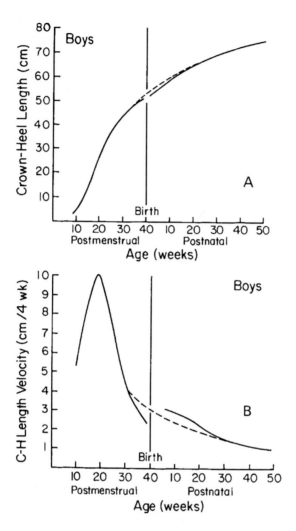

Figure 2.6. Distance (A) and velocity (B) curves for growth in body length during human prenatal and postnatal life. The figure is diagrammatic, as it is based on several sources of data. The interrupted lines depict the predicted curve of growth if no uterine restriction takes place. In fact, such restrictions does take place toward the end of pregnancy and this may impede the flow of oxygen or nutrients to the fetus. Consequently growth rate slows, but rebounds after birth and returns the infant to the size she or he would be without any restriction (from Tanner, 1990).

As for all mammals, human infancy is the period when the mother provides all or some nourishment to her offspring via lactation, or by some culturally derived imitation of lactation. One reason for this is that infants do not have teeth, and thus cannot eat solid food. During infancy the deciduous dentition (the so-called 'milk teeth') emerges through the gums. Infants will erupt five deciduous teeth in each quadrant of their mouth, the central incisor, lateral incisor, canine, first molar, and second molar (I_1, I_2, C, M_1, M_2). Table 2.4 provides the median age of emergence, and range in age, of these teeth for French-Canadian infants. The deciduous teeth of boys emerge about one month earlier than girls. This is noteworthy only because in most other aspects of physical growth and maturation girls are, on average, ahead of boys.

One surprising feature of human growth during infancy is the similarity that most infants show in both amount and rate of growth in length and weight during the first six months of life. One might expect that variation in hereditary and environmental factors between individual infants and populations would lead to marked differences in amounts and rates of growth. However, Habicht *et al.* (1974) and Van Loon *et al.* (1986) show that the growth of infants of normal birth weight, from a wide variety of ethnic and socioeconomic classes, in both the developed and developing nations, is remarkably similar during the first six months of life. Perhaps breast-feeding, which supplies the nutrient, immunity, and psychological needs of the infant, overrides the effect of variations in other aspects of the environment. After six months of age, when breast milk alone no longer meets the nutritional demands of the growing infant and other specially prepared infant foods must be supplemented, infants from the developed nations or higher socioeconomic classes may become significantly larger than their less privileged age-mates from poorer environments. If alleviated by improved nutrition and health status early on, the disadvantaged children may catch up in size (as shown by Pagliani in 1876). Otherwise, the differences in size between the well-off and the deprived become greater and greater, and by childhood the differences may have become irreversible.

Motor skills (what a baby can do physically) develop rapidly during infancy. At birth, states of wakefulness and sleep are not sharply differentiated and motor coordination is variable and transient. By one month the infant can lift its chin when prone and by two months lift its chest by doing a 'push-up'. By four months the infant can sit with support, by seven months sit without support, by eight months crawl, and by 12 months walk with support. By two years of age the infant can walk well and turn the pages of a book, one at a time. By three years of age, the end of the infancy

Table 2.4. *The median and range in age (in months) of eruption of the deciduous teeth for French-Canadian infants (after Demirjian, 1986)*

Boys	Median age (months)	Range (months)
Maxilla		
I^1	8.49	5.70–11.84
I^2	9.81	5.87–14.74
C	17.56	12.44–23.56
M^1	15.20	11.26–19.74
M^2	27.04	19.95–35.20
Mandible		
I_1	6.86	3.90–10.63
I_2	11.48	6.95–17.13
C	17.77	12.98–23.30
M_1	15.25	11.49–19.54
M_2	26.13	19.19–34.14
Girls	Median age (months)	Range (months)
Maxilla		
I^1	9.42	6.19–13.33
I^2	10.53	6.13–16.12
C	18.31	13.28–24.15
M^1	15.06	11.51–19.09
M^2	27.86	20.67–36.11
Mandible		
I_1	7.31	3.96–11.68
I_2	12.54	7.70–18.57
C	18.58	13.14–24.96
M_1	15.32	11.67–19.48
M_2	26.90	20.22–34.54

Abbreviations for the teeth are I_1, central incisor; I_2, lateral incisor; C, canine; M_1, first molar; M_2, second molar; the subscripts are for mandibular teeth and the superscripts are for maxillary teeth.

stage, the youngster can run smoothly, pour water from a pitcher, and manipulate small objects, such as blocks, well enough to control them. There is a similar progression of changes in the problem solving, or cognitive, abilities of the infant.

The development of the skeleton, musculature, and the nervous system account for all of these motor and cognitive advancements. The rapid growth of the brain, in particular, is important. The human brain grows more rapidly during infancy than almost any other tissue or organ of the body (Figure 2.7). All parts of the brain seem to take part in this fast pace of growth and maturation, including those structures that control the repro-

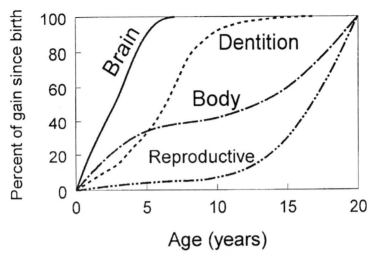

Figure 2.7. Growth curves for different body tissues. The 'Brain' curve is for total weight of the brain (Cabana *et al.*, 1993). The 'Dentition' curve is the median maturity score for girls based on the seven left mandibular teeth (I_1, I_2, C, PM_1, PM_2, M_1, M_2) using the reference data of Demirjian (1986). The 'Body' curve represents growth in stature or total body weight and the 'Reproductive' curve represents the weight of the gonads and primary reproductive organs (Scammon, 1930).

ductive system. The **hypothalamus**, a center of neurological and endocrine control, is one of these brain structures. During fetal life and early infancy the hypothalamus produces relatively high levels of gonadotropin-releasing hormone (GnRH). This hormone causes the release of luteinizing hormone (LH) and follicle stimulating hormone (FSH) from the **pituitary**. LH and FSH travel in the bloodstream to the gonads (**ovaries** or **testes**) where they stimulate the production and release of estrogen or androgen hormones (see Figure 7.4). These gonadal hormones are, in part, responsible for the rapid rate of growth during early infancy. By late infancy, however, the hypothalamus is inhibited for reasons as yet not completely known. GnRH secretion virtually stops and the levels of the sex hormones fall, which suspends reproductive maturation (see Figure 2.7 'Reproductive' curve). The hypothalamus is reactivated just before **puberty**, the event of development that marks the onset of sexual maturation.

The age of emergence of the last deciduous tooth, usually M_2, is important for this is one of the events that signals the end of infancy. Emergence of all of the deciduous teeth allows the infant to switch from dependence on breast-feeding (or formula/infant food feeding) to eating appropriate weaning foods. The strict definition of **weaning** used in this book is the termination of breast-feeding. In pre-industrialized societies, such as hunters and gatherers, horticulturists, and pastoralists, weaning occurs between 24 to 36 months of age. By this age all the deciduous teeth have erupted, even for very late maturing infants. Thus, by 36 months of age there occur both biological developments (tooth emergence) and behavioral changes in the mother–infant relationship (weaning) that signal the end of infancy.

Childhood stage

The **childhood** stage follows infancy, encompassing the ages of about three to seven years. Childhood may be defined by its own pattern of growth, feeding behavior, motor development and cognitive maturation. The rapid deceleration in rate of growth that characterizes infancy ends at the beginning of childhood, and the rate of growth levels off at about five centimeters per year. In terms of feeding, children are weaned from the breast, or bottle, but are still dependent on older people for food and protection. Most mammalian species move rapidly from infancy, and its association with dependence on nursing from the mother, to a stage of independent feeding. Post-weaning dependency is found in several species of social mammals, especially carnivores (such as lions, wild dogs, hyenas) and in some species of primates. Lion cubs, for example, are weaned at about 6–8 months old, but remain dependent on their mothers until about 24 months old. During that time the cubs must learn how to hunt for themselves. Many species of primates must also learn to hunt for high-quality foods, such as insects, and learn how to open fruits and seeds with tough skins. This learning also requires a period of post-weaning dependence on the mother and sometimes the father (as for marmosets and tamarins).

Post-weaning dependency is, by itself, not a sufficient criterion to define human childhood. Human children do, of course, learn how to find and prepare food, but there are a suite of features that define the childhood stage. Not all of these features are found for the social carnivores and non-human primates. Human children require specially prepared foods due to the immaturity of their dentition, the small size of their stomachs and intestines, and the rapid growth of their brain (Figure 2.7). The metabolic activity of the human brain is especially important. The newborn uses 87 percent of its resting metabolic rate (RMR) for brain growth and function. By age five years the RMR usage is still high, at 44 percent, while in the adult human the figure is between 20–25 percent of RMR. At the comparable stages of development the RMR values for the chimpanzee are about 45, 20, and 9 percent respectively (Leonard & Robertson, 1992).

The human constraints of immature dentition and small digestive system necessitate a childhood diet that is easy to chew and swallow and low in total volume. The child's relatively large and active brain, almost twice the size of an adult chimpanzee's brain, requires that the low volume diet be dense in energy, lipids, and proteins. Children do not yet have the motor and cognitive skills to prepare such a diet for themselves. Children are also especially vulnerable to disease and accidents and thus require protection. In past times, and in some areas of the world today, children were also targets for predatory birds and mammals. There is no society in which children survive if deprived of this care provided by older individuals. So-called 'wolf children' (referring to children reared by wolves or children living 'wild' on their own) and even 'street children', who are sometimes alleged to be living on their own, are either myths or, in fact, not children at all. A search of the literature finds no case of a human child, that is a youngster under the age of six, living alone either in the wild or on urban streets.

The physical growth changes in height, weight, and body composition that take place during the childhood years are reviewed by Johnston (1986). He points out that one of the more striking features of human growth at this time is its predictability, both within individuals and between populations. The distance and velocity curves for height depicted in Figure 1.4 are examples of the predictability of childhood growth. This French boy, the son of the Count Montbeillard, was raised in the country-side under near-optimal conditions for that time. Though this figure represents but a single child, the pattern of growth of all normal children follows a very similar course. For example, the boy gained 59.9 cm in height between his second and twelfth birthdays (Scammon, 1927). Children of generally middle socioeconomic class born in the United States during the 1960s and early 1970s, average a 61.6 cm gain in height between their second and twelfth birthdays (Hamill *et al.*, 1977). The difference between the gains in height of the French boy and the US sample are not statistically significant. The similarity in growth between a child and a sample of children, across time periods, and across geographic boundaries, emphasizes the common pattern of growth shared by all normal children and the predictability of this pattern. These features of human growth have important practical implications. For instance, they form the basis of epidemiological and clinical examinations that detect pediatric health disorders by searching for deviations in the expected trajectory of growth.

Though the pattern of childhood growth is predictable, there are several factors which may influence the amount and rate of growth of the individual child or groups of children. These factors include heredity, nutrition, illness, socioeconomic status, and psychological well-being. Chapters 5 and 6 of this book are devoted to a detailed discussion of the action of each of these factors, and their combined interactions, on growth and development. Here it may be briefly stated that, all other factors being equal, short or tall parents are likely to have children who achieve similar stature. However, malnutrition, chronic illness, poor living conditions, and chronic psychological stress are each capable of retarding the growth of a child. An example of all of these factors is the case of the 'Maya in Disneyland' described in the Introduction.

Two of the important physical developmental milestones of childhood are: (1) the replacement of the deciduous teeth by the emergence of the first permanent teeth, and (2) completion of growth of the brain in weight. First molar eruption (eruption is usually defined as first appearance of the tooth through the gingival, or gum, surface) takes place, on average, between the ages of 5.5 and 6.5 years in most human populations. Eruption of the central incisor quickly follows, or sometimes precedes, the eruption of the first molar. There is some variation between human populations in the age of eruption, but the number of permanent teeth in seven-year-olds varies from about eight teeth in European 'whites' to nine or ten teeth in African Zulu and Bantu groups (MacKay & Martin, 1952). By the end of childhood, usually at age seven years, most children have erupted the four first molars and, in addition, permanent incisors have begun to replace 'milk' incisors. Within another year the four lateral incisors will also erupt, replacing the 'milk teeth' that had been in that position Along with growth in size and strength of the jaws and the muscles for chewing, these new teeth provide sufficient capabilities to eat a diet similar to adults. The mean age of eruption of the permanent dentition of boys and girls is given in Table 2.5.

A close association between human dental development and other aspects of growth and maturation was noted many years ago by anatomists and anthropologists. More recently, Holly Smith (1991a, 1992) analyzed data from humans and 20 other primates species and found that age of eruption of the first molar is highly associated with brain weight and a host of other growth and maturation variables. The correlation coefficient between age of eruption of the first molar and adult brain weight is $r = 0.98$ ($r = 1.00$ is a perfect relationship). The big brain of humans predicts a late age of first molar eruption. Other research, based on direct measurements of victims of accidents and disease, shows that human brain growth in weight is complete at a mean age of seven years (Cabana *et al.*, 1993), confirming Smith's statistical analysis. At about age seven years, then, the child becomes capable dentally of processing an adult-type diet. The end of

Table 2.5. *The mean age and standard deviation (in years) for the eruption of the permanent teeth for North American boys and girls (Smith et al., 1994)*

Boys		
	Mean (years)	SD (years)
Maxilla		
I^1	7.34	0.77
I^2	8.39	1.01
C	11.29	1.39
P^3	10.64	1.41
P^4	11.21	1.48
M^1	6.40	0.79
M^2	10.52	1.34
M^3	20.50	—
Mandible		
I_1	6.30	0.81
I_2	7.47	0.78
C	10.52	1.14
P_3	10.70	1.37
P_4	11.43	1.61
M_1	6.33	0.79
M_2	12.00	1.38
M_3	19.80	—
Girls		
	Mean (years)	SD (years)
Maxilla		
I^1	6.98	0.75
I^2	7.97	0.91
C	10.62	1.40
P^3	10.17	1.38
P^4	10.88	1.56
M^1	6.35	0.74
M^2	11.95	1.22
M^3	20.50	—
Mandible		
I_1	6.18	0.79
I_2	7.13	0.82
C	9.78	1.26
P_3	10.17	1.28
P_4	10.97	1.50
M_1	6.15	0.76
M_2	11.49	1.23
M_3	20.40	—

Abbreviations for the teeth are I_1, central incisor; I_2, lateral incisor; C, canine; P_3, first premolar; P_4, second premolar; M_1, first molar; M_2, second molar; the subscripts are for mandibular teeth and the superscripts are for maxillary teeth.

brain growth in weight means that nutrient requirements for brain growth diminish.

During late infancy and childhood human locomotive skills develop and mature. Nakano & Kimura (1992) review previous research on human and non-human primate locomotive development and present some of their own new research. At age three years, the beginning of childhood, the human is still a 'toddler', that is, able to walk bipedally but without the efficiency and characteristic gait of the adult. Nakano & Kimura find that by age seven years, on average, humans are able to walk with the adult-type efficiency and gait. A study by Kramer (1998) examined the energy costs of locomotion in children and juveniles. Children use more energy per kilogram of body weight when walking than do adults. Five- to six-year-old children are about 85 percent efficient as adults. Seven- to eight-year-old juveniles have more than 90 percent the efficiency of adults. The onset of adult-style locomotion, along with the eruption of the first permanent teeth, and the end of brain growth, are all indicators that the physically dependent child is moving on to independence.

The end of childhood is also marked by a small increase in velocity, depicted in Figure 2.5, called the **mid-growth spurt** (Tanner, 1947). This small increase in growth velocity was first noted by Backman (1934), and then by Meredith (1935) and Count (1943). In a London-based study, Tanner & Cameron (1980) noted the presence of the mid-growth spurt in the average velocity curve of boys, but not girls (more than 150 children of each sex were measured). Molinari *et al.* (1980) found in a longitudinal study of children from Zurich, Switzerland (112 boys and 110 girls) that two-thirds of boys and girls had mid-growth spurts. Bock & Thissen (1980) found a mid-growth spurt in the longitudinal measurements of the 'average' child from a sample of 66 boys and 70 girls from Berkeley, California, and Meredith (1981) reported that 14 percent of the 70 boys and 70 girls measured for the Iowa City growth study sample had the spurt. Berkey *et al.* (1983) found that 17 of 67 boys and none of the 67 girls from a Boston, Massachusetts growth study had mid-growth spurts. Analysis of data from the Edinburgh Longitudinal Growth Study, 80 boys and 55 girls, finds that all the boys and all but one of the girls show a mid-growth spurt at age 7.0 years for the boys and 6.7 years for the girls (Butler *et al.*, 1990).

Varying methods of statistical analysis may explain the differences in findings between these studies. Especially important are the effect of data 'smoothing' techniques, which are often used in longitudinal analysis that would tend to obliterate the mid-growth spurt. Curve fitting is also used commonly in longitudinal analysis and many curve-fitting routines assume that the mid-growth spurt does not exist. Other factors that may influence the ability to detect the mid-growth spurt are the number of serial measurements of height available for analysis (few measurements make detection difficult), and the frequency of measurements (whether annual, semi-annual, or more frequent). Those studies finding the mid-growth spurt tend to avoid most of, or all, of these problems.

Juvenile stage

Juveniles may be defined as, '. . . prepubertal individuals that are no longer dependent on their mothers (parents) for survival' (Pereira & Altmann, 1985, p. 236). This definition is derived from ethological research with social mammals, especially non-human primates, but applies to the human species as well. In contrast to infant and human children, juvenile primates can survive the death of their adult caretaker. The human primate is no exception to this, as ethnographic research shows that juvenile humans have the physical and cognitive abilities to provide much of their own food and to protect themselves from accidents and disease (Weisner, 1987; Blurton-Jones, 1993). Remember those so-called 'street children' mentioned above – they are in fact 'street juveniles!' During the juvenile stage the rate of growth declines once more. This decline follows the mid-growth spurt in those children who experience it. But even in children without a detectable mid-growth spurt the rate of growth declines. Thus, juveniles grow at the slowest rates since birth. This slow rate of growth applies to global measures of size, such as stature and weight, and also to individual tissues, organs, and body systems.

Juvenile growth is also predictable, stable, and harmonious. The growth in height and weight of two groups of Guatemalan juveniles between the ages of 7 and 13 years, shown in Figures 2.8 and 2.9, is an example of a predictable and stabilized difference in size. Both groups live in Guatemala City and attend school. The larger juveniles are from high SES families, the smaller juveniles are from low SES families. The high SES group are about the same size as healthy, well-nourished juveniles from the United States (Johnston *et al.*, 1973). The families of the low SES group are known to suffer from inadequate nutrition, primarily a shortage of total food intake, poor living conditions in terms of health care, the supply of potable water, and education (Bogin & MacVean, 1978, 1981a, 1983). Though unequal in size, these children all display a similar regularity in their growth as may be seen from the mean values plotted in the figures. The differences in size between girls and boys is not of biological importance at these ages. Even so, the pattern of growth by sex is virtually identical in both high and low SES samples, which is further evidence of the predictability of juvenile

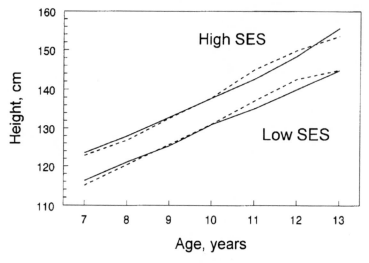

Figure 2.8. Mean heights of Guatemalan boys (solid lines) and girls (dashed lines) of high and low socioeconomic status (SES) (from Bogin & MacVean, 1978).

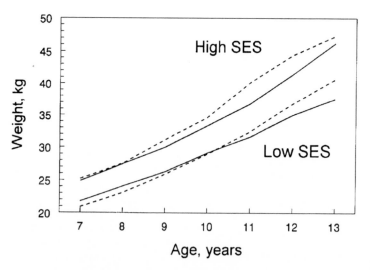

Figure 2.9. Mean weights of Guatemalan boys (solid lines) and girls (dashed lines) of high and low socioeconomic status (SES) (from Bogin & MacVean, 1978).

growth. Generally changes in size from age to age are similar for both of the SES groups. Indeed, a formal statistical comparison using the technique of tracking analysis (i.e., does the pattern of growth in one sample parallel, or track, the pattern in another sample) finds a high degree of tracking between the groups. The parallel tracking means that the differences in height and weight between these two groups were established prior to age seven years. Most likely this difference is the result of growth retardation in infancy and childhood due to the undernutrition of the low SES group. Growth during the juvenile stage maintains the stable and predictable differences between these two SES groups.

The harmony of juvenile growth is shown in Figure 2.10. Juveniles from both SES groups have the same height for weight proportionality regardless of absolute size. The 'regression line' drawn in Figure 2.10 represents the best-fitting straight line (estimated by the statistical method of least squares) drawn through the data points for the high SES juvenile boys and girls. The data points of the low SES juvenile boys and girls show no statistically significant deviations from the regression line. The maintenance of proportionality under the stress of low SES reflects the stability, predictability, and harmony of juvenile growth. Since both height and weight are equally affected in the low SES Guatemalan juveniles, it is likely that some common mechanism is regulating the growth of several different tissues (e.g, bone, muscle, adipose). The exact nature of this mechanism is not known.

Widdowson (1970) observed that to some extent a harmony of growth is displayed in the normal development of many body parts not only during the juvenile stage of life, but also during all stages of human growth and development. For example, from childhood to adulthood there is a coordinated growth of the teeth and the craniofacial complex of bones (mandible, maxilla, etc.) that maintains the functional integrity of the masticatory system. Without a harmony of growth of these bones and the teeth the growing individual would not be able to eat. Another example is the phenomenon of **catch-up growth** (Prader *et al.*, 1963). This is a rapid increase in growth velocity following a short-term period of starvation or illness which slowed or stopped growth. The increase in growth velocity usually restores a youngster to the size he or she would have achieved had there been no growth delay, and then the rate of growth slows and returns to its former level. Many segments of the body, such as the head, neck, truck and extremities, participate in this catch-up growth and each segment maintains its normal pattern of growth, that is, the affected individual does not end up with one leg shorter than the other or with arms stunted but legs of normal length. The nearly global nature of this harmony of

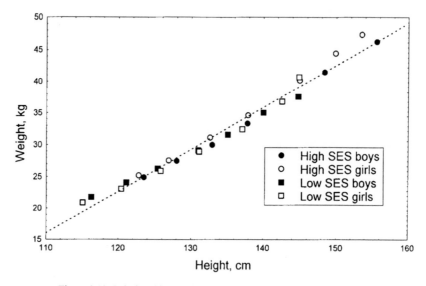

Figure 2.10. Relationship of height to weight for Guatemalan boys and girls, aged 7.00 to 13.99 years old, of high and low socioeconomic status (SES). The broken line is the linear regression, the 'best fitting' straight line running through the data points (from Bogin & MacVean, 1978).

growth has stimulated several researchers to propose a theoretical concept of growth as biologically self-regulating (Tanner, 1963; Goss, 1978; Bogin, 1980). This is a topic that is discussed in detail in Chapter 7 of this book.

In girls, the juvenile period ends, on average, at about the age of 10, two years before it usually ends in boys, the difference reflecting the earlier onset of adolescence in girls. The juvenile period is often accompanied by a pronounced, but short-lived decrease in rate of growth. The data for Montbeillard's son (Figure 1.4) show such a dip in velocity at about age eight to 11 years. Similar dips are known from recent longitudinal studies of growth (Stutzle *et al.*, 1980), although not all children display them. The cause of this decrease in growth rate is not known. The nadir of this dip marks the end of the juvenile stage, for once the rate of growth begins to again increase the growing individual has entered the adolescent stage of growth.

Adolescent stage

The transition from the juvenile to the adolescent stage of growth and development is signaled by a reversal in the rate of growth, from deceleration to acceleration (Figure 2.5). Human **adolescence** is also the stage of life when social and sexual maturation takes place. In fact, adolescence begins with **puberty**, or more technically with **gonadarche**, which is the reinitiation of activity of the hypothalamic–pituitary–gonadal system of hormone production. The current understanding of the control of gonadarche is that one, or perhaps a few, centers of the brain change their pattern of neurological activity and their influence on the hypothalamus. The hypothalamus, which has been basically inactive in terms of sexual development since about age two years, is again stimulated to produce GnRH. It is not known exactly how this change takes place. As stated above, the production of GnRH by the hypothalamus is inhibited by about the age of two years. The 'inhibitor' has not been identified, but probably is located in the brain, and certainly not in the gonads. Human children born without gonads, or rhesus monkeys and other primates whose gonads have been surgically removed, still undergo both the inhibition of the hypothalamus in infancy, and the reactivation of the hypothalamus at puberty (Figure 2.11a). The transition from juvenile to adolescent stages requires not only the renewed production of GnRH but also its secretion from the hypothalamus in pulses (Figure 2.11b). Gonadarche is triggered when the pulsatile secretion reaches both the necessary frequency (number of pulses in a given time period) and amplitude (peak amount of secretion during each pulse). Further discussion of the neuroendocrine control of growth and development is presented in Chapter 7.

None of these hormonal changes can be seen without sophisticated technology, but the effects of gonadarche can be noted easily since visible signs of sexual maturation appear. One such sign is a sudden increase in the density of pubic hair (indeed the term 'puberty' is derived from the Latin *pubescere*: to grow hairy). Another sign, for girls, is the development of the breast bud, the first stage of breast development. The pubescent boy or girl, his or her parents, and other relatives, friends, and sometimes everyone else in the social group, can observe these signs of early adolescence.

Sexual development

The adolescent stage also includes development of other **secondary sexual characteristics**, such as development of the external genitalia, sexual dimorphism in body size and composition (Figure 2.12), deepening of the voice in boys, as well as the onset of greater interest and practice of adult patterns of sociosexual behavior and food production. Some of these physical and behavioral changes occur with puberty in many species of social mammals. What makes human adolescence unique are two important differences. The first is the length of time between puberty and adulthood, that is, full

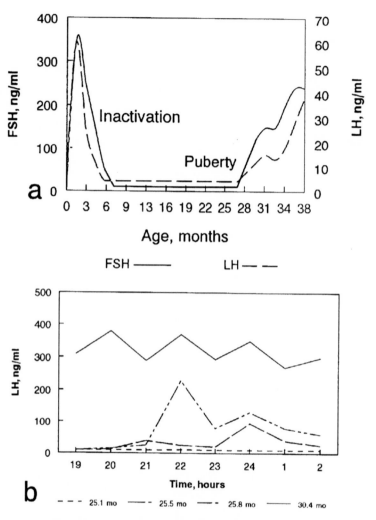

Figure 2.11. (a) Pattern of secretion of follicle stimulating hormone (FSH) and luteinizing hormone (LH) in a male rhesus monkey (genus *Macaca*). The testes of the monkey were removed surgically at birth. The curves for FSH and LH indicate the production and release of gonadotropin-releasing hormone (GnRH) from the hypothalamus. After age 3 months (i.e., during infancy) the hypothalamus is inactivated. Puberty takes place at about 27 months and the hypothalamus is reactivated. (b) Development of hypothalamic release of GnRH during puberty in a male rhesus monkey with testes surgically removed. At 25.1 months (mo) of age the hypothalamus remains inactivated. At 25.5 mo and 25.8 mo there is modest hypothalamic activity, indicating the onset of puberty. By 30.4 mo adult pattern of LH release is nearly achieved. The pattern of LH release shows both an increase in the number of pulses of release and an increase in the amplitude of release. In human beings a very similar pattern of infant inactivation and late juvenile reactivation of the hypothalamus takes place. This figure is redrawn, with some simplification, from Plant, 1994.

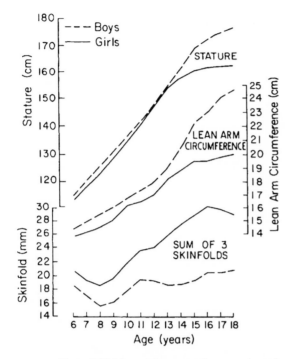

Figure 2.12. Mean stature, mean lean arm circumference, and median of the sum of three skinfolds for Montreal boys and girls (from Baughn *et al.*, 1980). Note that sexual dimorphism increases markedly after puberty, about age 12 to 13 years.

reproductive maturity. Humans take five to eight years for this transition. Monkeys and apes take less than three years. The second human difference is that during this life stage, both boys and girls experience a rapid acceleration in the growth velocity of virtually all skeletal tissue – the **adolescent growth spurt** (see Figure 2.5). The magnitude of this acceleration in growth was calculated for a sample of healthy Swiss boys and girls (112 boys and 110 girls) measured once a year, near their birthdays, between the ages of four and 18 years (Largo *et al.*, 1978). At the peak of their adolescent growth spurt the average velocity of growth in height was + 9.0 cm/year for boys and + 7.1 cm/year for girls. Similar average values are found for adolescents in most human populations. In contrast to humans other primate species either have no acceleration in skeletal growth or an increase in growth rate that is very small (the next chapter includes a detailed analysis of growth spurts in human and non-human primates).

Largo *et al.* (1978) analyzed several other features of the adolescent spurt in growth using the Swiss data. First, they ascertained that the age range of the sample, four to 18 years, is greater than the variation in the timing of the onset of adolescence. That onset occurred within the range of 6.6 to 13.5 years in the study sample, and thus the transition from juvenile to adolescent stage could be observed in every individual. This fact, and the relatively large number of children measured repeatedly from year to year, gives confidence to the growth statistics derived from the study. Using the data published by Largo *et al.* the change in acceleration of growth from childhood to adolescence can be calculated. During the juvenile stage the deceleration in the rate of height growth averages − 0.46 cm/year/year for boys and − 0.48 cm/year/year for girls. From the point of minimal juvenile velocity to the peak of the adolescent growth spurt the acceleration in height averages + 1.66 cm/year/year for boys and + 0.88 cm/year/year for girls.

The change in the velocity and acceleration of growth at adolescence affects almost all parts of the body, including the long bones, vertebrae, skull and facial bones, heart, lung and other visceral organs. Not all parts of the body experience the adolescent spurt at the same time. Different regions of the skeleton, for example, reach the peak rate of growth during adolescence at different ages (reviewed by Satake *et al.*, 1994). In a nationally representative sample of 18 004 girls from the former East Germany, measured by Holle Greil (1997), the average age at peak velocity of the spurt occurred at about nine years for foot length, 10 years for the hand length, 10.5 years for leg length, and 12 years for arm length, standing height, and trunk length. The sample of boys measured by Greil, totaling 18 123 individuals, followed the same general sequence of peak velocities, but the boys reached those peak values two to three years latter than the girls. A study of the skeletal sequence of adolescent spurts of Japanese boys ($n = 15$) and girls ($n = 20$) measured longitudinally from age 9.0 to 15.0 years was published by Satake *et al.* (1994). They found that, on average, the spurt in leg length preceded the spurt in trunk length, biacromial diameter (bony shoulder breadth), and biiliac diameter (bony hip width) in both boys and girls. Again, girls reached these spurts at an average earlier age than boys, but the sex difference was only about one year. Satake and colleagues noted that that there was considerable variation in this sequence for individual subjects, so much so that order of this sequence could, '. . . provide a unique method of clarifying individual growth variations' (p. 359).

Muscle mass of boys also undergoes a spurt at adolescence, and it is relatively greater than the spurt for growth in height. After the peak of the skeletal spurt, the rate of bone growth declines more steeply than the rate for muscle growth, meaning that adolescent boys continue to increase their muscle mass at a faster rate than they grow in stature. Expressed another way, the average healthy boy reaches 91 percent of his adult height by his age at peak height velocity (PHV) during adolescence, typically about 14 years of age. However, the same average boy achieves only 72 percent of his total muscle mass at the age of PHV and takes about four more years to reach 91 percent of his adult value (Buckler, 1990).

Some body parts and tissues do not evidence an adolescent growth spurt, for example, adipose tissue, both subcutaneous fat and the deep body fat, decreases in mass during adolescence in British and American boys, and perhaps in many girls as well (Tanner, 1965; Johnston *et al.*, 1974). Lymphatic tissues and the thymus show no adolescent increase in growth rates (Scammon, 1930). Another body part unaffected by the adolescent growth spurt is the female pelvis. Marquisa LaVelle (Moerman, 1982) measured the growth of the **pelvic inlet**, the bony opening of the birth canal, from X-rays taken on a sample of healthy, well-nourished American girls who achieved menarche between 12 and 13 years. These girls attained adult pelvic inlet size at 17 to 18 years of age. Quite unexpectedly, the adolescent growth spurt, which occurs before menarche, did not influence the size of the pelvis in the same way as the rest of the skeleton. Rather, the pelvis and pelvic inlet of these girls followed its own slow pattern of growth, which continued for several years even after adult stature is achieved.

One consequence of the pattern of growth of the female pelvis is that adolescent girls under age 17 who become pregnant risk their own health and the health of their fetus. This is due, in part, to the immaturity of their pelvic inlet. Other risks of teenage pregnancies are due to the fact that the mother is still growing in height and weight (bone, fat, muscle, etc.) and her own growth diverts nutrients away from the fetus. The result may be growth retardation of the fetus leading to a low birth weight infant. These risks, and the biological and social strategies that humans have developed to avoid them, are discussed in more detail in relation to the evolution of the human life cycle in Chapter 4.

Changes in stature, muscle mass, and fatness that typically occur from childhood through adolescence are illustrated in Figure 2.12 for a longitudinally measured sample of French-Canadian children (Baughan *et al.*, 1980). Muscle mass is an estimate of the amount of muscle at the mid-point of the arm. This estimate is derived from measurements of upper arm circumference and triceps skinfold. These measurements are usually taken at the mid-point of the upper arm, half way between the olecronon process (the 'elbow') of the ulna and the acromion process (the 'shoulder') of the scapula (Lohman *et al.*, 1988). The circumference measures the amount of skin, subcutaneous fat, muscle and bone in a cross-section of the arm. The triceps skinfold estimates the contribution of skin and subcutaneous fat to arm circumference. If it is assumed that the arm is cylindrical in shape, simple geometry may be used to calculate the lean arm circumference, which is the circumference of the muscle and bone at the midpoint of the upper arm. Gurney & Jelliffe (1973) give the following three formulae to make this calculation (readers with access to a skinfold caliper and tape measure may take the measurements and apply the formulae for themselves):

(1)   arm muscle diameter (mm) = (arm circumference in mm$/\pi$) − triceps skinfold in mm
(2)   arm muscle area (mm^2) = $(\pi/4) \times$ (arm muscle diameter)^2
(3)   arm fat area (mm^2) = ((arm circumference)$^2/4\pi$) − arm muscle area

If it is also assumed that the circumference of the humerus is equal for all individuals, variation in lean arm circumference represents differences in the amount of muscle at this site. Though the arm is not cylindrical in shape and the circumference of the humerus is not equal in all individuals, the differences between reality and the assumptions of the technique are small enough so that when applied to groups of individuals the estimate of average muscle mass at the mid-arm are reliable and accurate (Jelliffe, 1966; Martorell *et al.*, 1976; Frisancho, 1981).

In Figure 2.12 fatness is represented by the sum of three subcutaneous skinfolds – triceps, subscapular, and suprailiac. While skinfold measurements do provide a good estimate of the subcutaneous fat, their relationship to the deep body fat is questionable. Some early research found that the amount of fat in the subcutaneous and deep body reserves are positively correlated and that changes in both occur in a similar fashion (Hunt & Heald, 1963). However, more recent studies using computed tomography to measure deep fat show a lack of correlation in the amount of fat held in the two reserves (Borkan *et al.*, 1982; Davies *et al.*, 1986). Nevertheless, since during the childhood, juvenile, and adolescent stages most fat is subcutaneous, a measurement of the amount of subcutaneous fat is a fair estimate of total fat (Brozek, 1960).

It may be seen in Figure 2.12 that there is little difference in average stature between boys and girls until adolescence, after which boys are typically taller than girls. Girls usually begin their adolescent growth spurt about two years earlier than boys, which means that on average girls are taller than their male age-mates for a couple of years. Boys have greater average muscle mass at all ages, though the differences become absolutely greater, and biologically important, at adolescence. Conversely, girls tend to have more subcutaneous fat at all ages, and again, the difference in fatness increases during adolescence. On average, girls add fat mass continuously from age eight to 18, with a slowing or possible loss of fat at the time of the adolescent growth spurt (about age 11 to 12 years as shown in Figure 2.12). Most boys experience an absolute loss of total fat mass during adolescence, and may have no more fat at age 18 than they had at age six (Holliday, 1986). The adolescent spurt in muscle mass in boys is usually accompanied by an increase in bone density, an increase in cardio-pulmonary function, larger blood volume, and greater density of red blood cells. Increases in each of these also occur in girls, but at levels relatively and absolutely lower than for boys (Shock, 1966).

As indicated in Figures 1.4 and 2.5, the shape of the adolescent growth spurt is not symmetrical. The rise to peak height velocity is relatively slower than the fall after the peak. The size of the spurt is usually greater in boys than in girls, although there is much individual and population variation in this (see Chapter 5 for a discussion of population variation in growth and development). The size of the spurt and the age when peak velocity is reached are not related to final adult height. In fact some normal, but slow maturing, individuals and people with certain endocrine disorders do not have a growth spurt but may reach normal adult height (Prader, 1984). This fact makes the otherwise universal nature of the adolescent growth spurt an even more striking human characteristic. The evolutionary and biocultural significance of the human adolescent growth spurt are discussed in Chapters 3 and 4.

On average, adult men are taller and heavier than adult women. Alexander *et al.* (1979) surveyed 93 societies, including Western and non-Western cultures, and found that the stature of women averages between 88 and 95 percent the stature of men. In England, women average 93 percent of the height of men, and this average difference is identical for men and women in the tallest (97th percentile), median (50th percentile), and shortest (3rd percentile) height groupings (Marshall & Tanner, 1970). One study in Switzerland (Largo *et al.*, 1978) found that the difference between men and women in adult height is 12.6 cm. Since this study had followed the growth of the subjects longitudinally, from the age of four years, it was possible to calculate how much of the adult difference in height occurred in the various stages of postnatal growth. It was found that four factors accounted for the difference: the boys' greater amount of growth prior to adolescence added 1.6 cm, the boys' delay in the onset of adolescence added 6.4 cm, the boys' greater intensity of the spurt added 6.0 cm, and the girls' longer duration of growth following the spurt subtracted 1.4 cm from the final difference.

Due to the interplay of these factors the regulation of size may be more precisely controlled and the 'harmony of growth' evidenced during infancy and childhood is continued during adolescence. For instance, Boas (1930) discovered that the age at which adolescent growth begins is inversely correlated with the size of the spurt, meaning that early-maturing children have higher peak height velocities than late-maturing children (Figure 1.6). This observation has been confirmed for American children (Shuttleworth 1937, 1939), British children (Marshall & Tanner, 1969, 1970), and Swedish children (Lindgren, 1978). Another compensating mechanism, described for Swiss children by Largo *et al.* (1978), is that a child with slow growth prior to puberty will tend to have a longer-lasting growth spurt during adolescence than a child who achieves a greater prepubertal percentage of adult height. Where chronic undernutrition, disease, and child labor are endemic, such as in highland Peruvian Indian societies (Frisancho, 1977), East African pastoral societies (Little *et al.*, 1983) and Guatemala Maya villages (Bogin *et al.*, 1992), height at every age is reduced compared with less stressed populations. However, the total span of the growth period is prolonged, up to age 25 or 26 years, so that a greater adult height may be achieved than if growth stopped at 18 to 21 years, as it does for healthy individuals in the United States. Presumably, these growth adjusting mechanisms are present in all children.

Patterns of secondary sexual development

In both sexes, the onset of the puberty is followed within a few months by the appearance of the secondary sexual characteristics. In boys these include changes in size of the penis and scrotum, the growth of pubic, axillary, and facial hair, the 'breaking of the voice', and seminal emission. In girls the secondary sexual characteristic include the growth of the breasts, appearance of pubic and axillary hair, **menarche** (first menstruation) and development of the uterus, vagina, and vulva to their mature size and appearance. One of the common methods used to assess the secondary sexual development of boys and girls is the Tanner Puberty Stage classification system (Tanner, 1962). This system divides the development into five stages. For both boys and girls there are five stages of pubic hair development: PH1 indicates no visible pubic hair, PH2 the first appearance of pubic hair, and PH3 to PH5 the progressive growth of pubic hair to the adult stage. For boys there are five stages for the development of the penis, testes, and scrotum, which rate their changes in size and coloration. This genital rating proceeds from G1, the prepubertal stage, to G5, the adult stage. For girls there is a breast development scale that begins at B1, no breast development, and proceeds to B2, initiation of the breast bud, to B5, the adult form of the breast (including areola and nipple).

Details of the development of the primary and secondary sexual characteristics, interrelationships between these events during adolescence, and sex differences in the timing of these events are available for Americans (Simmons & Greulich, 1943; Reynolds & Wines, 1948; Nicolson & Hanley, 1953), for Chinese (Lee *et al.*, 1963; Chang *et al.*, 1966), for the English (Marshall & Tanner, 1969, 1970; Billewicz *et al.*, 1981), the Swedes (Taranger *et al.*, 1976), for Poles (Bielicki, 1975; Bielicki *et al.*, 1984), and for Turks (Onat & Ertem, 1974; Neyzi *et al.*, 1975a, b). Differences exist between these populations in the timing of onset of the stages of adolescent maturation. For instance, breast development in American (of European descent) and British girls begins at an average age of about 11 years, which is about one year later than for Turkish or Chinese girls. Turkish boys begin pubic hair development at, on average, 11.8 years, compared with Swedish boys at 12.5 years and Chinese boys at 13.0 years. Even so, the results from each of these samples are remarkably similar, despite the variation in ethnicity, geographic areas, and cultural practices of each population, and variation in the methods of measurement and analysis used by the authors. Indeed, the amount of variation in the age at which individual adolescents achieve any maturational stage is greater within the samples studied than between the samples. Thus, there may be average differences between samples, but they are of little biological importance.

All of these studies report findings for, generally, healthy adolescents of middle to upper socioeconomic class. There are also studies of the sexual development of South African adolescents who had been severely malnourished (requiring hospitalization) when they were between 5 months and 4 years 4 months old (Cameron *et al.*, 1988, 1990). These former patients were followed-up at 10 years and 15 years after their hospitalization and compared with a control sample of adolescents from similar low socioeconomic and poor nutritional circumstances who had not developed clinical signs of severe malnutrition. Amazingly, both the former patients and the controls show an identical sequence in the order of appearance and timing of the secondary sexual characteristics. From this one can conclude that severe malnutrition at an early age does not seem to disrupt the basic human pattern of secondary sexual development. In Chapter 5 the population variation in sexual development, and the environmental causes of this variation, are explored in more detail.

The common control of adolescent development

As the sequence of appearance of secondary sexual characteristics is so highly predictable in people, some researchers have attempted to find a common control mechanism for adolescent development. Two such attempts were by Nicolson & Hanley (1953), who used factor analysis on their American sample, and by Bielicki *et al.* (1984), who used principal component analysis on their Polish sample. These statistical techniques are similar in that they divide the total variance in a set of data into discrete sources of variation, called 'factors' or 'components'. In both studies the component accounting for the largest percent of the total variance in maturation was a general maturity factor. Clustered on this general maturity factor were growth velocities for height and other linear dimensions, stages of sexual maturation (e.g., breast development in girls or pubic hair growth in boys), and skeletal maturation. For the data of Nicolson & Hanley, the general maturity factor accounted for an average of 71 percent of the variance in maturation in the boys, and an average 73 percent of the variance in girls. Bielicki *et al.*, found that the general maturity factor accounted for 77 percent of the variance in boys and 68 percent in girls. This statistical finding, along with the similarities in adolescent maturation found in different populations, supports the idea that adolescent maturation is controlled by some central organ or system within the body. These data also demonstrate that there is a human pattern of adolescent growth and development which is shared by all people. The search for the central control of maturation is a subject discussed in Chapter 7 (which treats genetic and endocrine regulation of growth). The reasons for the universal pattern of adolescent maturation is the subject of Chapters 3 and 4, which discuss the evolution and ecology of human growth.

Adult stage

The attainment of adult stature is one of the hallmarks used to mark the transition from adolescence to **adulthood**. In the United States, young women and men of middle to upper socioeconomic status reach adult height at about 18 years of age and 21 years of age respectively (Roche & Davila, 1972). These ages may be close to the lower limit for onset of adulthood. In other populations growth continues to later ages. Hulanicka & Kotlarz (1983) studied a sample of 221 young men from Wroclaw, Poland, an industrial city of 600 000 people, and found that only 54 percent of the subjects reached final adult height by age 19 years. The other 46 percent added an average of 2.13 cm in height between the ages of 19 and 27 years. It is well known that individuals suffering from undernutrition, chronic diseases, and certain drug therapies may continue to grow in height until reaching about 26 years of age. Though these individuals may grow for a longer period of time, they usually achieve less total growth and end up shorter than their more privileged or healthier age mates.

Height growth stops when the long bones of the skeleton (e.g., the femur, tibia, etc.) lose their ability to increase in length. Usually this occurs when the epiphysis, the growing end of the bone, fuses with the diaphysis, the shaft of the bone. As shown in Figure 2.4, the process of epiphyseal union can be observed from radiographs of the skeleton. In their study of Polish men, Hulanicka & Kotlarz (1983) found that the amount of growth that occurred after age 19 was a function of **skeletal maturation,** late maturers grew more than average or early maturers. This fact has been known for many years, and an estimate of skeletal maturation, often called **skeletal age,** is incorporated into equations used to predict the adult height of children (Bayley & Pinneau, 1952; Roche *et al.,* 1975b; Tanner *et al.,* 1983a). The fusion of **epiphysis** and **diaphysis** is stimulated by the gonadal hormones, the androgens and estrogens. However, it is not the fusion of epiphysis and diaphysis that stops growth, for children without gonads, or whose gonads are not functional, never have epiphysial fusion, but they do stop growing (Tanner, 1978). Rather, it is a change in the sensitivity to growth stimuli of cartilage and bone tissue in the **growth plate region** (Figure 2.13) that causes these cells to lose their hyperplastic growth potential.

Reproductive maturity is another hallmark of adulthood. The production of viable spermatozoa in boys, and viable oocytes in girls, is achieved during adolescence, but these events mark only the early stages, not the completion, of reproductive maturation. For girls, menarche is usually followed by a period of one to three years of adolescent sterility. That is, there are menstrual cycles, which are often irregular in length, but there is no ovulation. So, the average girl is not fertile until 14 years of age or older. Fertility, of course, does not indicate reproductive maturation. Becoming pregnant is only a part of reproduction: maintaining the pregnancy to term and raising offspring to adulthood are equally important aspects of the total reproductive process. Adolescent girls who become pregnant have a high percentage of spontaneous abortions and complications of pregnancy. This is true for girls in developed nations such as the United States (Taffel, 1980) and developing nations such as Peru (Frisancho *et al.,* 1985). Teenage mothers also have higher rates of low birth weight infants than older mothers and, consequently, these infants suffer high rates of mortality (Taffel, 1980; Garn & Petzold, 1983). There are many reasons for the reproductive difficulties faced by teenage girls, ranging from physiological immaturity of the reproductive system to socioeconomic and psychological trauma induced by the pregnancy. The fact that the mother is still growing means that the nutritional needs and

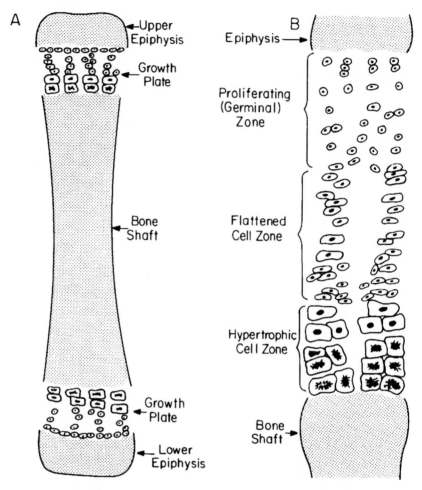

Figure 2.13. (A) Diagram of a limb bone with its upper and lower epiphyses. (B) Diagrammatic enlargement of the growth plate region: new cells are formed in the proliferation zone and pass to the hypertrophic zone to add to the bone cells accumulating on top of the bone shaft (from Tanner, 1990).

hormonal activity of her body may compete with and interfere with the growth and development of the fetus. This problem was suggested by Pagliani over 100 years ago and confirmed by Frisancho *et al.* (1985). For all these reasons, most researchers agree that female reproductive maturity is reached at the end of the adolescent stage of life, which occurs, on average, at 19 years.

Analysis of urine samples from boys age 11 to 16 years old show that they begin producing sperm at median ages that cluster between 13.4 and 14.5 years (Richardson & Short, 1978; Hirsch *et al.*, 1979; Laron *et al.*, 1980; Muller *et al.*, 1989). Whether this event marks the onset of fertility is not known. The quality of viable sperm from teenage boys is also unknown, though one may speculate that pubertal endocrine activity in the boy may have some effect on his sperm cells. Even if fertile, the average boy of 14 years is less than half-way through his adolescent growth spurt and, therefore, his developmental status is incomplete. In terms of physical appearance, physiological status, and psychosocial development, he is still more of a child than an adult. The cross-cultural evidence shows that few boys successfully father children until they are into their third decade of life (Bogin, 1993, 1994a). In the United States, for example, only 3.09 percent of infants live-born in 1990 were fathered by men under 20 years of age (National Center for Health Statistics, 1994). A notorious exception occurred in 1997 when a 13-year-old schoolboy in the state of Washington fathered a child with his teacher, a married women who had already given birth to four children.

The transition to adulthood is marked by dramatic events, such as the cessation of height growth and full reproductive maturity. In contrast the course of growth and development during the prime reproductive years of adulthood are relatively uneventful. Most tissues of the body lose the ability to grow by hyperplasia (cell division), but many may grow by hypertrophy (enlargement of existing cells). Exercise training can increase the size of skeletal muscles and caloric over-sufficiency will certainly increase the size of adipose tissue. However, the most striking feature of the prime adult stage of life is its stability, or **homeostasis**, and its resistance to pathological influences, such as infectious disease and psychological stress (Dubos, 1965; Timiras, 1972).

Senescence and old age

Following the prime reproductive years of adulthood the process of aging becomes more noticeable. Aging, or **senescence**, is a process of decline in the ability to adapt to environmental stress. The pattern of decline varies greatly between individuals. Though specific molecular, cellular, and organismic changes can be measured and described, not all of these occur in all people and rarely do they follow a well-established sequence. **Menopause** may be the only event of the later adult years that is experienced universally by women who live past 50 years of age; men have no similar event. The biology and possible value of menopause is described in Chapter 4.

There are many hypotheses about the aging process and about why we must age at all. One explanation for why we must age is called the 'pleiotropy hypothesis'. In now classic works on the biology of senescence Medwar (1952) and Williams (1957) argued that aging is '. . . due to an accumulation of harmful age-specific genes . . . [or] . . . pleiotropic genes which have good effects early in life, but have bad effects later . . .' (Kirkwood & Holiday, 1986, p. 371). Kirkwood (1977), Charlesworth (1980) and others refined this hypothesis further in terms of a general theory of aging. Some of the empirical experimental support for the pleiotropic hypothesis links aging with the limited mitotic (cell duplication) ability of hyperplastic cells. Hayflick (1980) found that when raised in tissue cultures, human embryo hyperplastic cell lines double in number by mitotic division only 50 (± 10) times and then die. Tissue cultures of cells from adult humans have an even more limited mitotic potential, doubling only 14 to 29 times before dying. This doubling limit of hyperplastic cells provides a theoretical limit to life. In practice, few cells and few people ever reach this limit (Austad, 1997). Many cells die as part of normal physiology, not senescence. These cell deaths appear to be programmed by gene–environment interactions, a process called **apoptosis**. In many cases cell death is required to bring about the mature form of a tissue or organ. An excess of brain cells are produced during prenatal growth and after birth, and apparently in response to many environmental stimuli, groups of brain cells die in order to create the functional architecture of the brain. Another example of apoptosis occurs during the growth of the mammalian skeleton. To maintain the form and efficient function of long bones as they grow, existing bone cells must die so that as the bone lengthens it does not also become wider (this process of bone remodeling is detailed in Chapter 3). People, as whole organisms, do not reach an age that would place them at or near Hayflick's theoretical limit. Rather, death occurs due to the inability of one or more cell types, including nerve, muscle and other non-replicating cells, to use nutrients and repair damage.

Other candidates for 'agents of aging' are: (1) free radicals, chemical by-products of metabolic activity, that accumulate with time and can damage DNA, proteins, and cell membrane; and (2) the amassed burden of DNA mutations caused by ionizing radiation or chemical pollutants. Undoubtedly, aging is a multi-causal process, but unlike the biological self-regulation of growth prior to adulthood, there may not be a biological plan for the aging process. There may be no biological reason to age in any particular way. It is only recently, in the evolutionary history of our species, that an appreciable number of individuals have come to live past the prime adult years. Throughout prehistory, death by predation, disease, and trauma caused by violence and accidents was probably more common than death due to old age. Death is inevitable, but nature did not have the time or the selection pressures to mold our manner of death into a predictable pattern.

In contrast to the process of aging and death, growth and development from conception to adulthood follow a predictable pattern. It was during the evolutionary history of our species, and those species ancestral to ours, that selective pressures operated to shape our pattern of growth. To understand why we grow the way we do, we must examine some of the events that occurred during human evolution. The next chapter describes the evolution of the human pattern of growth.

3   The evolution of human growth

Biologists and anthropologists have proposed a number of taxonomic schemes for classifying *Homo sapiens*. Lovejoy (1981) suggested five characteristics of humans as defining features: bipedality, a large neocortex, reduced anterior dentition with molar dominance, material culture and unusual sexual and reproductive behavior. The development of each of these characteristics can be seen in the ontogenetic unfolding of the human pattern of growth and development. For instance, bipedality is made possible by differential growth of the legs and pelvis, including the bone, muscle, tendons, and ligaments, versus the arms and shoulder girdle (Simpson *et al.*, 1996). Our unusual reproductive behavior results, in part, from our prolonged childhood, delayed maturation, and species-specific neuroendocrine physiology. The human pattern of endocrine physiology results in the menstrual cycle of women, the continuous sexual receptivity of both sexes, the development of our secondary sexual characteristics, e.g., patches of hair in the groin and armpits rather than fur all over, and menopause in women. These are only some of the features of *Homo sapiens* and human growth. Though these characteristics set us apart from all other species, they have their origins in evolutionary history. In this sense we share many basic growth patterns with other species, but we differ in other ways due to some special evolutionary developments. To better understand both the shared and special features of human growth, this chapter explores the phylogeny of growth of lower vertebrates, the mammals, and the primates – the group that includes monkeys, apes and humans.

Vertebrate and mammalian foundations for human growth

The growth and development of any organism may be viewed as movement through space or time (Thompson, 1917, 1942). It moves through time in two ways, first in the conventional sense of the passage of days, months, and years, and secondly in the sense of development and maturation from earlier to later stages of life. These temporal movements can be

98

represented in mathematical form, for instance as the distance and velocity curves of growth illustrated in Figures 2.5 and 2.6, or as the curves for maturation of body systems illustrated in Figure 2.7.

Most non-human organisms share the same basic curve of growth. It is an S-shaped, or sigmoid, curve (Figure 3.1, A). The growth of chickens, rats, and cattle follow this curve (Brody, 1945; Bertalanffy, 1960; Laird, 1967; Timiras, 1972). The growth of parts of organisms and colonies of cells, such as bacteria and tumors in animals, also conform to this sigmoid curve. Figures 3.1 B and C illustrate some of the other mathematical features of the general growth curve. In B the velocity, or rate of growth, is given; only a single peak, or maximum rate of growth occurs, meaning that there is an initial acceleration and then a period of deceleration in growth rate. In C the changes in acceleration are more clearly revealed; the point of zero acceleration corresponds to the inflection point in the velocity curve where the rate of growth stops increasing and begins to slow. Growth rate at any subsequent point on the curve is decreasing with time.

In mathematical terms, parts B and C of this figure are related to A as its first and second derivatives, respectively. That the curve of general growth is completely differentiable means, mathematically, that the growth process represented by this curve is smooth and continuous. That continuity means that we can predict changes in amounts and rates of growth during the course of development with precision. Such predictions allow us to make quantitative and qualitative assessments of the growth of any individual organism and to make comparisons between different individuals, groups of individuals, and even different species of animals in terms of the mathematical properties of their growth curves.

An example of this type of smooth and predictable growth pattern is given in Figure 3.2 for the chicken. Only the physical constraints of the egg, around the time of hatching, interfere with a smooth growth trajectory. The rigid shell and the depletion of nutrients from the yolk sac of the egg slow growth before hatching. After hatching the growth rate 'rebounds', but only to the point where an averaging of the prenatal and postnatal growth rates would yield a smoothly decelerating curve. A similar pattern of growth occurs for humans just before and after birth (see Figure 2.6). During the last part of the third trimester of human pregnancy the fetus is large enough to press against the inner surface of the uterus and the placenta, which probably constricts blood vessels and inhibits the fetal–maternal exchange of nutrients, gases, and wastes. Fetal growth slows, but rebounds several days following parturition so that the child 'catches-up' to the size he or she would have achieved if there had been no prenatal decrease in growth rate.

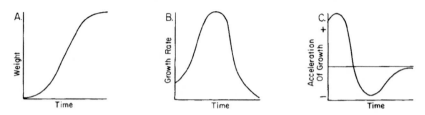

Figure 3.1. General growth curves: (A) weight versus time, (B) rate of growth (velocity) versus time, and (C) acceleration of growth rate versus time (after Medawar, 1945).

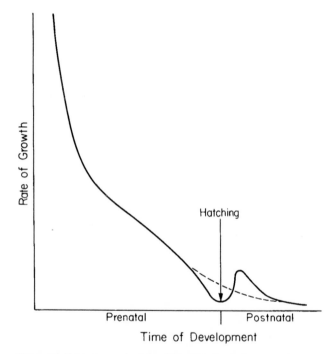

Figure 3.2. Rate of growth of the chick before and after hatching. The interrupted line is the theoretical curve if no growth restriction prior to hatching takes place (from Timiras, 1972).

Brody (1945) and Bertalanffy (1960) showed that the growth of mice and Brahma cattle may be modeled with the same curve used for the chick. Thus the pattern of growth of these animals of very different absolute size, distinct evolutionary history, and ecologically diverse life styles is qualitatively identical. Some important exceptions to the general pattern of animal growth are found when plotting the growth of highly social mammals, including humans. The human pattern was illustrated and its basic aspects discussed in the previous chapter (Figure 2.5). The distance curves in Figure 2.5 are, at first glance, not markedly different from the general sigmoid curve of Figure 3.1 A. However, the human velocity curves illustrated in Figure 2.5 are different from the velocity curves for other animals (Figure 3.1 B). Humans have a series of both rapid and gentle decreases in growth rate after birth punctuated by both small and large increases in growth rate (the mid-growth spurt and adolescent growth spurt). Even a casual inspection of the human and non-human curves presented so far finds both qualitative and quantitative differences exist in the pattern of growth.

Unlike the non-human curve of growth, the human growth curve cannot be modeled with a single mathematical function: that is, it is not completely differentiable over its length. The distance curve requires at least two functions, one for the pre-pubertal segment and one for the post-pubertal, or adolescent, segment (Shohoji & Sasaki, 1984). The velocity curve requires at least three mathematical functions for adequate description, one for the infancy stage (birth to about three years), a second for the childhood and juvenile stages (from about three years to the beginning of the adolescent spurt) and a third for the adolescent stage (Laird, 1967; Bock & Thissen, 1976; Bogin, 1980; Karlberg, 1987). These mathematical features of human growth are unusual for mammals, but they have their origins in the patterns of growth followed by other species of social mammals, especially the primates.

Mammalian growth

The growth of mammals differs from that of other vertebrates for two basic reasons. One reason relates to mammalian locomotion and the other relates to mammalian reproduction. Animals, both invertebrates and vertebrates, evolved the capabilities for rapid and flexible movement. This requires muscle tissue and something for it to work against. Most vertebrates (such as fish, amphibians, reptiles, birds, and mammals) utilize bone for this purpose, a tissue that provides support and protection due to its rigidity, but also the developmental flexibility that allows for growth (Goss, 1978). Most fish and amphibians make poor use of bone compared to reptiles, mammals and birds (the bones of birds are not discussed here). The fish bone grows by **periosteal** deposition, the addition of tissue on all of the external surfaces of the bone. As they grow, the bones of fish not only elongate but also widen becoming heavier over time. The buoyancy of water compensates for the additional weight. The skeleton of higher vertebrates does not grow this way. Goss states that 'amphibians evolved marrow cavities . . . [and] also acquired cartilaginous epiphyses . . .' (1978, p. 65). The epiphysis is the growing end of a long bone, and by maintaining the epiphysis in a cartilaginous form growth can continue. If the epiphysis develops into hardened bone tissue, growth ceases. The marrow cavity allows for the transport of nutrients to both ends of a growing bone. With these new structures amphibians were able to grow skeletons that were longer without also growing wider.

The reptiles also took the next step in bone evolution and developed the cartilaginous **growth plate**, which allows for several improvements in skeletal efficiency. The growth plate of a typical mammalian long bone, including the diaphysis (bone shaft), epiphysis, and growth plate, was shown in Figure 2.14. The growth plate separates the growing part of the bone from the rigid part. A photomicrograph of the growth plate region of a rat is given in Figure 3.3. The figure is labeled to indicate four regions described by Horton & Machado (1992), beginning with the reserve zone, which is farthest from the surface of the growing bone. In this zone are the reserve chondrocytes, which are cells that form the cartilage precursor of mature bone cells. When these reserve cells begin to form bone they first migrate into the proliferative zone where they flatten and undergo mitosis to form clusters of cells. In the hypertrophic zone the cells in these clusters increase in size, especially in vertical diameter. These cells also begin to mineralize, but they do not become true bone cells. That happens when the enlarged clusters finally reach the zone of primary ossification where they are invaded by blood vessels and then receive deposits of bone matrix from osteocytes (bone-making cells). As these ossified cells undergo remodeling during growth the original cartilage core is completely replaced by bone matrix. The growth plate is located behind the bony articular surface, which allows for efficient skeletal operation while growth is taking place. The amphibians have to make do with an entirely cartilaginous epiphysis, which is not as hard or durable as bone and limits the range of work and the size of the organism that the skeleton can accommodate. The new structure of reptiles has obvious use in the prodigious size attained by dinosaurs, as well as giant Galapagos tortoises and other large-bodied reptiles alive today.

The evolution of the growth plate overcame many of these problems and allowed reptilian and mammalian bones to perform their functions more efficiently throughout life. The growth plate system also allows for both relatively rapid growth in early life and then the cessation of growth. In fish, amphibians, and some reptiles bone growth never completely stops, although growth continues very slowly after sexual maturity is reached. This

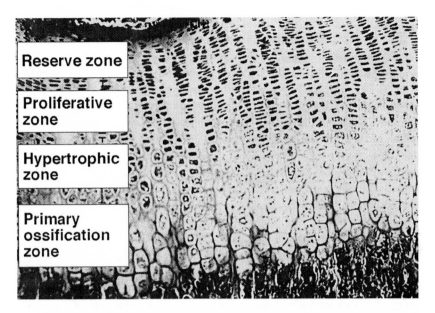

Figure 3.3. Photomicrograph of the growth plate region of the rat. The tissue sample comes from an animal at about 30 days post-conception. The sample was Azan stained and photographed at 6.3 × 20 magnification by Dr. A. A. Missankov, Department of Anatomical Sciences, University of the Witswatersrand, South Africa. See the text for explanation of the four zones.

pattern of growth is unsuitable for mammals, who must limit growth, for the most part, to the pre-reproductive stages of life. Mammalian reproduction, especially for females, demands so much energy and body resources that it precludes the possibility for significant body growth of the mother and, at the same time, the fetus. One of the dangers of human teenage pregnancy is that the mother's own growth may divert resources from the fetus, resulting in fetal growth retardation (as mentioned before). Mammals and birds make up for a finite period of growth by being able to grow very rapidly. The maximum rate of bone growth, measured in birds, is 3.0 mm/day per growth plate, or 6 mm/day for bones with growth plates at both ends (Kember, 1992). Human infants can achieve growth rates of slightly more than 1.0 mm/day in crown–heel length (Heinrichs et al., 1995). By limiting growth to an early stage of life an organism can develop a strategy of investing first in its own development, and then, after ending growth, become sexually mature and begin to invest in the development of its offspring.

An end to growth is also necessary for terrestrial animals that must support their own body weight without the help of water or any other buoyant medium. The largest terrestrial mammals, the Proboscideans (elephants and the allied extinct mammoths and mastodons), may have reached the limits of size for land mammals. The limbs of these animals are used almost entirely for support of the body and locomotion. The evolution of a flexible muscular appendage, the trunk, serves the function of a limb for food gathering and environmental manipulation. Some smaller mammals, such as rats and other rodents, never fuse their growth plates with the diaphysis and never stop growing. However, because of their relatively small size they are capable of flexible and rapid movement throughout their lives. These small, ever-growing rodents increase in size so slowly and die so soon after sexual maturity that they never attain the sizes suggested by certain second rate horror films, though one rodent, *Hydrochoerus hydrochaeris*, the capybara of South America, may attain 1.3 meters in length and 50 kg in weight.

To maintain efficiency of function, a pattern of limited growth and **remodeling of bone** evolved to meet mammalian needs for movement and diet. The physiological characteristics of mammals, including **homoiothermy** (self-regulation of a relatively constant body temperature), efficient placentation by which the fetus continuously benefits from maternal blood circulating in the uterus, rapid bodily movement and other features associated with a relatively high metabolic rate, require a diet rich in energy and other nutrients. In the long term, mammalian metabolism requires a constant and high quality dietary intake rather than episodes of abundant food, like that of a snake gorging itself at a single meal. Mammals must be able to move rapidly and efficiently to find and capture quality foods on a regular basis. This requires an efficient musculo-skeletal system during both the early and the later phases of growth. Bone remodeling helps to maintain such efficiency. As a bone grows in length or size its surfaces must be reworked so that its characteristic shape and function can be retained. In a long bone this remodeling is achieved by removing old ossified bone tissue from the **periosteal** (outer) surface and adding new bone tissue to the **endosteal** (inner) surface (Enlow, 1963, 1976). This process is schematically illustrated in Figure 3.4. Via bone remodeling, the evolution of the growth plate, the cessation of growth of the skeleton, and other evolutionary alterations in the function of limb bones (see Romer, 1966), mammals were able to achieve the efficient, rapid and flexible mobility that they require.

Mammalian reproduction

Reproduction is the second aspect of mammalian biology that influences growth. Mammalian reproduction is based upon a high degree of invest-

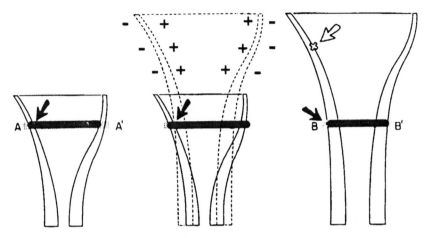

Figure 3.4. Diagrammatic representation of remodeling in a limb bone. As the bone grows in length, the level indicated by the line AA' becomes repositioned into the level indicated by the line BB'. The relative level of AA' in the larger bone is indicated by an X. The structural remodeling of the bone occurs by the process of resorbtion (− signs) and deposition (+ signs). One result of the subtraction and addition of bone is that the point indicated by the black arrow at level AA' has been relocated from the inner side of the cortex of the bone to the outer side of the cortex in BB'. As the bone continues to grow in length, and new bone tissue replaces the older bone tissue, the point will eventually be lost by continued resorbtion. From Enlow (1963), with permission of the author and publisher.

ment in offspring, both before and after they are born. That investment takes the form of energy and time that mammalian parents, especially mothers, provide to their offspring. Mammals were not the first class of vertebrates to evolve parental care and investment in their offspring, indeed such care is found in all classes (fish, amphibians, reptiles, birds), but mammals have carried it to a level of physiology and behavior exceeding that of other vertebrates. The high quality of mammalian parental investment may be measured by the efficient internal fertilization and placentation of most living mammals, lactation by the mother during her offspring's infancy, and by the capacity of each individual offspring to help ensure its own survival to reproductive age. Each of these mammalian features has a direct relationship to growth.

Placentas

Mossman (1937) described the **placenta** as '. . . any intimate apposition or fusion of fetal organs to the maternal tissues for physiologic exchange'. Though the contact between tissues may be 'intimate' there is never any direct connection between mother and fetus and the exchange of substances always occurs through a tissue boundary. The evolution of the placenta removed some of the limitations to prenatal growth, including both growth in size and length of gestation. The prenatal growth and gestation of non-placental animals such as most reptiles, birds, monotremes (the platypus and echidna) and most marsupials (e.g., opossums and kangaroos) is limited by the need to 'package' fetal nutrients in the **yolk sack** (YS) and fetal waste products in a separate compartment called the **allantosis** (AL) (Figure 3.5). In contrast, the placenta provides for fetal nutrition, respiration and the removal of metabolic wastes continuously throughout gestation.

Internal gestation and live birth, or **viviparity**, evolved in all classes of living vertebrates, except the birds. Even placenta-like structures evolved independently in many non-mammalian species, such as snakes. Mammals, however, developed the most efficient types of placentas. Amoroso (1961) described three basic kinds of placentas for mammals (the first and third are depicted in Figure 3.5). The first is the **yolk sac placenta**, found in some marsupials and rabbits, in which blood vessels connect the yolk sac with the uterine wall. Nutrients from the mother's circulation may be transferred to the yolk sac to replenish the fetal nutrient supplies. Wastes are still confined to the allantosis and cannot be removed via the maternal circulation. The second type is the **chorionic–allantoic** placenta, characteristic of higher mammals, in which parts of the surface of the allantosis fuse with the chorion (a membrane surrounding the fetus composed of maternal and embryonic tissues). This apposition allows for a more efficient exchange of substances between mother and fetus than in the yolk sac placenta, because a greater surface area of fetal and maternal tissues are in contact. In the egg-laying birds, reptiles, and monotremes, and in the mammals with yolk sac placentas, the allantosis serves only as a receptacle for embryonic and fetal wastes. The conversion of this waste-sac function to a system for the exchange for nutrients, gases and wastes in some placental mammals is an example of the conservative nature of evolution. Existing organs are often 'retooled' for new functions rather than having organisms develop totally new organs.

The third type Amoroso described is the **chorionic placenta**, in which there is a more direct connection between the chorion and the fetus via the umbilical cord. The chorionic placenta is found in rodents, monkeys, apes and humans. There is generally a greater amount of surface area for the exchange of substances between fetus and mother in this type of placenta. The chorionic placenta also presents fewer boundaries between the maternal and fetal blood circulation. The chorionic placenta type may be

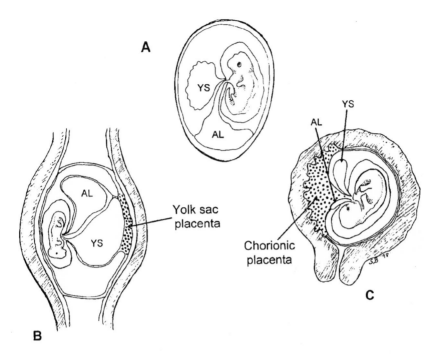

Figure 3.5. Schematic diagrams of the reptilian egg (A), the yolk sac placenta of the rabbit (B), and the chorionic placenta of haplorhine primates (C). See text for details (after Hamilton & Mossman, 1972 and Zihlman, 1982).

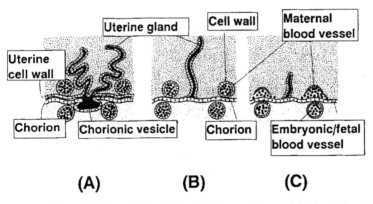

Figure 3.6. Schematic drawings of the three sub-types of chorionic placentae: (A) epitheliochorial, (B) endotheliochorial, (C) hemochorial. Embryonic tissues are in light stippling and maternal tissues are in heavy stippling (after Martin, 1990).

classified into three sub-types, which are illustrated in Figure 3.6. A brief discussion of these sub-types is important because human beings possess the most invasive and intimate type of placenta of any mammal. The nature of the human placenta helps us to better understand how the mother's environment and her behavior presents both benefits and risks to her embryo and fetus. Benefits include the efficient transport of needed nutrients and oxygen to the fetus and the efficient removal of wastes from the fetus. Behavioral risks of the mother include the use of tobacco and alcohol consumption – these behaviors may interfere with the operation of the placenta and may also introduce harmful substances to the embryo and fetus.   The first sub-type, called the **epitheliochorial placenta**, maintains the separation between the chorion and the epithelial lining of the uterus. Maternal blood vessels are separated from fetal blood vessels by four barriers – the two cell walls of the blood vessels themselves and the cell walls of the chorion and uterus. The development of a profuse network of uterine glands and chorionic vesicles increases the area of contact and exchange across epitheliochorial placenta and partially overcomes the barrier to nutrient exchange imposed by the four cell walls. The epitheliochorial sub-type is found in the lemurs and lorises (the strepsirhine primates). The second sub-type is called the **endotheliochorial placenta**, and it eliminates the uterine epithelium barrier. This seems to allow for more direct exchange between the maternal and fetal blood vessels, without the need for the network of chorionic vesicles. The endotheliochorial sub-type is found in rodents. The third sub-type is called the **hemochorial placenta**. This is the most invasive kind of placenta because the cell wall of '. . . the maternal blood vessels has itself broken down so that the chorion is directly bathed by maternal blood . . .' (Martin, 1990 p. 446). The hemochorial sub-type is found in tarsiers, New World monkeys, Old World monkeys, apes, and humans (the haplorhine primates). In human beings, the fetal capillaries fuse with the chorion and protrude through to the maternal side of the placenta where these capillaries sit in pools of maternal blood. The maternal and fetal circulatory systems are never joined directly, but they share a single wall of tissue between them. Such close contact over a relatively large surface area – gross external dimensions of the human placenta average 16 to 20 cm in diameter and 3 to 4 cm in thickness with a total surface area of contact of between 10 and 11 square meters (Timiras, 1972) – allows for the greatest ease of diffusion and active transport of substances across the placenta of any mammal.

The critical advance in the biology of placental mammals is that the fetus can develop and grow to an advanced stage protected and well-nourished in the uterus. In general, the length of gestation and the growth in size of the whole fetus (that is, the total body weight) of placental mammals are constrained primarily by the mechanical limitations of the mother's uterus and the size of the birth canal. However, in proportion to the weight of the mother, the primates with hemochorial placentae (monkeys, apes, humans) give birth to heavier newborns than the lorises and lemurs with their epitheliochorial placentae. The 'higher' primates also develop offspring with proportionally larger brains than the prosimian primates (Martin, 1990). Placental type is not the only reason for these differences in growth of the fetal body and brain, but it is a significant correlate.

Lactation

All mammals, even the egg-laying monotremes, nurse their young. **Lactation** continues to supply high-quality nutrients to the newborn. However, the evolution of lactation required behavioral changes in both mother and offspring, particularly in mother–infant bonding, which maintains the infant in contact and communication with the mother so that it can be suckled when hungry. The mother–offspring contact ensuing from this feeding method establishes a period of dependency in the young and a reciprocal period of parental investment by the mother. This time of life for the newborn is called infancy (as defined in the previous chapter), and it has become a stage in the life cycle and growth curve of all mammals. A similar period of dependency occurs for birds, but is of shorter duration and less physical intimacy than that for most mammals.

Lactation and infancy prolong the period of dependency, but allow for rapid and high-quality growth and greater physical and behavioral adaptability (Pond, 1977). Compared with other animals, infant mammals may be better able to adjust total rates of growth, or the rates for specific body parts, to adapt to environmental stress. For example, in cold environments, growth rates may be adjusted to produce adult mammals with relatively larger bodies, shorter extremities, or both, compared with mammals living in warmer climates. Large bodies with short extremities conserve heat better than smaller bodies with relatively long extremities; the former body type has relatively less surface area for heat dissipation. The body size and body proportions of polar bears (*Thalarctos maritimus*) and Malayan sun bears (*Helarctos malayanus*) conform to these growth adaptations. Infancy may also increase behavioral adaptability by allowing young mammals the time to practice and improve innate behaviors, such as the stalking of prey in carnivores. The mother–infant bond increases the opportunity for young mammals to acquire learned behaviors by observing and imitating their mothers or other adult animals with whom the mother socially interacts.

Caroline Pond (1977) provides an original and highly useful discussion of the evolution of lactation. She points out that viviparity and lactation evolved independently in many species of vertebrates. Only in the marsupial and placental mammals do these both occur regularly as a package. Together, viviparity and lactation protect the mother's investment in ova, embryos, and newborns. Viviparity allows the mother to take the embryo and fetus with her during the whole gestation, rather than leaving an egg unprotected at a nest. Lactation supplies high-quality nutrients to the infant, even if this requires cannibalizing the mother's body reserves of fat, protein, vitamins, and minerals. Non-lactating species must often confine the birth of their young to places and times of the year when appropriate foods are available, but lactation provides appropriate food at all times and places. The high-quality nutrition provided via lactation allows for rapid postnatal growth, especially of the skull, jaws, and teeth during infancy. By the end of infancy these developments mean that most mammalian species are able to eat the adult type diet. As a consequence, states Pond, sexual maturation and reproduction may follow quickly after the end of lactation/infancy, and indeed they do for most species of mammals. Pond observes that the investments in offspring and patterns of growth and maturation associated with the viviparity–lactation package result in high levels of reproductive success for the mammals.

Brains and learning

There is more, however, to successful reproduction than what one's mother provides via viviparity and lactation. After **weaning** (defined here are the *cessation* of lactation, not the process that leads to cessation) the young mammal must find food and shelter, avoid predators, find a mate, and care for its own offspring until they reach reproductive age. The way mammals accomplish this is through the growth of relatively large brains and the flexibility in behavior, that is, the capacity for learning, that these large brains allow. The evolutionary record shows that the mammalian brain has undergone repeated selection for increases in size and complexity. Jerison (1973) compared **endocasts** (molds of the interior of the skull which may be used to estimate brain size) of fossil skulls of mammals and found that the brains they contained were, in proportion to body size, smaller in earlier times and have increased in size steadily over the last 60 million years. In contrast, Jerison found that reptiles have not undergone this selection; the brain size to body size ratio of reptiles has not changed appreciably during the last 200 million years.

Mammals have also evolved more complex and functionally diversified brain structures. The mammalian neocortex and its neurologically distinct regions (the motor-sensory region, the auditory and visual regions, etc.) are examples. Jerison (1976) showed that mammalian brains have a system of neurological pathways that bring together, at various locations, information from the visual, auditory and olfactory senses. The 'integrative neocortical system', as Jerison called it, joins neurological regions of the paleocortex (including the olfactory bulb and the limbic system) and the neocortex (including the visual, auditory and somatic systems) of the brain (Figure 3.7). More recent research shows that there may be several integrative systems in the human brain which allow for more detailed memory and task performance than in other mammals (D'Esposito *et al.*, 1995). Lower vertebrates, such as reptiles, rely mostly on the paleocortex for the control of behaviors which Jerison characterized as '. . . fixed-action patterns . . . with few requirements for plasticity or flexibility' (1976, p. 101). Higher vertebrates, the birds and mammals, rely on both the paleocortex and the neocortex for the control of behaviors which are plastic and flexible in all species. Birds do not have the integrative neocortical system and, according to Jerison, their behavior displays to perfection the fixed-action pattern of response to stimuli, but he did not address the fact that birds can learn quite complex and lengthy behavior routines.

Mammals have the integrative system, allowing '. . . sensory information from various modalities [to be combined] as information about objects in time and space' (Jerison, 1976, p. 101). Mammals do not just react to environmental stimuli, they perceive, store, retrieve and evaluate information and adjust behavior responses according to the present situation and past experience. More neurological tissue is required to accomplish these sensory, brain, and behavioral tasks and mammals do have brains that are larger, in proportion to body size, than the brains of reptiles and most birds. Larger, more complex brains allow for a greater capacity for learning and more flexible behavior, because learned behavior may be constantly modified by further learning.

The evolution of learning as an adaptive strategy is associated, in a classic feedback manner, with the series of changes in mammalian biology and behavior that have just been described. The tissues of the central nervous system have relatively high metabolic activity, requiring a regular supply of nutrients and oxygen for maintenance and growth. The evolution of the placenta is thus directly related to the evolution of larger, more complex brains and greater learning abilities. The placenta is the organ that constantly supplies oxygen and nutrients to the developing fetal brain and allows that brain to develop to an advanced stage before birth, and the

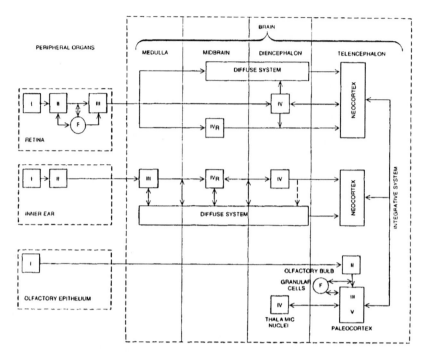

Figure 3.7. Schematic view of the neurological connections between the visual, auditory, and olfactory systems of living mammals. The arrows show the general direction of the flow of information through successive orders of nerve cells (I to V; IV_R indicates parts of the reflex control systems; F indicates feedback loops). The integrative system (right) is a mammalian characteristic (from Jerison, 1976).

evolution of the placenta and of the brain are clearly correlated in mammals. Yolk sac placentas are found in the lower mammals, chorionic–allantoic placentas are found in higher mammals and chorionic placentas are found in the primates. In turn, each higher mammalian group has a brain that is relatively larger than expected for its body size (Jerison, 1973). It is no coincidence that humans have one of the largest brain to body weight ratios, perhaps the most complex and active brain of any mammal, and the most efficient placental system (the hemochorial variety of chorionic placenta) of any primate.

Other biological changes related to brain evolution occur during postnatal life. One example is a correlation of lactational behavior of the mother in relation to the brain size and learning capacity of the infant. Martin (1968) described such a case for the tree shrew (*Tupaia belangeri*). Relative to all mammals, tree shrews have a moderate brain to body weight ratio and average learning ability. A female tree shrew may cache her infants (two or three are born per litter) in a nest and leave them for up to 48 hours while she searches for food. The infant tree shrews are virtually silent and unmoving during their mother's absence, which may be a behavioral adaptation to avoid attracting predators. The seclusion and immobility also limit the variety of sensory stimuli that the infants experience. Upon the mother's return the infants are nursed with a milk that is concentrated in calories and other nutrients. This pattern of periodic feeding coupled with sensory deprivation during infancy works well for a species with limited brain growth after birth and a limited learning potential. This feeding style would not work for a species with rapid postnatal brain growth, requiring a constant nutrient supply during infancy, and a greater dependence on learning in later life.

In neurologically more advanced mammals, especially primates, mother and infant remain in virtually constant physical contact for several weeks or months after birth. Most primate females usually give birth to one infant per pregnancy, but marmosets and tamarins regularly give birth to twins. Singleton births facilitate intimate physical contact since there is no competition between siblings for the mother. Suckling is done 'on demand', 24 hours per day. The concentration of nutrients in the milk of primates is lower than that of the 'primitive' mammals, but the efficiency, constancy and quality of nutrient supply is superior (Widdowson, 1976). The newborn primate is highly active compared with the tree shrew infant. The primate infant travels with its mother, clinging to her body, sensing many of the things that the mother experiences and developing motor and sensory skills in the process. This type of early sensory stimulation is known to be conducive toward further learning (Jolly, 1985). The infant primate grows more slowly than most other mammalian newborns and is, therefore, dependent for a longer time on this intimate relationship with its mother. Infant dependency extends the period of growth, development and protection and also increases the opportunity for the infant to learn survival skills by observing successful maternal behaviors.

Infancy, dependency, and learning are advantageous to both mother and infant since they lead to a greater probability that the young will survive to reach reproductive age. The drawback of infant dependency is that it is incompatible with some adult behaviors, particularly reproductive behavior. Competition for nesting space, breeding territories, aggressive encounters with conspecifics for mates, and mating itself are often precluded behaviors for mothers with dependent young, but the limits to offspring production by females are partially offset by the higher quality of mammalian reproduction. That is, some non-mammalian species, such as many kinds of insects, fish, and reptiles, rely on prodigious egg production to assure the survival of some of their offspring to reproductive age. In contrast, mammals maximize the probability of survival of each individual offspring to achieve a high degree of reproductive success.

Stages of mammalian growth

All placental mammals share the basic skeletal and reproductive adaptations related to rapid and flexible locomotion and to efficient feeding of both the fetus, via some type of placenta, and the infant, via lactation. All placental mammals also share the neurological systems, especially the brain, that allow for intense parental investment and high levels of learning by the offspring. In contrast to these universals of mammalian biology and behavior, mammals can however be grouped by patterns of postnatal growth. The velocity growth of the mouse (Figure 3.8) and the Holstein cow (Figure 3.9) shows the pattern that is typical for most mammalian species.

The shape of these velocity curves are fundamentally different from the human growth curve (Figures 2.5B, 2.6B) in several ways. First, humans achieve their maximum rate of growth in length and weight during gestation, and after birth growth rates decelerate during infancy. Humans are able to achieve such rapid rates of growth during gestation by having only one fetus per pregnancy (the most common case) and the most efficient placentation of all mammals, which provides an optimal environment for rapid growth. The mouse and cow reach their maximum rate of growth after birth during the infancy stage. For the mouse this is probably due to the competition between the multiple fetuses for placental resources (litters are the norm). For both mouse and cow, the improved nutrition provided by lactation, over that provided via the placenta, may also promote a surge in growth rate. For both mouse and cow, growth rates begin to decline very soon after weaning.

The second difference is that for the mouse and cow, sexual maturation occurs relatively soon after weaning and just after the maximal rate of growth. Within a few days of weaning, female mice have vaginal opening and can become pregnant. Male mice produce sperm about two weeks after weaning. Human growth and maturation stand in sharp contrast to the mouse and cow in that humans delay puberty and sexual maturation for many more years after the end of infancy (weaning). The delay is even longer if measured from the time of maximal rate of growth after conception. The mouse and cow achieve peak velocity after birth, but the human peak velocity is before birth, meaning that humans experience an average

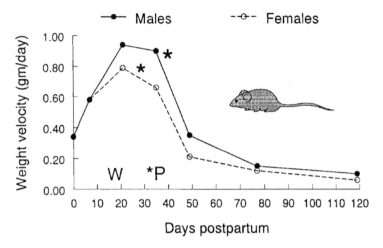

Figure 3.8. Velocity curves for weight growth in the mouse. Weaning (W) takes place between days 15 and 20. In both sexes puberty (P), meaning vaginal opening for females or spermatocytes in testes of males, occurs just after weaning and maximal growth rate (redrawn from data reported in Tanner, 1962).

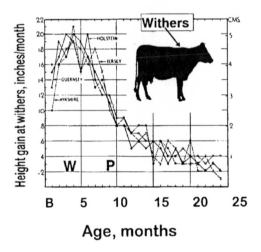

Figure 3.9. Monthly gains in height at the withers for the several varieties of cows. The data are from Brody (1945), who reports monthly values for height as the mean value for many animals (sample size varies from 67 to 239 at each age) measured monthly from birth to 24 months.

delay of more than a decade between peak velocity during gestation and puberty.

A third difference is that for the mouse and cow, puberty occurs while growth rates are in decline but still near the maximal rate. Human puberty occurs when the postnatal rate of growth in height and weight are at their lowest points since birth. The fourth difference is that human puberty, or gonadarche (defined in Chapter 2), initiates a series of changes in endocrine function that lead to the adolescent growth spurt. In contrast, mouse and cow growth rates continue to decline after puberty. The fifth, and final, difference is that soon after puberty, mice are fertile and begin to reproduce, but humans have a delay of years between puberty and the onset of the adult reproductive stage of life.

Many of these differences between human and non-human mammalian growth were recognized by other researchers. In a classic analysis of mammalian growth, Samuel Brody (1945) demonstrated that most mammals have a pattern of growth similar to the mouse. Brody's analysis was confirmed by Bertalanffy (1960) and Tanner (1962). Tanner presents many growth curves for mice, rats, rabbits, sheep, and cattle showing that the size of the mammal makes little difference in the pattern of its growth. The growth curve of Holstein cattle, animals with a mean adult weight of 2640 kg and shoulder height of 135 cm (Figure 3.9), is virtually identical to the mouse (Figure 3.8). Brody, Bertalanffy, and Tanner all agree that the majority of mammals progress from infancy to adulthood seamlessly, without any intervening stages of development. Human beings, in contrast, follow a very different path from conception to maturity. Brody states this most clearly by writing that his analysis,

> demonstrates the close similarity between the age curves of different animal species. The human age curve, however, differs from the others in having a very long *juvenile* period, a long interval between weaning and puberty (approximately 3 to 13 years); this period is almost absent in laboratory and farm animals. In these animals, weaning merges into adolescence without the intervention of the juvenile phase found in man (p. 495, emphasis added).

Brody's terminology is out of date by today's standards, for example he makes no distinction between puberty and adolescence, and equates both of these with sexual maturation. We know today that puberty is an event of the neuroendocrine systems in the brain that control sexual maturation. In humans this event occurs at the end of the juvenile period and before the start of the adolescent growth spurt and the adolescent life stage. Nevertheless, Brody correctly identifies the fact that of all the animals he studied, only humans and the chimpanzee have a juvenile growth period. Despite this very important discovery, Brody still tried to show that in essence all animals, parts of animals, colonies of cells, and even populations of animals followed a common pattern of growth. Brody's own words, again, present this idea most concisely,

> The general similarity between the curves of growth of individuals and of populations is not surprising, since ultimately both are collections of individuals. Our bodies are made of cells, and our bodies, in turn, are cells in a social body. Individuals are organisms and also units of a larger organism, an epiorganism (p. 495).

Brody was searching for a type of 'grand unification theory' of growth. His specific hypothesis was that all growing organisms and populations or organisms could be modeled with one curve of growth. Brody tried to prove his ideas by aligning the maximum rate of growth after birth for many species, including mouse, cow, and human, and then showing that they all have two phases of growth velocity. The first is a self-accelerating phase, from birth to peak velocity, and the second is a self-inhibiting phase, from peak velocity to adult size. Humans, even with their juvenile growth period and adolescent growth spurt, would have to fit into this unification of growth curves. Accordingly, Brody equated the adolescent growth spurt of humans with the velocity curve of growth for animals such as the mouse, the rat, and many domesticated farm animals (cow, pig, sheep). Brody believed that all mammals followed the same curve of growth, but that some mammals had their growth spurt soon after birth (e.g., the mouse and the Holstein cow) while others delayed the spurt until later in life (e.g., humans).

In many ways, Brody's methodology and speculative ideas are ingenious and stimulating. His analyses and findings do apply to the laboratory and farm animals in his data base, but they do not apply to humans. The main problem with Brody's unification hypothesis is that some animals, especially humans, have more than one phase of self-acceleration and self-inhibition of growth rate. Humans, in fact, have three such phases; the first occurs during gestation, the second is the mid-growth spurt, and the third is the adolescent growth spurt. Brody only knew about the last of these, the adolescent growth spurt, and incorrectly equated it with the solitary growth spurt of his laboratory and farm animals.

Brody's work is mentioned here because it had a strong influence on the study of human growth for many years. Even today, it is common to read about 'adolescent growth spurts' or an 'adolescent phase' in animals such as rodents or farm animals which have nothing like human adolescence. While these ideas seem preposterous today, bear in mind that the nature of the human postnatal growth curve, especially the existence of the adolescent growth spurt, was still being debated during the first third of the twentieth century (see Chapter 1), the time when Brody carried out most of his original research. The human postnatal stages of childhood and juvenile growth, as well as the human prenatal curve of growth (Figure 2.6), were completely unknown to most researchers. Moreover, the neuroendocrine control of growth were poorly understood, and the control of puberty was completely unknown.

Juvenile mammals

Brody's discovery of the human juvenile growth period is one of his lasting contributions to the field. Since Brody's time, juvenile growth stages have been discovered for several other mammalian species. The highly social mammals, such as social carnivores (wolves, lions, hyenas), elephants, many cetaceans (porpoises, whales), and most primates all evolved a new stage of development between infancy and adulthood – the juvenile stage (Bekoff & Byers, 1985; Pereira & Fairbanks, 1993). Juveniles were defined in the previous chapter as offspring who are no longer dependent on maternal lactation, but who are still prepubertal. Juveniles are largely responsible for their own care and feeding. They must find their own food, avoid predators on their own, and compete with adults for food and space. Juveniles may even compete with their own mothers, who may be encumbered with another pregnancy or nursing infant. Clearly, the addition of the juvenile stage adds several new risks along the path of growth and development toward reproductive maturity. In fact, the highest rates of post-neonatal mortality (i.e., deaths after the first month following birth) for social mammals occur during the juvenile stage (Pereira & Fairbanks, 1993).

Juvenility must have added some benefits to the life of social mammals to have evolved, but there is some debate as to the function of the juvenile growth stage. A stimulating review and analysis of the evolution of the juvenile growth stage is offered by Janson & van Schaik (1993). They propose two benefits. The first is the traditional 'learning hypothesis' explanation, that is, the juvenile period allows for the extended period of brain growth and learning necessary for reproductive success in various species of social mammals. Social carnivores, elephants, and primates must all learn how to live within the social hierarchy of the group. They must also learn complex feeding skills such as hunting animal prey, opening fruits or seeds with protective coverings, and where and when to find food. Reproductive skills must also be learned, including competition for mates and care of offspring. Johnson (1982) and Lancaster (1985) show that the selective benefit of learning is that it permits adaptation to ecological changes that are not predictable. Included in these changes are common problems faced by all living and extinct hunting and gathering societies, such as seasonal variability in climate, plant growth and animal migrations. There are also the rare crises that exemplify the unique capacity for learning in humans, such as the 1943 drought in Central Australia. Birdsell (1979) relates that during this time an old Aborigine man, Paralji, led a band of people on a 600 kilometer trek in search of water. After passing 25 dry waterholes he led them to a fallback well that the old man had not visited for more than 50 years. That well was also dry, forcing Paralji to trek 350 kilometers on ancient trails, locating water holes by place names learned from initiation rites and ceremonial songs he memorized as a juvenile.

Even the way humans forage for food requires extensive learning during the childhood and juvenile growth periods. Much of the food humans utilize is hidden from view or is encased in protective coatings; tools are usually needed to extract and process these foods. The costs of tool manufacture, the time and energy needed to find and process raw materials, are outweighed by the benefits. Tool-using human gatherers extract twice as many calories from savanna-woodland environments as non-tool-using primates. Other foods are poisonous before processing by washing, leaching, drying or cooking. For example, acorns and horse chestnuts, eaten by many North American Indians, and manioc, a staple food of many tropical living cultures, are toxic if eaten raw, and must be leached by boiling in water and dried before consumption. Furthermore, knowledge of the location of these foods and their methods of processing, and the location of raw materials and their manufacture into tools requires learning. An example is the Arunta, a hunting and gathering people of Central Australia. They are compelled to live in self-sufficient nuclear families by the widely dispersed nature of food resources in their habitat (Service, 1978). As soon as they are able, the children follow their mothers and fathers on the daily rounds of food collection and preparation. Elkin (1964) observed boys and girls as young as five years being taken by their fathers on hunting trips and being shown how to collect raw materials and prepare them for spears, points and other tools. Since it takes more than a decade to become proficient in the manufacture and use of these tools, early learning and slow growth and maturation during the juvenile stage are mutually beneficial.

Learning seems like a great idea, but the learning hypothesis does not take into consideration the high juvenile mortality. After all, evolution works by differential mortality and reproductive success, and mortality during the pre-reproductive juvenile stage will not lead to evolutionary success. The second benefit of juvenility, and a complement to the learning hypothesis, is the 'ecological risk aversion hypothesis'. Janson & van Schaik develop this hypothesis to deal with the risks for juvenile mortality. The reasoning for this hypothesis is as follows. When the risk of predation is high for the individuals of a species, natural selection often favors the formation of social groups – there is relative safety in numbers. The formation of social groups comes at a cost however, and that cost is an increase in feeding competition within the group – there are more mouths to feed. According to Janson & van Schaik, newly weaned individuals are most likely to be adversely affected by this competition as they have less-developed foraging skills than adults. To reproduce, the young must develop and grow to mature size, and there are two basic ways to get from infancy to adulthood. One is to develop fast and minimize the duration of the non-reproductive period between weaning and sexual maturation – this is the strategy followed by the mouse, the bison, and most mammals. This strategy places a premium on quantity of reproduction rather than on quality, as there is little time to learn. The other strategy is to develop towards adulthood slowly, increase the time for learning, and produce higher-quality adults. A corollary of each strategy for mammals of medium to large body size is that fast development requires fast growth, but slower development allows for slow growth. Janson & van Schaik's hypothesis is that slow juvenile growth reduces the risk of death while allowing juveniles to learn.

The risk of death from feeding competition is an example. Juveniles must forage for their own food, a skill that must be practiced until mature levels of success are achieved. Much of the foraging of juveniles is in competition with adults. This competition becomes clear during times of food scarcity, when juvenile primates die in greater numbers than infants or adults. But, not all juveniles die, and there is a relationship between growth and mortality. A small, slow-growing juvenile requires less food input than a larger, fast-growing individual and may survive periods of food scarcity, and there is evidence for this from both primates (Janson & van Schaik, 1993) and elephants (Lee & Moss, 1995). Moreover, slow growth and smaller size allow juveniles to practice feeding skills with less risk of starvation during all seasons of the learning period. According to this risk-aversion hypothesis, the juvenile stage evolved by slowing the rate of growth and prolonging the duration of growth. In the next chapter this type of evolutionary change, which is part of a process called heterochrony, is discussed in greater detail.

An example of the juvenile pattern of growth is given in Figure 3.10 for the elephant. The distance curves are based on cross-sectional means of

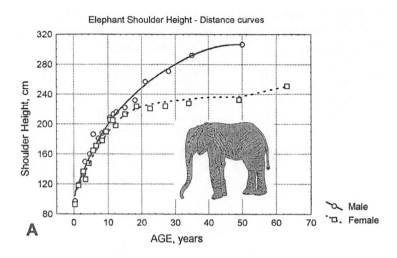

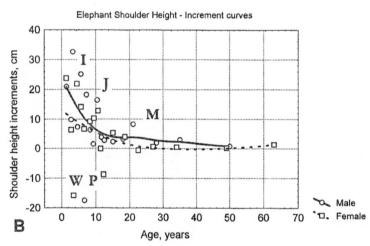

Figure 3.10. Growth curves for elephants illustrating the pattern of growth in social mammals with infant, juvenile, and adult stages of growth. (A) Distance curves, (B) increment curves. Weaning (W) takes place at a mean age of 40 months. Puberty (P), meaning reproductive maturity, occurs at about age nine years in females. Other abbreviations are: I, infancy; J, juvenile; M, mature adult. The mean values of shoulder height were kindly supplied by Dr. P. C. Lee, Department of Biological Anthropology, University of Cambridge, UK and published by Lee & Moss (1995).

shoulder height, a common measure of skeletal growth for elephants (see Lee & Moss, 1995 for details of the methodology of measurement). The original means were plotted and fit with a 'distance weighted least squares regression'. The curve produced by this statistical technique represents a type of 'average' growth and the curve bends to more closely approximate the values of mean shoulder height at each age. The distance curve for both male and female elephants shows the typical juvenile pattern of prolonged and relatively slow growth for many decades. The increments in growth between successive data points were calculated and, as was done for the distance curves, the increments between means were plotted and fit with a distance-weighted least-squares regression. The shape of this fitted increment curve reveals the stages of postnatal growth more clearly than the distance curves. The approximate duration of infancy (I), juvenile (J), and mature adult (M) stages are indicated.

The increment curves must be interpreted with a great deal of caution since they are calculated from means rather than individual data. Increments are presented only to provide a very rough approximation of growth velocity. Overall, the fitted curve shows a fairly rapid decline in rate of growth during the first decade of life and then a slower, prolonged decline to age 25 years for females and age 50 for males. The difference between the sexes in shoulder height seen in the distance curve is explained by the additional 25 years of growth for males. The increment estimates fluctuate widely, and some are even negative. This is due to the cross-sectional nature of the data, that is, the adjacent measurements are from different individual elephants, and the unequal spacing between increments for each individual. The fluctuations also may be due to very different ecological conditions for growth during the life of these individuals. It is possible that negative increments represent the difference in size between individuals who experienced good ecological conditions and those growing up under harsh conditions. Notice that the negative increments, and most of the fluctuations, are confined to the first 20 years of elephant life. That time includes the infancy and juvenile stages – in this sample of elephants, from the Amboseli reserve in Kenya, males do not reproduce until after 25 years of age, females reach puberty at about age nine years and begin reproduction after ten years of age. The fluctuations in these increment data, despite their limitations, do add some support to the ecological risk aversion hypothesis as an explanation for the juvenile stage of growth. In other words, the environment in which these elephants live is unpredictable, and a strategy of slow growth to adulthood may well offer the best chance to avoid starvation in bad years.

Another possible explanation for a juvenile stage for social mammals may be called the 'dominance hypothesis'. Research with wild and captive primates, with elephants, and with social carnivores (wolves, lions, hyenas) shows that high-ranking individuals in the social hierarchy can suppress and inhibit the reproductive maturation of low-ranking individuals (Pereira & Fairbanks, 1993). The inhibition may be due to the stress of social intimidation acting directly on the endocrine system, or may be secondary to inadequate nutrition due to feeding competition. Juveniles are almost always low-ranking members of primate social systems. In the past, individuals with slow growth and delayed reproductive maturation following infancy may have survived to adulthood more often than individuals with rapid growth and maturation, and thus juvenile stage may have evolved. Whatever the cause of the juvenile growth stage, be it learning, risk aversion, or social inhibition, one scholar points out that in broad perspective, '. . . juvenile life has two main functions: to get to the adult stage without dying and to become the best possible adult' (Alexander, 1990). Adding a juvenile stage must have served this purpose well for it to evolve independently in so many social mammals.

Primate growth patterns

Most primates are highly social mammals and, as expected, have a juvenile stage of growth and development. The primates, however, are a highly diverse group including the prosimians (such as lemurs, lorises, and tarsiers), New World monkeys (ranging from marmosets and tamarins to cacajos and howlers), Old World monkeys (such as baboons, rhesus and colobus), the apes (gibbons and siamangs, orang-utans, gorillas, common chimpanzee, and bonobo), and humans. In all there are more than 50 genera and more than 200 species of living primates (Napier & Napier, 1967). Primates live in environments that range from semi-arid scrub, to savanna, to woodland, and rain forest, and from sea level to more than 2000 meters. Adult sizes range from the male gorilla, at 160 kg, to the mouse lemur (*Microcebus murinus*) at 0.08 kg. Maximum life span ranges from 8.8 years for the dwarf lemur (*Cheirogaleous major*) to about 120 years for humans (Harvey *et al.*, 1986). Given this variation in world distribution, general environments, local ecologies, adult body size, and life span it is reasonable to expect that primates have a variety of patterns of growth and development, and indeed they do.

Three patterns of primate postnatal growth in weight are illustrated in Figure 3.11, based on the work of Steven Leigh (1992, 1994a). There are differences in growth patterns between the three species (marmoset,

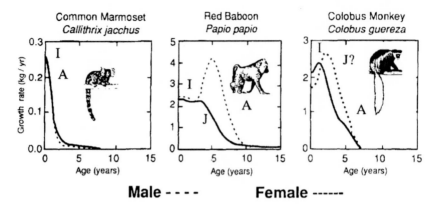

Figure 3.11. Three primate species showing different patterns of postnatal growth in weight. Abbreviations are: I, infancy; J, juvenile; A, Adult. Marmosets are weaned at 63 days and can breed at one year of age (Harvey *et al.*, 1986). Marmoset growth rate shows no postnatal growth spurt and no change in growth rate, typical of a juvenile growth stage. Baboons are weaned by 18 months and begin puberty at about 3.5 years for females and 4.5 years for males (ibid). Baboons have a juvenile growth stage for both males and females, but only males have clear weight growth spurt. Colobus monkeys are weaned at about 13 months and females have their first birth at about 4.6 years of age. Colobus monkeys show a post-natal spurt for both sexes, but it is not clear if there is a juvenile growth stage. The velocity curves are fit to cross-sectional data collected from captive animals. The curves are fit using the statistical estimates based on lowess regression – see the text for details (after Leigh, 1994b).

baboon, colobus) and between males and females within species. Further-more, none of these species follows a pattern of weight growth similar to the human species. The marmoset seems to follow the type of growth curve Brody described for most mammals, that is an acceleration phase followed immediately by a deceleration phase. The switch between phases seems to occur just before the time of birth, and the infancy stage merges seamlessly into the reproductively mature stage. Based on the velocity curves of weight growth there is no evidence of a juvenile stage. The lack of a juvenile growth stage may relate to marmoset social behavior. Marmosets, and the closely related tamarins, live in a type of 'family' group, in which females form a social bond with one or a few males. These polyandrous 'families' establish feeding and breeding territories and exclude other marmosets. The young, usually born in pairs or triplets, are raised cooperatively by the mothers and their male consorts (there may be two males in a breeding group). After weaning, which takes place as early as six months for some species and by 12 months for other species, the young are protected and provisioned with food by both adult males and the mother (Goldizen,

1987) until reaching sexual maturity at about age 15 to 24 months. Following Janson & van Schaik's risk aversion hypothesis, a juvenile stage of growth would not be needed in this type of territorial 'family' social group (see Garber & Leigh, 1997, for an alternative interpretation). Leigh (1996) analyzed growth data for five other species of marmosets or tamarins and all of these follow a pattern of growth essentially the same as for the common marmoset.

The baboon velocity curves show evidence for infancy, juvenile and adult stages of growth. Baboon infants are weaned by about 18 months after birth, but growth rates remain fairly stable until about age four years – a clear juvenile pattern of growth. Baboons reach a peak in growth rate before birth, just as humans do, but by the time of birth baboons are already growing at the slower, steady rate that characterizes late infancy for humans. Both male and female baboons follow this pattern during infancy, but during the juvenile phase males show a pronounced acceleration in weight growth – a spurt. Baboons live in large social groups composed of both males and females of all ages, and therefore their growth pattern conforms to the prediction that they will have a juvenile growth stage. Why males, and not females, have a juvenile growth spurt in weight is not known. It is known that the spurt is due mostly to an increase in muscle mass, which males seem to use to compete with each other for scarce resources such as food and mates.

Colobus monkeys show yet a third pattern of postnatal growth. Both males and females have a clear acceleration in weight growth after birth. What is not clear is whether these accelerations are juvenile growth spurts. It is possible that colobus monkeys grow more like marmosets than like baboons. Like the marmoset, the colobus may have only a single peak rate of growth, but that peak is reached soon after birth. The colobus, then, would be following a growth pattern like that for the mouse and the cow (Figures 3.8 and 3.9). Colobus social behavior and diet do not provide a clear case for the prediction of a juvenile stage. Colobus monkeys live in troops of 3 to 15 animals, with a typical troop composing a single adult male, three to four adult females, and their offspring. These troops are highly social, there is little evidence for a female dominance hierarchy or female aggression, and infants seem to be groomed and cared for by troop members other than the mother. This mutual infant care, as well as intense allo-grooming between adult females, is believed to maintain a highly cohesive social group (Struhsaker & Leyland, 1987). Colobus monkeys eat mostly young leaves from the hackberry tree (*Celtis durandii*) and two other tree species, which together comprise 69% of the total diet. The colobus, and related leaf-eating monkeys of the subfamily Colobinae,

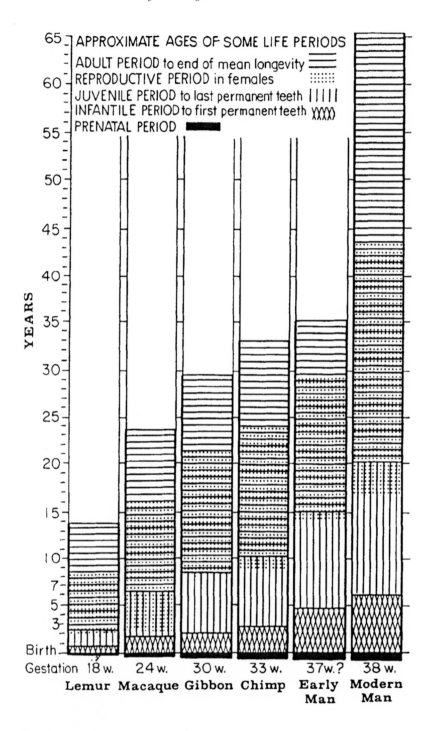

APPROXIMATE AGES OF SOME LIFE PERIODS

ADULT PERIOD to end of mean longevity

REPRODUCTIVE PERIOD in females

JUVENILE PERIOD to last permanent teeth

INFANTILE PERIOD to first permanent teeth

PRENATAL PERIOD

YEARS

Birth

Gestation 18 w.   24 w.   30 w.   33 w.   37 w.?   38 w.

Lemur Macaque Gibbon Chimp Early   Modern
                                      Man   Man

digest this large quantity of leaves in a specialized stomach comprising three or four subcompartments. Their digestive system is very similar to the ruminant digestive system of the cow. Given the wide distribution and availability of their food, and given their cohesive and non-competitive social organization, it is possible that young colobus monkeys are exposed to very low risk for mortality. If so, then the risk aversion hypothesis would predict the absence of a juvenile growth stage, or the presence of a very brief stage. There is some empirical support in favor of this prediction. In an analysis of 42 primate species, Leigh (1994a) found a significant correlation between diet and rate of growth. Folivores, such as the colobus monkey, have faster rates of growth than species of non-folivore monkeys and apes, such as the rhesus monkey and the gibbon. So, perhaps it is not so far-fetched to state that the colobus has both a cow-like stomach and a cow-like pattern of growth.

A more detailed discussion of primate juvenile growth, especially postnatal growth spurts, is presented later in this chapter. Before that review, some of the history and findings of research into primate growth is discussed so as to better appreciate what primates share with other mammals and what is special about primate growth. The origin of comparative studies of primate growth begin with the work of Adolph Schultz (1924), and from inception, Schultz's studies were aimed at '. . . the relation of the growth of primates to man's evolution . . .' (1924, p. 163). Perhaps Schultz's most lasting contributions were summarized in his 1960 illustration of the 'Approximate ages of some life periods' of the primates (Figure 3.12). Despite the grandeur and vision of Schultz's research on primate growth and primate evolution, very few primate species were actually studied in detail by 1960. In fact, until the 1980s details of skeletal, dental, and somatic growth were known from only three species, the rhesus monkey (*Macaca mulatta*), the chimpanzee (*Pan troglodytes*), and humans.

Ana Laird continued Schultz's interest in the 'Evolution of the human growth curve' – the title of Laird's 1967 paper. In that paper she reviewed

Figure 3.12 (left). Schultz's diagram of the proportional increase in the length of life stages across the *scala naturae* of living primates. Note that Schultz used eruption of the permanent teeth to mark the boundary between life periods. Also Schultz did not recognize the childhood or adolescent stages for modern humans. Indeed, all primate species have the same life stages, which just increase in length from prosimian to human. The estimates for total length of life are based on average expectations rather than theoretical maxima. The data for 'Early Man' are entirely speculative as no species is given and very little data were available when Schultz prepared this figure. From Schultz (1960).

studies of the growth of the three well-studied species; rhesus monkey, chimpanzee, and human. Laird took a mathematical approach to the study of growth. By fitting mathematical functions to the growth data, Laird hoped to reveal more precisely the stages in the evolutionary development of the human growth curve. The curves were fitted to the monthly weight data points by the method of least squares regression. Essentially, this method minimizes the sum of the squared deviations of each data point from the fitted curve, that is, it produces a type of 'average' line between the data points.

Laird found that monthly weight increases in the rhesus monkey and the chimpanzee followed two separate growth curves. For the male rhesus, the first curve fit the data from birth to 22 months and the second from 23 months onwards (Figure 3.13). A change in growth rate that occurred between months 22 and 23 necessitated the use of different mathematical curves to model growth in the two periods. Sexual development in the male rhesus takes place during the second growth phase, after month 40. The deviations of weight growth above the fitted curve between months 48 and 54 corresponded to the time of reproductive maturation and the beginning of adult levels of gonadal hormone secretions. The curves of growth for the female rhesus were similar to the male, except that the time of sexual maturation was earlier, occurring at about 42 months. For the chimpanzee, the first curve fit the early phase of growth, from birth to six years. This early phase is essentially the infancy period, which lasts to age five years in chimpanzees (Teleki *et al.*, 1976; Goodall, 1983; Nishida *et al.*, 1990). The second curve was fit to what Laird called the 'adolescent phase'. Male and female chimpanzees followed the same curve of growth from birth to six years. During the 'adolescent phase' males grew in weight at a faster rate than females and this required separate mathematical functions for each sex. Due to the different rates of weight growth sexual dimorphism in weight became well-marked. The sexual dimorphism reached its greatest level at the age when male chimpanzees begin to sire offspring. Laird's analysis confirmed the work of Tanner (1962), who had shown that the sexual dimorphism in adult chimpanzee weight was largely due to a weight growth spurt for the males (Figure 3.14).

Laird (1967) found that the velocity curve of human growth, as depicted in Figure 2.5B, required three mathematical functions to model its course. This conclusion was confirmed, independently, by Bock & Thissen (1976), Bogin (1980) and Karlberg (1987). The need for the third function is one aspect of human growth that makes it different from the growth of the other primates. Laird described the similarities of growth between the rhesus monkey and chimpanzee and the distinct pattern of human growth as follows:

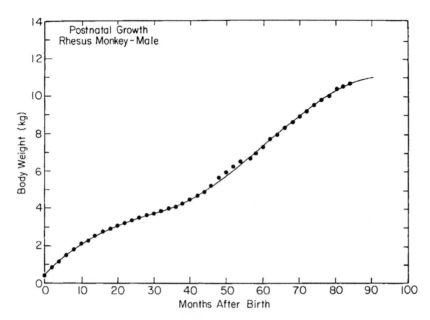

Figure 3.13. Growth in body weight of the rhesus monkey (from Laird, 1967).

... the curvilinear growth by which the body weight of an organism approaches its mature value and during which sexual maturation characteristically occurs, starts at birth in sub-primate mammals and birds, but is deferred in monkeys ... the preliminary growth occupying about $\frac{1}{3}$ and the adolescent growth about $\frac{2}{3}$ of the time required to reach fully mature size. In the chimpanzee, adolescent growth is deferred to the last $\frac{1}{2}$ of the total period ... In the human a further delay has occurred so that adolescent growth with its concomitant development of sexual maturity occupies only the last $\frac{1}{3}$ of a prolonged growth period. The delay in the human can be interpreted as being due to the *insertion*, between birth and adolescence, of two growth phases, rather than the single phase identifiable in the monkey and the chimpanzee ...

(1967, pp. 351–2, emphasis added).

It seems that Laird considered all three primates to be alike in having a juvenile phase between infancy and what she calls 'adolescence'. The second growth phase that humans add between infancy and puberty is childhood. The insertion of childhood into the human life cycle not only requires a third mathematical function to model growth, but, more importantly, changes the biological and social ecology of the human species. These biosocial changes are discussed in detail in the next chapter.

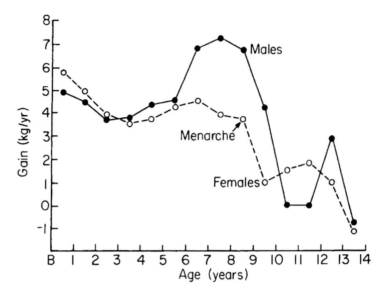

Figure 3.14. Weight velocity curve for the male and female chimpanzee (from Tanner, 1962).

Of brains and bodies

There is another key difference between non-primate and primate growth, and this involves the relative rates of growth of the body, the brain, and the reproductive system. Most mammals, including the rat (depicted in Figure 3.15) and the social mammals, show an advancement of brain growth relative to body growth. In the rat reproductive maturation occurs before the brain or the body achieve final adult size (Donaldson, 1895). Primates delay body growth and reproductive development, but do not delay brain growth. Figure 2.7 illustrated these relationships in humans. The weight of the human brain reaches 80 percent of adult size by age four and virtually 100 percent of adult size by age seven. Yet body growth continues to age 18 and beyond. Even more to the point, brain growth is finished before reproductive maturity even reaches 10 percent of the adult value. This pattern of relative growth of the brain, body and reproductive system is also found in the rhesus monkey and the chimpanzee (Laird, 1967). Other organs of the primate body (e.g. heart, lungs, liver) follow the body growth curve. The primate brain, then, is most unusual in its pattern of accelerated growth in comparison to other organs and the body as a whole.

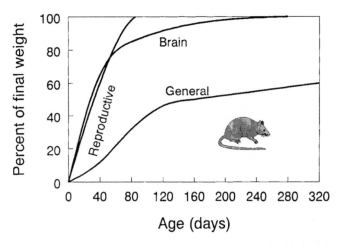

Figure 3.15. Growth of different types of tissue in the rat (after Timiras, 1972).

In one obvious way, this pattern of growth relates to the fact that primates are learning creatures *par excellence*. There are other consequences of the primate pattern of growth that are not as obvious as learning. Some of these are dealt with in the next chapter. At this point it is worth noting that rapid brain growth, deferred body growth, and progressively delayed sexual maturation greatly enhance the quality and quantity of reproductive efficiency in primates. Not only is a given offspring endowed with great flexibility of learned behavior, but also the older but still pre-reproductive juvenile can help its mother provide care for new infants. In humans, with our childhood and adolescent periods of growth, reproductive efficiency is further enhanced, and because of this there was strong selection at the genetic level of primate biology to develop and maintain these patterns of growth.

Is the human adolescent growth spurt unique?

One of the hallmarks of human growth is the adolescent growth spurt, the acceleration in rate of growth in height and weight. Many individual organs and most body dimensions experience the adolescent spurt (Shock, 1966; Cameron *et al.*, 1982; Satake *et al.*, 1993; Dasgupta & Das, 1997; Greil, 1997). Moreover, the adolescent spurt is detectable in both boys and girls, and in virtually every human population so far examined. The adolescent spurt, which is easily visualized from Figure 2.5, is a regular and normal feature of human growth. Only chronic and severe illness, malnutrition or physiological stress can obliterate the growth spurt. For instance, Quechua Indian boys and girls living at high altitude in the Peruvian Andes have a late and poorly defined adolescent growth spurt. The reason for this is that these children suffer from the combined stress of hypoxia (insufficient delivery of oxygen to the tissues of the body), energy malnutrition, heavy workload, and cold. Instead of a clear spurt, the Quechua experience a prolonged adolescent growth period, lasting until age 22 and beyond (Frisancho, 1977).

There is some controversy as to whether the adolescent growth spurt is a uniquely human feature. On the basis of empirical observations and evolutionary considerations, I conclude that the human adolescent growth spurt in stature is a species specific characteristic, that is, a skeletal growth spurt of the human type is not found in any other primate species. This conclusion, which I have developed and modified in several publications (Bogin, 1993, 1994a; Bogin & Smith, 1996), stands in contrast to that of previous research. As discussed above, Brody (1945) believed that the human adolescent growth spurt, both for height and weight, was homologous with the peak velocity of growth shown just after birth by most laboratory and farm animals. Bertalanffy (1960) and Laird (1967) also believed this to be the case. For these researchers, the special features of primate growth were the prolongation of the juvenile period, found in all haplorhine primates, and the evolution of the childhood period, found only in humans. The addition of the juvenile, and then the childhood, growth phases, just delayed the ubiquitous postnatal growth spurt of all animals to the adolescent phase of humans. As shown above, this interpretation is incorrect.

Tanner (1962) was, perhaps, the first to reject the 'grand unification hypothesis' proposed by Brody. Both a pediatrician and a human biologist, Tanner was able to understand that primates follow a pattern of growth that is different in many ways from that of other mammals. One such difference, according to Tanner's analysis, is that the Old World monkeys, the apes, and humans, show two growth spurts. The first is the universal spurt shared by all mammals that occurs just before or after birth, and the second is a spurt at the time of puberty. Confirmation and further clarification of the unusual pattern of primate growth was provided by Laird (1967) and then by Timiras & Valcana (1972), but Tanner's analysis is clear enough on this point. Tanner analyzed growth curves for male and female rhesus monkeys and chimpanzees. The weight velocity curves for these primates show that growth rates peak at or just before birth and then decline after birth. The period of decline lasts for more than a year for the rhesus and for more than three years for the chimpanzee. Both species then show an acceleration in weight velocity that has all of the characteristics of the human adolescent spurt in weight; including the earlier onset of the spurt in females and the greater intensity of the spurt in males. Tanner presents velocity curves for crown–rump length for the rhesus monkey and finds, again, that both males and females have an adolescent spurt. This skeletal spurt, however, is both much smaller than the spurt in weight and of much shorter duration. The rhesus weight spurt spans almost three years in males, but the skeletal spurt lasts at most six months. Based on other skeletal measurements Tanner concludes that the rhesus growth, '. . . spurt is largely one of shoulder breadth and, above all, of muscle mass' (1962, p. 235).

Based on the evidence available today, there is no question that some monkeys and apes have pubertal growth spurts in weight. The work of Leigh (1996) is the most comprehensive study to date that is directed at the question of the primate growth spurts in weight. Some of Leigh's findings were presented above in Figure 3.11. Leigh collected data for chronological age and body weight for 2395 captive primates, housed at both zoos and primate laboratories. There are 35 non-human species in his sample, including representatives of the New World monkeys, Old World monkeys, and apes. He also included a data set for healthy humans (English boys and girls measured in the 1970s and 1980s). Although there are many longitudinal records for individual animals in the data base, the data are analyzed in a cross-sectional fashion. Leigh explains this approach by stating that cross-sectional analysis '. . . leads to artificially depressed estimates of growth spurt magnitudes . . . [and] should lead to a conservative diagnosis of the presence of growth spurts' (p. 457). Leigh also used the nonparametric mathematical technique called lowess regression to fit smoothed curves to the growth data for each primate species. There are many advantages and disadvantages to the use of mathematical curve fitting, and I suggest that interested readers consult books on the theory and application of regression for details. I can state here that Leigh's decision to use the mathematical smoothing technique of lowess regression is useful, since it helps to reduce the unimportant variability (often called the 'noise') in time-series data sets and helps to reveal the important main trends that occur over time. Lowess curves were fit to the distance data and then the first derivative of the distance curve was estimated to produce a velocity curve. This follows exactly the procedure used at the beginning of this chapter to construct Figure 3.1.

A summary of Leigh's results for the 35 non-human species he analyzed is given in Table 3.1. Only one species of New World monkey shows a

Table 3.1. *Summary results of Leigh's (1996) analysis of weight growth in primates. Species evaluated, sample sizes, and presence or absence of a postnatal growth spurt by sex*

Species	Sample size (Male/F)	Growth spurt (by sex)
Ceboidea (New World Monkeys)		
Cebuella pygmaea	36/51	None
Callithrix jacchus	48/71	None
Cailirnico goeldi	42/47	None
Saguinus fuscicollis	18/19	None
Saguinus geoffroyi	9/10	None
Saguinus imperator	11/14	None
Saguinus oedipus	46/18	None
Leontopithecus rosalia rosalia	26/31	None
Saimiri sciureus	32/28	None
Cebus apella	26/28	Male
Callicebus moloch	30/23	None
Aotus trivirgatus	25/23	None
Cercopithecoidea (Old World Monkeys)		
Cercopithecus aethiops	30/30	Both
Cercopithecus mitis	27/37	Male
Cercopithecus neglectus	29/23	Male
Erythrocebus patas	41/52	Both
Cercocebus atys	38/71	Male
Macaca arctoides	52/58	Male
Macaca fascicularis	13/13	Both
Macaca fuscata	64/71	Both
Macaca mulatta	52/58	Both
Macaca nemestrina	39/64	Both
Macaca silenus	39/41	Male
Papio hamadryas	33/53	Male
Mandrillus sphinx	49/59	Both
Colobus guereza	46/49	Both
Presbyns entellus	29/24	Both
Presbytis obscura	19/17	Male
Hominoidea (Apes and Humans)		
Hylobates lar	25/25	None
Hylobates syndactylus	19/21	None
Pongo pygmaeus	42/42	Male ?
Gorilla gorilla	77/64	Both
Pan paniscus	13/23	Both
Pan troglodytes	22/23	Male
Homo sapiens	Literature data	Both

spurt, *Cebus apella*, and only for males. All of the Old World monkeys studied to date show a spurt for males, and many species also show a spurt for females. Among the ape species, gibbons (*Hylobates spp.*) show no spurt, orangutans (*Pongo*) show a possible male spurt, the common chimpanzee (*Pan troglodytes*) shows a male spurt but no female spurt, and both the bonobo (*P. paniscus*) and the gorilla (*Gorilla*) have male and female weight spurts. The variation in presence or absence of a postnatal weight spurt among primate species, and between sexes within a species, is in keeping with the diversity of primate biology and ecology. Leigh's data show clearly that a postnatal weight spurt is not a primate characteristic, rather it is a variable trait found in one, or both, sexes of some species.

The velocity curves for the apes are illustrated in Figure 3.16. The rate of weight growth of the gibbon generally decelerates from birth to sexual maturity in a fashion similar to the marmosets of Figure 3.11. Gibbons are also somewhat like marmosets socially in that both kinds of primates live in family groups. But, unlike the marmosets, gibbons have a juvenile stage of growth. Gibbon infants are weaned at about age three years and the female's age at first birth is about nine years, which means that gibbons spend about six years as juveniles. The gibbon data show that postnatal growth spurts, especially spurts around the time of puberty, and the juvenile growth stage are not causally associated with each other. The great apes, orang-utan, gorilla, bonobo, and chimpanzee, all have infant, juvenile and adult stages of growth, but show varying patterns of stage duration and varying patterns of growth spurts. Weaning takes place by four years of age for gorillas, by five years for the chimpanzee and bonobo, and by about age six years for the orang-utan. First birth for adult females occurs at age 10 years for gorillas, 12 years for orang-utans, and between 13 to 15 years for chimpanzee and bonobo. Each species, then, has varying amounts of time spent in the juvenile stage. Even though gorillas have the shortest infancy and juvenile stages, they achieve the largest body weights of all the apes. This is due to the fact that at any age gorillas are always growing faster than any of the other great ape species, as may be seen by reading the *y*-axis scale in Figure 3.16.

In contrast to Leigh's comprehensive survey of primate weight growth, there are only a few studies of skeletal growth. Coelho (1985) measured gains in crown–rump length (CRL) and weight in a mixed-longitudinal sample of 250 male and 452 female olive baboons (*Papio cynocephalus anubis*). The animals were part of a laboratory colony living under naturalistic conditions in terms of the physical environment and social group composition. All animals were healthy and well-nourished, and none showed signs of clinical obesity. The date of birth for all individuals was

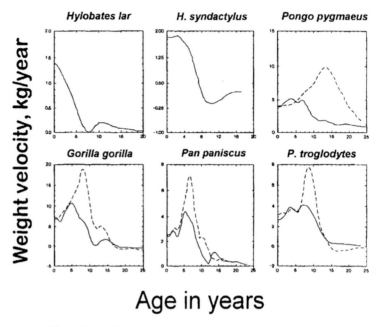

Figure 3.16. Velocity curves for weight growth in six species of apes (from Leigh, 1996).

known and the animals were measured once a year. The mixed-longitudinal design of the study provided data on growth between birth and eight
years of age, which for this species is the total span of the growing years.
Velocity curves for CRL are presented in Figure 3.17 (weight velocity is
essentially identical to the data presented in Figure 3.11). The crown–rump
velocity data are presented in a cross-sectional fashion, which would tend
to reduce the apparent size of any growth spurt (Boas, 1892). Even so, the
crown–rump velocity curves stand in sharp contrast to the weight velocity
curves, which show a large spurt for males, for neither male nor female
baboons show a spurt in crown–rump length. Baboons do show three
phases of declining growth rate in crown–rump length: an infancy phase
until two years, a juvenile phase from two to six years, and a post-pubertal
phase until eight years. Similar patterns of decreasing velocity in skeletal
growth from birth to maturity were found in studies of three other monkey
species, *Macaca nemestrina* (Orlosky, 1982), *Papio cynocephalus* (Sirianni
et al., 1982), and *Macaca sinica* (Cheverud *et al.*, 1992). Based on these
studies, there is no evidence for the presence of human-like adolescent
growth spurts in the skeleton of any of these species of Old World monkeys. Another mixed-longitudinal study of growth in weight and CRL with

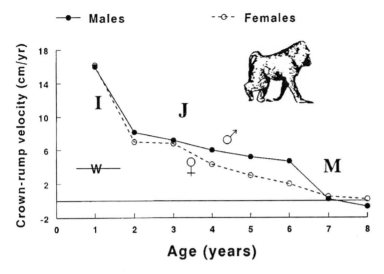

Figure 3.17. Velocity curves for crown–rump length in the baboon (after Coelho, 1985).

baboons offers the suggestion of a pubertal spurt in skeletal growth. The authors of this study report '. . . the presence of a pubertal growth spurt in body weight, and possibly CRL, in male but not in female baboons' (Crawford *et al.*, 1997). The male CRL 'spurt' amounts to about a two centimeter increase in growth rate from pre-pubertal values. However, the analysis of the spurt is based on a varying number of individual baboons, with as few as two and as many as eight individuals measured on different occasions. It is difficult, if not impossible, to derive an accurate and reliable estimate of growth velocity from data of this type.

Cross-sectional studies of both weight and skeletal growth in Japanese macaques (*Macaca fuscata*) and chimpanzees (*Pan troglodytes*) have been published by Hamada (1994) and Hamada *et al.* (1996). The macaque group consisted of both laboratory colonies and semi-natural colonies reared in 'monkey parks' in Japan. The data were cross-sectional, representing 2886 animals (1288 males, 1598 females) between the age of birth to 15 years. The chimpanzee data, which were also cross-sectional, came from animals reared at a primate park owned by a Japanese corporation. There were a total of 172 chimpanzees (81 male, 91 female) between the ages of birth and 15 years. Hamada and colleagues analyzed the data by fitting spline curves to the median values of growth for each age class, e.g. 3 months, 6 months, 12 months, 2 years, etc. The first derivative of these distance curves was used to produce a velocity curve. The spline curve method is similar to the lowess regression method used by Leigh, in that both are designed to fit a smooth continuous curve to a set of data points. One difference between the methods is that spline curves are forced to pass through each data point, while lowess curves may pass between data points.

Hamada found that the postnatal growth of the Japanese macaque, in terms of both distance and velocity, had, in general, three distinct phases: '. . . [1] from birth to two years when the growth velocity decreases rapidly; [2] from two to three-and-a-half (in females) or four-and-a-half (in males) years of age when the velocity stays constant or growth accelerates; [3] from those ages to eight years of age when velocity decreases again to zero . . .' (p. 57). This overall pattern is a textbook case of an animal with infant, juvenile, and adult growth stages. Note that during the juvenile stage, Hamada's phase 2, growth rate 'stays constant' or accelerates. Weight growth for Hamada's male and female monkeys provide an example of both juvenile patterns. Male monkeys show a clear spurt in weight, but female monkeys show no spurt (similar to the baboons in Figure 3.11). Hamada's analysis of the macaques also includes the calculation of distance and velocity curves of growth for 15 different body segments or cranial dimensions, such as trunk length, tail length, upper arm and forearm length, thigh length and head breadth. Some of these dimensions show what might be a growth spurt, for example hand length and leg length, but other dimensions do not, such as forearm length and head breadth. Generally, male monkeys show spurts more often in these 15 dimensions than do female monkeys. The chimpanzees data analyzed by Hamada *et al.* (1996) included measurements of body weight and trunk length (measured from the cranial end of the sternum to the cranial end of the pubic symphysis). Velocity curves for these two measurements show that the chimpanzees have three stages of growth: infancy, juvenile, and adult. During the juvenile stage male chimpanzees experience a clear spurt in weight, but not for trunk length. Female chimpanzees do not show a postnatal growth spurt for either weight or length.

These studies by Leigh and by Hamada and colleagues are valuable because they include fairly large samples of non-human primates and many species of primates. Taken together, these studies show that some primate species have postnatal growth spurts in weight that occur around the time of sexual maturation. About half of the species studied so far show a weight spurt for the males only, and when a female weight spurt is present it is always smaller than the spurt for the males. These accelerations in weight growth are the result of an increase in size of several tissues, particularly muscle mass and body fat. Behavioral studies of free-ranging animals show that adult males use this muscle mass for reproductive competition with each other. Female primates may also be choosing to mate with larger males, who may be healthier and socially more successful than smaller males. In contrast to the males, the absence of a female growth spurt, or the smaller size of a spurt, allows females to maintain smaller adult body size compared with males. This may be advantageous during pregnancy and lactation, since smaller bodies require less energy and other nutrients, and this permits the mother to divert more of the food she eats to her developing offspring. Viewed in this context of reproductive strategies, one may begin to understand how evolutionary pressures and natural selection have shaped the type and intensity of growth spurts that non-human and human primates experience. Human boys and girls experience a growth spurt in weight and in height during adolescence, and the male spurts are more intense than the female spurts. The evolutionary pressures that shaped human growth at adolescence deserve more detailed treatment than can be provided at this point. Before moving on to this topic, which is treated at length in the next chapter, we must turn our attention to a more detailed consideration of skeletal growth spurts. For it is in the growth of the skeleton that we see the clearest difference between the patterns of growth for human and non-human primates.

A serious limitation of the otherwise excellent studies by Leigh and by Hamada and colleagues is that the growth velocity curves were computed from cross-sectional data, which tends to minimize the size of any actual growth spurt. However, cross-sectional data may also produce a growth spurt when none is really present. Differences in the care, feeding, or health of animals providing information for adjacent data points may result in discordant patterns of growth that produce spurious growth spurts. Only a pure longitudinal analysis can faithfully represent the growth of a primate, or group of primates, over time. One of the first longitudinal analyses of non-human primate growth was published by James Gavan (1953), who analyzed the skeletal growth of nine male and seven female chimpanzees. The chimpanzee data were collected at the Yerkes Primate Laboratories beginning in 1939. The chimpanzees were measured once a year from birth to age 12. Influenced by Brody's speculations, and based on the mathematical methods and devices (i.e., adding machines) available at that time, Gavan concluded that chimpanzees had a human-like skeletal growth spurt at the time of puberty. In 1971, however, after using the newer mathematical methods and the computer to reanalyze his chimpanzee data, Gavan found that no spurt in linear growth could be detected. 'After all, a smoothly decelerating curve gave the best fit to most of my data with a very small residual variance' (Gavan, 1982, p. 3).

Watts & Gavan (1982) reanalyzed the same chimpanzee data and new growth data for the rhesus monkey. The rhesus data were collected by Gavan between 1960 and 1967 from a laboratory colony. Longitudinal assessments of growth were made at least once a year from birth to five years of age. Watts & Gavan found that simple plots of height or weight for age did not reveal a growth spurt in either the rhesus monkey or the chimpanzee. This stands in contrast to the human case where simple graphical methods of analysis reveal the adolescent growth spurt in most individuals. To search for non-human primate growth spurts, they fitted the longitudinal data for each chimpanzee or rhesus with an exponential regression formula. 'This model depicts growth as a process that is gradually and constantly decelerating as size increases. Its use, therefore, assumes that an adolescent spurt does not exist' (Watts & Gavan, 1982, p. 56). This is the same model Gavan (1971) used when he concluded that the chimpanzee had no growth spurt in trunk length nor other linear dimensions.

Watts & Gavan found that deviations in skeletal growth from the regression model followed a consistent pattern in all 16 chimpanzees and in all the rhesus monkeys. An example is depicted in Figure 3.18 for the growth of the thigh of one chimpanzee. Deviations are positive, above the curve, in early infancy, negative in later infancy and just prior to puberty and then positive again following puberty. The authors noted that the differences between the observed and predicted values at each age are small, less than a centimeter and often only a few millimeters. It was impossible for them to depict these differences graphically as two distinct curves. 'Therefore, to exaggerate the differences for drawing this figure [Figure 3.18] the actual deviations were multiplied by a factor of three . . .' (Watts & Gavan, p. 58). Watts & Gavan emphasized their consistent finding of growth deviations from the predicted curve as evidence for the presence of adolescent growth spurts in non-human primates. No statistical test of the mathematical significance of the deviations was made. Watts does make the following qualitative assessment: '. . . the magnitude of the change is *very small*' (emphasis added, 1985, p. 56).

In 1990, Tanner, Wilson & Rudman published an analysis of skeletal growth in the female rhesus monkey (*Macaca mulatta*) and found evidence for 'clear pubertal growth spurts' (p. 101) for tibia length, crown–rump length (CRL), and weight. Their data consist of measurements taken every three months on nine monkeys housed indoors and six monkeys housed outdoors at the Yerkes Regional Primate Center in Atlanta, Georgia. Reproduced here as Figure 3.19 are the original illustrations from the article, as these illustrations are the essence of Tanner *et al.*'s analysis. The velocity curves shown in Figure 3.19 are mean values that were calculated

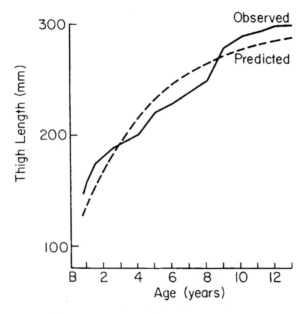

Figure 3.18. Observed and predicted growth curves for thigh length in a male chimpanzee (from Watts & Gavan, 1982).

by combining the data from all nine indoor-housed monkeys by aligning their curves on the age at maximum velocity and producing peak length velocity curves. This method is identical to the procedure invented by Boas to study human adolescent curves (Figure 1.6), and, '. . . an exact copy of Shuttleworth's (1937) famous analysis of human growth and produces essentially a mean-constant velocity curve' (Tanner *et al.*, 1990, p. 102).

Tanner *et al.* state that these figures of rhesus monkey growth '. . . establish the occurrence of a pubertal growth spurt beyond any reasonable doubt, at least for this species' (p. 101). The mean curve for tibia length certainly seems to show this, as there is a single peak velocity of about 2.25 cm/3 month (on an annual basis this peak velocity equals about 0.6 cm/year). The mean curves for CRL and weight are less easy to interpret. Both CRL and weight show two or three spurts, albeit the one preceding menarche being the largest. Tanner and colleagues offer only these figures as evidence for the primacy of the pre-menarchial spurt. They do not carry out any statistical analysis to demonstrate that the pre-menarchial spurt represents a significant increase in growth rate compared with any other period of growth. Nor do they analyze the duration of the spurt, which is important if we want to understand the contribution of this

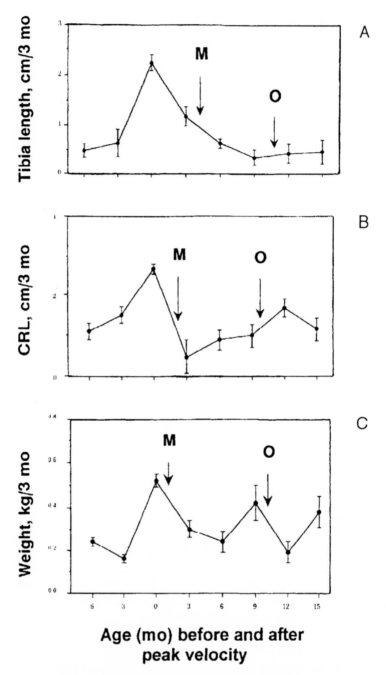

Figure 3.19. Mean-constant velocity curves for growth of female rhesus monkeys. (A) Tibia length, (B) Crown–rump length, and (C) weight (after Tanner *et al.*, 1990).

spurt to the total pattern of growth of these monkeys. Another complication is that the pattern of growth and growth spurts seen for the nine indoor-housed monkeys differs significantly from the six outdoor-housed monkeys. The outdoor group show a spurt for tibia length that occurs just before menarche, as is the case for the indoor-housed monkeys. The outdoor-housed females show either no spurt for CRL, or a small post-menarchial spurt, and two spurts for weight (Wilson *et al.*, 1988). The first weight spurt occurs before menarche and the second spurt well after menarche, which is unlike the situation for the indoor-housed monkeys.

Tanner *et al.* provide illustrations of the velocity curves for the tibia and CRL for, '. . . three individual indoor-housed monkeys . . . selected at random from the total group' (p. 103). That illustration is reproduced here as Figure 3.20. The data show much variability, with skeletal growth rates oscillating between 0.00 cm/3 mo and more than 2.5 cm/3 mo for adjacent measurements. Maximum individual peak velocities occur both before and after menarche (CRL for monkey Nos. 1 and 4) and smaller spurts occur both before and after first ovulation (CRL for monkeys Nos. 4 and 7, note the variation in 'spurts' for tibia length for all three monkeys).

Longitudinal data for primate skeletal growth are very rare, and it is even less common to see this data analyzed in a proper longitudinal fashion. This paper by Tanner, Wilson & Rudman is valuable because it presents and analyzes such longitudinal data. The quality of their data is unquestionable, the authors' analysis is meticulous. Nevertheless, human adolescents do not follow the patterns of growth seen in the rhesus monkey. Figure 3.21 is a mean-constant curve of human velocity growth in height and weight produced from the longitudinal records of growth of 12 Guatemalan girls. The data are derived from a study of children, juveniles, and adolescents attending a private school in Guatemala City (some data from this study were presented in Chapter 2 as Figures 2.8 and 2.9). The girls are from families of high socioeconomic status (SES), and all were healthy and well-nourished at the times of measurement – the girls were measured once per year. The same methods were used to construct Figure 3.21 as were used by Tanner *et al.*, (1990) to construct the mean-constant velocity curves for the nine rhesus monkey females of Figure 3.19. One indicator of human puberty is the change of growth velocity, from decreasing to increasing, that marks the transition from the juvenile stage to the adolescence stage. That point is labeled on the figure. The mean age at PHV, and peak weight velocity (PWV), for these 12 girls is 12.5 ± 1.07 years and the standard error of the mean is 0.31.

There are at least five differences between humans and rhesus monkeys in the pattern of velocity growth: (1) humans have one relatively large and

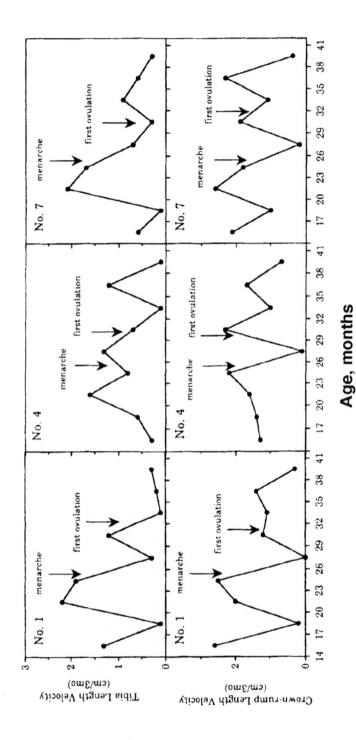

Figure 3.20. Velocity curves for tibia length and crown–rump length for three individual rhesus monkeys chosen at random from the sample used to construct the curves in Figure 3.19 (from Tanner *et al.*, 1990).

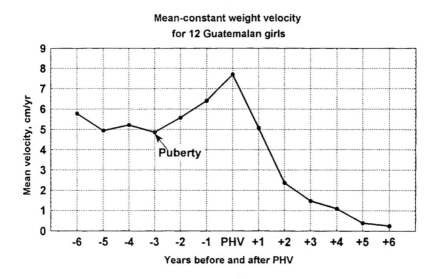

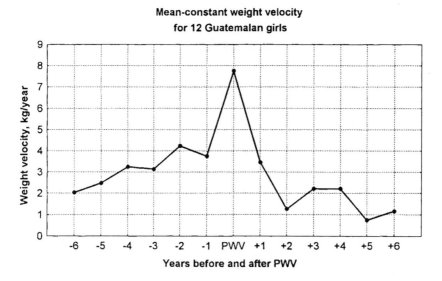

Figure 3.21. Mean-constant height velocity and weight velocity curves for 12 Guatemalan girls from a large longitudinal study. The change from negative to positive growth velocity at the time of puberty marks the transition from juvenile to adolescent stages. See the text for further discussion.

unambiguous growth spurt for both height and weight during adolescence, (2) the magnitude of the human adolescent spurt in the rate of skeletal growth is absolutely large compared to that of the non-human primates, (3) human growth velocities do not oscillate from near maximum velocity to near minimum velocity between adjacent measurements, (4) the human adolescent spurt lasts for years, the rhesus spurts last only a few months, which is a relatively short amount of time even considering that the monkey's life-span is only one-third that of humans, and (5) the human adolescent growth spurt is found in virtually all human populations, including both 'indoor-housed' and 'out-door housed' (that is, societies ranging from urban/industrialized to rural/hunting and gathering groups). An exception to each of these six points can be seen in the rhesus data.

The Guatemalan girls were measured once per year, so it is not possible to show with this particular data set that humans maintain a fairly smooth trajectory of adolescent growth even over shorter time periods. At monthly intervals there is a good deal of variation in stature increments for boys and girls, as was shown by Bogin (1977, 1978) and Togo & Togo (1982). Even so, during the time of the adolescent spurt in stature, human growth increments never approach zero as they do for the rhesus monkey. At six month intervals between measurements most of this human variability disappears. Six month intervals would be a fair comparison with the rhesus monkey three month intervals, as humans take more than double the time to grow up. The growth data for Montbeillard's son (Figure 1.4) was one famous example of human growth measured every six months. Togo (1995) analyzed the boy's growth based on the full data set of six month intervals between measurements, and showed that during this boy's adolescent growth spurt in height he increased smoothly to PHV and decreased smoothly for the next year. Togo did not find any of the oscillations that typify rhesus monkey growth after puberty.

Some important differences between human and non-human primate growth

There is no doubt that some non-human primates have small increases in the rate of skeletal growth for some body dimensions at the time of puberty. There is considerable evidence that these small increases in skeletal growth represent real and important biological processes. Most likely, the perturbations in growth around the time of puberty are due to the onset of adult levels of gonadal hormone secretions. It is now well-established that testosterone in male primates and estrogens, such as estradiol, in female primates can increase growth rates (Martin *et al.*, 1977; Bercu *et al.*, 1983; Prader, 1984). However, more is involved since roughly equivalent amounts of hormone production in both the chimpanzee and humans results in strikingly different rates of growth. In the male chimpanzee, the concentration of testosterone in blood serum prior to puberty (from one to six years of age) averages 13 nanograms per deciliter (ng/dl) (Martin *et al.*, 1977). For the human male, the pre-pubertal serum testosterone concentration (from ages one to 12 years) averages 9 ng/dl (Winter, 1978). The peak velocity in long bone growth of the eight male chimpanzees studied by Watts & Gavan (1982) occurred at a mean age of 10.96 years, with a standard deviation of 1.31 years. At this age, serum testosterone averages about 400 ng/dl (Martin *et al.*, 1977). Peak height velocity in human boys from western Europe occurs at a mean age of 14.06 years, with a standard deviation of 0.92 years (Marshall, 1978), when serum testosterone levels average about 340 ng/dl (Winter, 1978). Based on these data, the serum testosterone concentration of the chimpanzee increases about 31-fold from the pre-pubertal to pubertal state. In the human male, serum testosterone concentration increases about 38-fold, or 1.23 times the increase for the chimpanzee.

Clearly, both chimpanzee males and human boys have large increases in testosterone production after puberty, but the effects on skeletal growth are not so similar. According to Watts & Gavan (1982), chimpanzees have a relatively small increase in the velocity of growth of individual long bones during puberty, '. . . usually less than a centimeter' (p. 58). Similar findings are reported by Copeland *et al.* (1985), who analyzed growth in weight and crown–rump length (CRL) in 86 male and female chimpanzees. In males there was a marked increase in serum testosterone between the ages of six to eight years. This was followed by a slight increase in the rate of weight gain but no detectable spurt in CRL. Females also showed a rise in serum testosterone between the ages of six to eight years and rise in estradiol (one of the estrogen hormones) after ten years of age. In contrast to the males, female chimpanzees had a much smaller increase in the rate of weight gain in relation to testosterone levels. Females had no detectable spurt in CRL at any age. Hamada *et al.* (1996) confirm these findings when they report no increase in the rate of skeletal growth at the time of puberty for their samples of male or female chimpanzees.

In sharp contrast to the chimpanzee research are the findings for human boys and girls, who show relatively large and easily detectable growth spurts for both height and weight during adolescence. As chimpanzees are not bipedal it is not possible to take a measurement that is exactly equivalent to human height. It is more appropriate, therefore, to compare chimpanzee and human growth in terms of body segments. Cameron *et al.* (1982) performed a longitudinal analysis of the growth of individual limb segments in British boys. It was found that the peak value in velocity during the adolescent growth spurt ranged between 1.34 cm per year for the forearm and 2.44 cm per year for the tibia. Satake *et al.* (1993) report that during human adolescence, peak velocity for sitting height (roughly equivalent to crown–rump length) equals 7.5 cm/year for Japanese boys and 6.2 cm/years for Japanese girls. From these findings one may propose that there are differences in the effect of testosterone and estrogens on the skeletal growth of chimpanzees and human beings. The growth response of the human skeleton to rising testosterone levels is greater than that of the chimpanzee skeleton, as the change in the serum hormone levels for males of the two primates differs by a factor of about 1.23, but the change in the velocity of growth of various body segments is two to four times greater for the human arm or leg, and more than ten times greater for sitting height.

The velocity of growth in weight of the male chimpanzee (and, presumably, male baboons and some female primates) is better correlated with endocrine changes at puberty, as may be seen in Figures 3.11 and Figure 3.16. This observation suggests that there are differences in the growth response of skeletal and non-skeletal tissue, such as muscle and adipose. There are also differences in the effect of hormones on tissue growth between males and females within the same species. Female chimpanzees experience a much smaller weight spurt than males and female baboons show no weight spurt. It is very important to contrast this to the human case; most girls experience significant growth spurts in both height and weight. Thus, the growth effects of pubertal endocrine changes differ between individuals of the same sex, but different species (e.g., male chimpanzees and human boys), and between the sexes within the same species (e.g., male and female chimpanzees).

Three major differences between human and non-human primate growth may be highlighted at this point. These are: (1) the residual growth potential of the non-human versus the human primate at adolescence, (2) the sensitivity of different body tissues to growth-promoting stimuli, and (3) sex differences in the expression of growth spurts at adolescence. Monkeys, apes, and humans all experience a prolongation in the time for growth and a delay in the age at onset of sexual maturation. In humans, the delay is both relatively and absolutely greater than in the monkey or ape (Laird, 1967). In addition to this, humans also have a markedly increased potential, compared with monkeys and apes, for growth in height and weight during adolescence. This growth potential is more likely to be regulated by the sensitivity of neuroendocrine receptors and post-receptors (i.e., biological tissues) to growth stimuli than by the rate or amount of production of the stimuli (e.g., hormones) themselves. The lack of linear associations between testosterone concentrations and growth velocities in skeletal and non-skeletal tissue of chimpanzee and human show this. The differences in cellular sensitivity to growth stimuli between non-human and human primate growth are probably controlled at the genetic level. While it is possible that the different patterns of growth seen in the various primate species are the result of the evolution of new **structural genes** (genes that code for specific proteins), it is more likely that the variation in growth control lies in the **regulatory genes** that initiate and terminate each of the distinct periods of growth and control their duration (Britten & Davidson, 1969; King & Wilson, 1975; and Chapter 7).

A philosophy of human growth

Non-human primate models are used to study human growth because of similarities in anatomy and physiology between the species. When Schultz started his studies of primate growth he assumed explicitly that there was an evolutionary continuum between the living primate species. His famous illustration of primate life history (Figure 3.12) shows this assumed continuum very nicely. There is an evolutionary connection relating all primate species, but the living monkeys, apes, and humans each have a separate evolutionary history. Cercopithecoids (Old World monkeys) and hominoids (apes) separated some 20 million years ago and the hominoid–hominid (ape–human) split occurred about five to six million years ago. There is no evolutionary reason to expect that the patterns of growth of these three divergent and ecologically distinct species should be identical or even similar. As Gavan (1971, p. 54) observed, chimpanzee postnatal growth begins '. . . with an initially high rate which decelerates smoothly as size increases, but it is well known that human growth is characterized by a growth spurt . . . Some change must have occurred in human growth since we and the chimpanzee have had a common ancestor'.

The notion of an evolutionary continuum is one legacy of the philosophy of a Great Chain of Being (Lovejoy, 1936). The Great Chain is a popular cultural construct in western society that has historical roots going back at least to ancient Greek writers. In biology, the Great Chain takes the form of the *scala naturae*. In its original usage this implies (erroneously) that all living creatures, from amoeba to human, form a living evolutionary sequence from the simplest to the most complex creature. We now understand that humans are not the culmination or the goal of evolutionary history. We are just one of more than two million animal species alive today, each the end-product of its own history and each with its own unique place in nature. Yet, the *scala naturae* is sometimes misapplied to the connection between human and non-human primates.

Some observers tend to see monkeys and apes more as models for human biology and behavior than as creatures in their own right. This point was cogently argued by Scott (1967, p. 72): 'Subhuman primates are not small human beings with fur coats and (sometimes) tails. Rather they are a group which has diversified in many ways, so that they are as different from each other [and humans] as are bears, dogs and raccoons in the order Carnivora'. Though Scott referred specifically to psychological attributes of species within the orders Carnivora and Primates, his cautionary remarks apply equally to morphology, physiology and, in the present context, patterns of growth. As mentioned above, within the Carnivora, certain social species (dogs, wolves, lions) experience a prolonged period of relatively slow growth between infancy and adulthood, which corresponds to the juvenile growth phase. Non-social species of carnivores mature from infancy to adulthood without a juvenile stage of growth. There is no reason to expect, a priori, that the Primates, as an order of mammals, would be any more uniform in growth patterns than the Carnivora. Newell-Morris & Fahrenbach (1985) reviewed the use of non-human primates as models for human development and growth and concluded, '. . . there are problems with the extrapolation from the non-human primate model to the human condition because of intergeneric differences in size, growth and development rates, and timing. Although investigators justify direct extrapolation of their findings on the basis of the close genetic relationships of all primates, this assumption in many cases may be little more than absolute faith in the evolutionary argument from which it stems' (p. 35).

Watts (1990) points out that the patterns of growth of the New World monkey *Cebus*, the Old World rhesus monkey, the chimpanzee, and humans are all derived from some ancestor or ancestors. Each of the living species is likely to be derived from its ancestor in ways that are independent and unequal, due to 'different ecological and adaptive circumstances' (p. 99). Watts argues that the rhesus monkey makes a poor model for human growth due to an advanced state of skeletal development at birth and an early onset of puberty and menarche compared with humans. The rhesus is also a seasonal breeder in nature, with ovulation taking place in the fall or winter months and birth occurring in the spring or summer. Female rhesus monkeys raised outdoors may reach menarche in any season but have their first ovulation only in the fall (Wilson *et al.*, 1988). When housed under artificial conditions, such as for the indoor-housed monkeys studied by Tanner *et al.* (1990), the sexual maturation of the female rhesus is very much altered as they can reach first ovulation in any season. Several studies show that when experimental animals, including primates, are housed indoors and exposed to 12 hours of light followed by 12 hours of darkness (a standard procedure) many of their normal endocrine system functions are disturbed, or even obliterated (Bogin, 1977). As both sexual maturation and growth of the body are influenced by many of the same hormones it seems most unreasonable to expect non-human primates reared indoors to serve as models for human growth and development.

My colleague Steve Leigh offered the following comment after reading a draft of this chapter: 'The presence of a spurt in tibia or in other anatomical units certainly does not, as your review indicates, imply homology with the human statural spurt'. In biology, the term **homology** is used to describe anatomical structures, physiological processes, or behaviors that are found in different species due to a common evolutionary origin. The five fingers and toes, and the intimate mother–infant relationship, of almost all primate species are due to homology. Leigh's point is that growth spurts for various parts of the body of different species of primates are not necessarily homologous. Continuing with Leigh's comment, he adds:

> For example, my recent studies of baboons demonstrate unambiguous growth spurts in the snout, probably comparable in magnitude to tibial spurts observed by Tanner *et al.* in macaques . . . The presence of these spurts, however, does not imply homology with human growth spurts in stature. CRL and sitting height would be the closest homologues, but even then, locomotor differences among these primates would greatly complicate assessments of homology. What I derive from the ideas you are presenting here, and in the last section, is that it might be best to think of growth spurts as modular and highly evolvable features of ontogeny. Natural selection (or sexual selection) can effectively 'put' spurts where (anatomically) and when they are needed to increase fitness. I think that the field has labored a little too long under Brody's desire to find a universal pattern. What may be happening here is more akin to a universal process. Perhaps we could propose that the universal process is modularity and evolvability of growth spurts (or decelerations), and that this process need not produce a uniform pattern between any two species or within a particular clade.

An acceptable philosophy of human growth must acknowledge the mammalian and primate foundations for the human pattern of growth. However, that philosophy must also allow for the evolution of variations on common themes, and the evolution of new stages of growth that may be unique to the human species. A robust philosophy of human growth must also account for the ecology to which the human species – indeed any species – is adapted. With this in mind, we may now proceed to a discussion of the evolutionary and ecological pressures that shaped the pattern of human growth.

4   *Evolution of the human life cycle*

The previous chapter described the evolution of the human pattern of growth in terms of its mammalian and primate foundations. The emphasis was on form and function of anatomical structures such as the placenta, the brain, and the skeletal system and how these structures are developed during the stages of human growth. In this chapter the emphasis is placed on how the human **life cycle** evolved. The life cycle of any organism includes all the stages of growth, development, and maturation from conception to death. In the following discussion the focus is on the stages of postnatal life because the most profound changes in the evolution of the human life cycle have occurred to the stages of life after birth.

The pattern of human growth evolved in the context of the biological and social **ecology** of our ancestors. The term 'ecology' is used here to refer to the relationship that an individual organism, or group of individuals of a species, has with its physical, biological, and social environment. At the core of any ecological system are two sets of behaviors; the first is directed toward how an organism acquires food and the second is directed toward how the organism reproduces. All organisms are alike in that they share behaviors related to what may be called simply 'food and sex'. Social mammals, including most primates, satisfy their needs for food and reproduction through a complex ecology of biological and social relationships with their environment. Human beings also share in this biosocial ecology, and add to it a significant cultural component. Human beings are **cultural animals**, meaning that we possess all the potentials and limitations of any living creature, to which we add a cultural trilogy of: (1) dependence on technology, (2) codified social institutions, such as kinship and marriage, and (3) ideology. In its anthropological sense, ideology refers to a set of symbolic meanings and representations particular to any society, through which its members view and interpret nature. Elements of the human capacity for culture may be found in many other species of animals, but only in the human species do all three aspects of the cultural trio become so intensified, elaborated, and universal. Because of this, human growth may be best understood by using a **biocultural** and anthropological perspective.

Anthropologists have become increasingly interested in explaining the

153

significance of life cycle characteristics of the human species. This is because the human life cycle stands in sharp contrast to other species of social mammals, even other primates. Several of these contrasts, such as the mid-growth spurt and the adolescent growth spurt, both of which are experienced almost universally by boys and girls, were described in previous chapters. Any theory of human growth needs to explain how and why humans successfully combined these growth spurts along with relatively helpless newborns, a short duration of breast-feeding coupled with a vastly extended period of offspring dependency, delayed onset of reproduction, relatively short birth intervals, unusual secondary sexual characteristics, such as the peculiar distribution of both hair and fat in women and men, and menopause. A central question is, did these characteristics evolve as a package or as a mosaic? The present evidence suggests that the stages and events of the human life cycle evolved as a mosaic and may have taken form over more than a million years.

Life history and stages of the life cycle

Life history theory may be defined as the study of the strategy an organism uses to allocate its energy toward growth, maintenance, reproduction, raising offspring to independence, and avoiding death. For a mammal, it is the strategy of when to be born, when to be weaned, how many and what type of pre-reproductive stages of development to pass through, when to reproduce, and when to die. Living things on earth have greatly different life history strategies, and understanding what shapes these histories is one of the most active areas of research in whole-organism biology.

Human life history, with nearly two decades of infant dependency, extended childhood, juvenile and adolescent stages prior to social and sexual maturation, has long been considered to be advantageous for our species because it provides:

1. An extended period for brain development.
2. Time for the acquisition of technical skills, e.g., tool making and food processing
3. Time for socialization, play, and the development of complex social roles and cultural behavior.

These statements are standard 'textbook' rationalizations for the value of the pattern of human growth. They emphasize the value of learning, an idea that Spencer (1886) popularized, but which actually goes back to the dawn of written history (Boyd, 1980). Learning as the reason for the evolution of several prolonged life stages prior to maturation was nicely summarized by Dobzhansky (1962): 'Although a prolonged period of juvenile helplessness and dependency would, by itself, be disadvantageous to a species because it endangers the young and handicaps their parents, it is a help to man because the slow development provides time for learning and training, which are far more extensive in man than in any other animal' (p. 58). The learning hypothesis for human ontogeny was also invoked by Allison Jolly, author of *The Evolution of Primate Behavior*. She writes that '. . . human evolution is a paradox. We have become larger, with long life and immaturity, and few, much loved offspring, and yet we are more, not less adaptable'. In an attempt to resolve the paradox of human evolution and our peculiar life history, Jolly concludes in the next sentence that 'mental agility buffers environmental change and has replaced reproductive agility' (1985, p. 44).

The reference to reproductive agility means that we are a reproductively frugal species compared with those that lavish dozens, hundreds, or thousands of offspring on each brood or litter. It is fairly easy to argue that humans, with relatively low wastage of offspring, are somehow more 'efficient' than other species. But a paradox still remains, for the learning hypothesis does not explain how the pattern of human growth evolved. It does not provide a causal mechanism for the evolution of human growth. Rather, it is a tautological argument for the benefits of the simultaneous possession of brains that are large relative to body size, complex technology, and cultural behavior. Specifically, the learning hypothesis does not answer the following questions: (1) Why not produce more offspring, instead of few mentally agile offspring? (2) Why do our offspring take so long to reach reproductive age? (3) Why is our path of growth and development from birth to maturity so sinuous, meandering through alternating periods of rapid and relatively slow growth and development? (4) Why do human women often live for many years past the age of reproduction? A theory of human growth must answer these and other questions about the presence of human life history events and their timing during the life cycle.

The evolution of ontogeny

Ontogeny refers to the process of growth, development, and maturation of the individual organism from conception to death. It is virtually axiomatic that every species has its own unique pattern of ontogeny (Bonner, 1965; Gould, 1977). Behind this truism lies what may be the 'secret' to life, hidden in the process that converts the fertilized ovum, with its full complement of **deoxyribonucleic acid (DNA)**, into a multicellular organism composed of hundreds of different tissues, organs, behavioral capabilities, and emotions. During human evolution, the form and function of our ancestors' structural and regulatory DNA was reworked to produce the genetic basis for the ontogeny of the human species. The literature is replete with proposals for how the reworking occurred. One tradition in the study of human evolution looks for a single major cause or process. It has been argued that humans evolved when we became big-brained apes, terrestrial apes, killer apes, hunting apes, aquatic apes, tool-making apes, symbolic apes, monogamous apes, food-sharing apes, and, even, apes with ventral–ventral copulatory behavior. None of these, or any other single factor hypothesis, proves to be helpful to understand human evolution, for a non-human primate exception can always be found. Another tradition looks instead at the pattern of ontogeny. In the book *Size and Cycle*, J. T. Bonner (1965) develops the idea that the stages of the life cycle of an individual organism, a colony, or a society are '. . . the basic unit of natural selection'. Bonner's focus on life cycle stages follows from the research of several nineteenth and twentieth century embryologists who proposed that speciation is often achieved by altering rates of growth of existing life stages and by adding or deleting stages. Bonner states that we should not think in terms of organisms *with* a life cycle, but rather think of organisms *as* life cycles. 'The great lesson that comes from thinking of organisms as life cycles is that it is the life cycle, not just the adult, that evolves. In particular, it is the building period of the life cycle – the period of development – that is altered over time by natural selection. It is obvious that the only way to change the characters of an adult is to change its development' (1993, p. 93).

A history of research on life cycle evolution was published by S. J. Gould in the book *Ontogeny and Phylogeny* (1977). Gould handily summarizes the mechanisms for biological change over time by stating, 'Evolution occurs when ontogeny is altered in one of two ways: when new characters are introduced at any stage of development with varying effects upon subsequent stages, or when characters already present undergo changes in developmental timing. Together, these two processes exhaust the formal content of phyletic change . . .' (p. 4). Gould contends that it is the second process that accounts for human evolution. This process is called **heterochrony**. Quoting Gould again, '. . . this book is primarily a long argument for the evolutionary importance of *heterochrony* – changes in the relative time of appearance and rate of development for characters already present in ancestors' (p. 2, author's italics). In the discussion that follows, the focus will be on whether the human life cycle evolved by altering 'characters already present' in our ancestors or whether it evolved by introducing new characters. The evidence argues strongly in favor of childhood and adolescence as new characters in the human life cycle.

Gould explains that there are several types of heterochronic processes, but only one accounts for human evolution. This is **neoteny**, defined in the glossary of Gould's book as 'Paedomorphosis (retention of formally juvenile characters by adult descendants) produced by retardation of somatic development'. In a subsequent publication, Gould provides a somewhat more readable definition: 'In neoteny rates of development slow down and juvenile stages of ancestors become adult features of descendants. Many central features of our anatomy link us with the fetal and juvenile stages of [non-human] primates . . .' (Gould, 1981, p. 333). Following Gould, we must add neoteny to the list of other single-cause hypotheses as the reason, or at least the mechanism, for human evolution.

Another heterochronic process, called **hypermorphosis**, is favored by other researchers as the mechanism for human evolution. Hypermorphosis may be defined, at this point, as an extension of the growth and development period of the descendant beyond that of the ancestor. The differences between neoteny and hypermorphosis may be summarized as follows. Neoteny is a slowing down of the *rate* of development. Neoteny produces an adult descendant that retains the immature body shape, and even the immature behavioral characteristics, of its ancestor. Hypermorphosis is a prolongation of the *time* for development, and this extra time for ontogeny produces a descendant with features that are hypermature compared with the ancestor. Both neoteny and hypermorphosis are evaluated critically in this chapter.

One point to stress at this juncture, however, is that Gould defined another way for life stages to evolve. This is '. . . when new characters are introduced at any stage of development' (1977). Some stages of the human life cycle, namely childhood and adolescence, may well be such new life cycle stages. If this is true, then these new stages were inserted *de novo* into human life history sometime in the past. The next sections of this chapter show that some types of heterochrony can account for the evolution of the juvenile life stage. It is also shown that neither the Peter Pan scenario of neoteny nor the Methuselah-like development of hypermorphosis are adequate to account for the evolution of the entire human life cycle. Moreover, neither neoteny nor hypermorphosis can unravel the paradox of human evolution. To resolve the puzzle of why we have so few offspring, why they take so long to develop and reproduce, and why our rate of growth takes a serpentine path to adulthood, requires a multi-causal, and more 'mature', view of human ontogeny and human evolution.

Neoteny and human evolution

That human beings of all ages are essentially child-like in morphology, behavior, and cognitive potential is the essence of the concept of neoteny. Notions of human neoteny may be traced back as far as Biblical writings (Montagu, 1989), and the concept has dominated popular and scientific attitudes toward the evolution of human growth for hundreds of years. William Wordsworth, for example, praised the concept of neoteny in 1802 with words of innocence and hope:

> My heart leaps up when I behold
>   A rainbow in the sky:
> So was it when my life began;
>   So is it now I am a man;
> So be it when I shall grow old,
>   Or let me die!
> The child is father of the man;
> And I could wish my days to be
> Bound each to each by natural piety.

The term 'neoteny' was coined by Julius Kollman in 1885 to describe the sexual maturation of the axolotl, a urodele amphibian (salamander), while still in its aquatic, gill breathing stage of development. Ashley Montagu (1989) states that Kollman intended neoteny to mean 'retaining youth' but that Kollman confused the Greek word *teinein* (meaning to stretch) with the Latin word *tenere* (to retain). Figure 4.1 presents contemporary dictionary definitions, as well as another current medical usage of the Latin root of the term. Despite the scatological humor that may be found in Kollman's etymological error, Montagu and Gould believe that Kollman had the right idea, that is, neoteny is the process for human evolution.

The idea that neoteny is the primary process for humanization was first formalized scientifically by Louis Bolk in 1926 (see Montagu, 1989 for a concise review of historical sources). Gould acknowledges that much of Bolk's neoteny is really an argument for scientific racism and sexism. Nevertheless, Gould tries to retain the baby of neoteny while discarding the bath water of racist and sexist science. To do so, Gould suggests that the major difference between human and non-human primate growth is that humans mature sexually while still in an infantile or juvenile stage of physical development.

Gould expressed his ideas on heterochrony via neoteny in terms of a 'clock model' (Figure 4.2). The clock has two hands, one that calibrates a size arc and the other that measures a shape arc. The clock sits above a type of bar-shaped calendar that records time in the form of biological age, that

Neoteny

ne·ot·e·ny -- noun

1. Retention of juvenile characteristics in the adults of a species, as among certain amphibians.

2. The attainment of sexual maturity by an organism still in its larval stage

— ne´o·ten´ic or ne·ot´e·nous adjective

[New Latin neotenia : NEO- + Greek teinein, ten-, to extend. See tenesmus.]

te·nes·mus -- noun

A painfully urgent but ineffectual attempt to urinate or defecate.

[Medieval Latin tenesmus, variant of Latin tenesmos, from Greek teinesmos, from teinein, to strain, stretch.]

Figure 4.1. Definitions and etymology of the term 'neoteny'. The sketch of the amphibian represents the axolotl, which often becomes sexual mature while still retaining the external gills typical of larval amphibians. The cartoon of Betty Boop illustrates some human features which are sometimes labeled as neotenous, such as a large head, short arms and legs relative to total height, and clumsy, child-like movements. Definitions from *The American Heritage Dictionary*, 3rd edition. Betty Boop © King Features Syndicate, used with permission.

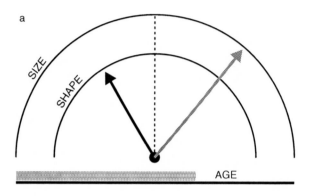

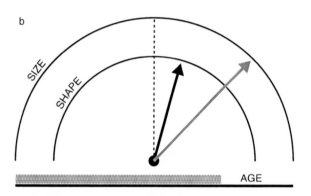

Figure 4.2. (a) Clock model for human evolution by neoteny; (b) clock model for human evolution by hypermorphosis. From Godfrey & Sutherland (1996) with permission.

is, age at sexual maturation. The clock may be used to compare an ancestor with its descendants. The ancestor's clock is fixed with the two hands at the zenith of the arcs and age of sexual maturation at the mid-point of the calendar. The descendant's clock measures changes in size, shape and age at maturation by differences in these three vectors. The major point that Gould emphasizes with the clock model is that size, shape, and maturation are disassociable from each other and can evolve independently. Figure 4.2(a) is Gould's clock for human evolution via neoteny. The 'size' hand of the clock extends to the right of the ancestral condition meaning that humans have larger-sized body, brain, and other organs than our ancestor. The 'sexual maturation' bar also extends to the right of the ancestral condition, indicating that humans mature at a later age, compared with our ancestor. Finally, the 'shape' hand of the clock is shifted to the left of the ancestral condition, meaning that when sexually mature, humans retain the shape of an immature ancestor. That shape includes a relatively large, bulbous cranium, small jaws and a flat face.

Gould makes very clear his interpretation of the consequences of neoteny. 'If humans evolved, as I believe, by neoteny, . . . then we are, in a more than metaphorical sense, permanent children' (1981, p. 333). Could it be true that humans are 'permanent children?' Some eminent scholars believe it. Benjamin Franklin wrote, 'Our whole life is but a greater and longer childhood'. Sigmund Freud was more circumspect; 'In our inner-most soul we are children and remain so for the rest of our lives'. Montagu is the most emphatic: 'We are intended to remain in many ways childlike, we were never intended to grow up into the kinds of adults we have become . . . Our uniqueness lies in always remaining in a state of development' (all quotes from Montagu, 1989).

Given the scientific respectability of works on neoteny by Gould and Montagu, the term and the concept have been popularized in Western society and adopted into many other arenas. The following are some recent examples. Philosophers posit neoteny as the process leading to the human capacity for language (Goldsmith, 1993). Psychiatrists allege it is the reason humans are so playful (Brown, 1995). Some types of neurological dementia are described as 'a failure of neoteny' (Bemprad, 1991). Human female sexual attractiveness is claimed to be a function of neoteny (Jones, 1995). Taking this cue, the advertising industry employs human infants, children, and neotenous adult women to 'sell' all sorts of products. Moreover, the product designers make the inanimate objects (everything from tea pots to automobiles) more appealing and purchasable by making them seem neotenous, that is shaped like a human infant or child (Boym, 1994 and see Box 4.1). Finally, a prominent social anthropologist claims that human evolution via neoteny allows for the development of human culture, especially religion (La Barre, 1991)!

Hypermorphosis and human evolution

Despite the intellectual weight that biology (Gould), social philosophy (Franklin), psychoanalysis (Freud), and anthropology (Montagu) bring to this issue, as well as the economic import of advertising, and despite the popularity of neoteny as a *cause célèbre* for so many human traits and behaviors, the proposition that adult human beings are permanent children is not accepted by all scholars. Michael Mckinney and Kenneth McNamara (1991), in their book *Heterochrony: The Evolution of Ontogeny*, provide a detailed case against neoteny in human evolution (see also McKinney, 1998). McKinney & McNamara argue instead for another type of heterochronic process to account for human growth and evolution, namely hypermorphosis. They state their position as follows, 'Neoteny is the process of growing slower. Yet humans do not grow particularly slow (relative to either chimp or our ancestors . . .). What we do is delay the offset of virtually all developmental events (growth phases) so that each phase is longer. This is hypermorphosis . . .' (p. xi). Figure 4.2(b) illustrates evolution by hypermorphosis using Gould's clock model. Both the neoteny and hypermorphosis clocks are in agreement that human beings have larger size and later sexual maturation than our putative ancestor. The difference between the two clocks is that hypermorphosis posits that adult humans have a more 'mature' shape than our ancestor. That shape includes a large, bulbous cranium, small jaws and a flat face, exactly the same as for neoteny! The difference between neoteny and hypermorphosis is not to be found in what evolved, but rather in how it evolved. McKinney & McNamara do admit that there are a number of non-hypermorphic features of human beings, but they state that '. . . hypermorphosis seems to best explain most of those traits that make us human: large body size, large brain, long learning stage and life span . . .' (p. xi).

According to the hypermorphosis hypothesis, humans are not permanent children, rather we are developmentally delayed and growth-prolonged apes. Recall from the previous chapter that Adolph Schultz (1960) proposed a hypothesis along these same lines slightly more than three decades before McKinney & McNamara's book was published. In his classic illustration, reproduced here as Figure 3.12, Schultz proposes that primate evolution occurred by the progressive delay in the onset and offset of primate life stages. McKinney & McNamara refine Schultz's idea by calling this process of delayed offset and onset of human development sequential hypermorphosis, '. . . because a number of sequential phases are affected' (1991 p. xi). Hypermorphosis, therefore, is hardly a new idea, but is it the right idea?

Sue Taylor Parker (1996) thinks that sequential hypermorphosis is the right idea, and she uses it to account for the evolution of human cognitive capacities. Parker applies Piaget's stage theory (1954; Piaget & Inhelder, 1969) to analyze the cognitive development of human and non-human primates. Based on many years of innovative and careful research, she finds that all primates share the first Piagetian stage, called the sensorimotor stage. At this stage a monkey, ape, or human infant can make logical relationships among objects and can perform rudimentary imitations. At the highest level of this stage, rhesus monkeys perform cognitively at the level of a two-year-old human infant. Chimpanzees complete the sensorimotor stage at about three years of age and move on to Piaget's preoperational stage. At this level of cognitive development, chimpanzees and humans understand cause-and-effect relationships and can engage in pretend play. According to Parker, chimpanzees do not complete the preoperational stage and remain at a cognitive level equivalent to a four-year-old human for the rest of their lives.

Human children remain in the preoperational stage from about age two to seven years. Humans share with chimpanzees the ability to discern cause-and-effect relationships and engage in pretend play during this stage, but humans also become preoccupied with learning and practicing verbal skills. During the preoperational stage the human infant and child begins to learn the names for objects and to reason intuitively. The next Piagetian stage, called the concrete operational stage, lasts from about age seven to 12 years (note that this stage corresponds to the shift from childhood to juvenile stages as defined in the previous chapter). During the concrete operational stage the juvenile begins to deal with abstract concepts such as numbers and relationships. Finally, humans progress to the formal operational stage, lasting from age 12 to 15 years. This stage, which corresponds with human adolescence, marks the beginning of reasoning in a logical and systematic fashion. Parker explains that the uniquely human stages were added by hypermorphosis, that is, by extending the period of cognitive development past that of the monkeys and apes. This negates neoteny, which in Parker's view would mean that human cognitive capacities evolved by halting mental development at an early or intermediate stage of non-human primate development. To quote Parker, 'Cognitively, humans are overdeveloped rather than underdeveloped apes' (p. 377).

Elizabeth Vrba (1996) agrees that hypermorphosis is the key process in human evolution, but she adds a twist to McKinney & McNamara's argument. Vrba's model of human evolution is set in the more general context of mammalian evolution in Africa. Vrba tries to show that climate change, specifically global cooling after 2.9 MYA (million years ago), resulted in an enlargement of body size along with a relative decrease in limb length for several species of African mammals. Such morphological change in response to a cooler climate accords with Bergmann's and Allen's ecological 'rules' concerning climate and mammalian body shape. According to Vrba, the African **hominids** alive at that time also conform to these ecological rules (the term 'hominid' refers to living humans and fossil species that walked about primarily on two legs). However, the hominids did not just increase in body size, rather they also increased brain size relative to body size. Vrba explains that this peculiar hominid morphological change resulted from '. . . the same evolutionary event of growth prolongation, or time hypermorphosis, as it acts on characters with different ancestral growth profiles in the same body plan'. Vrba continues by stating that this '. . . can result in a major reorganization – or "shuffling" – of body proportions such that some characters become larger and others smaller, some hyperadult and others more juvenilized' (1996, p. 1). Many of the 'ancestral growth profiles' of the hominids are, according to Vrba, still to be found in the great apes. She states 'I do not imply that the chimpanzee itself is ancestral, but only that its growth profile resembles that of the common ancestor' (1996, p. 17). Vrba predicts that by maintaining chimpanzee growth rates for legs, arms, torso, skull, brain, etc. and prolonging the total time for growth it is possible to derive a modern human morphology from an African pongid morphology. As will be shown in the next section of this chapter, this prediction has been tested and found to be invalid.

Critiques of neoteny and hypermorphosis

The concepts of neoteny and hypermorphosis posit that modern humans evolved by either a delay or an extension of ancestral patterns of growth. Human growth and development, anatomy, physiology, and behavior cannot be explained by either of these heterochronic processes. Humans display new patterns of growth not seen in any other species of mammal, especially other primates. Two such new patterns are the mid-growth spurt that marks the end of childhood and the adolescent growth spurt in skeletal growth. As discussed in the previous chapter these two novel growth spurts result from patterns of endocrine action and regulation of tissue sensitivity that are found only in human beings. Proponents of heterochrony argue that these growth spurts are not really 'new' because they result from modifications of genetic, hormonal, and physiological systems that were present in the primate ancestors of humans. At this point any critique of heterochrony slips into the murky depths of semantic meanings of words such as 'new' and 'novel'. One way to avoid this semantic morass is to think about the pattern of human growth as one might think about the elephant's trunk. In one sense the elephant's trunk is a nose, and it is possible to model the evolutionary development of the trunk as a heterochronic change in the growth of any mammalian nose. But, the trunk is not like most other noses since the trunk is a muscular

appendage that manipulates both the physical and social world of elephants. At what point in elephant evolution did the nose become a trunk? If
we had all the elephant ancestors that ever lived, then maybe we could see a
smooth and continuous evolution of the trunk from a nose. But, maybe we
would not see this sort of smooth continuity, rather at some point what was
a nose would take on the functions of a trunk. In a similar fashion, the
human stages of childhood and adolescence are not just heterochronic
extensions of life stages found in other mammals. Childhood and adolescence, like the trunk, serve new functions and have their own peculiar
growth characteristics and functional implications.

   Criticisms of the application of neoteny to human evolution are not
new. Kummer (1953), Bertalanffy (1960), and Starck & Kummer (1962)
argued against neoteny, or what they called fetalization, on the basis of
human craniofacial and postcranial growth; '. . . the concept of "fetalization" is to be refused with respect to the ontogeny and evolution of the
human . . . for this is not the result of an arrest of growth at an early phase
but of a differentiation in changed direction . . . Hence we may speak of
"retardation" but not of fetalization in human development' (Bertalanffy,
1960, p. 250). The phrase to emphasize in this quote is 'differentiation in
changed direction'. More recent analysis confirms this conclusion and
provides more details. One analysis by Dean & Wood (1984) compares the
growth of the bones of the cranial base of juvenile humans, orang-utans,
gorillas, and chimpanzees. Ten measurements of growth were taken from
each primate skull. The authors find that the growth changes within and
across species are so complex that a relatively simple process, such as
neoteny, is 'insufficient to explain the observed morphological differences
between the cranial base of modern *Homo sapiens* and the great apes' (p.
157). The proponents of neoteny believe that only by retarding development could the human brain, along with childhood learning and culture,
have evolved. However, another detailed study of brain development
shows this not to be the case. Armstrong *et al.* (1995) studied the ontogeny
of human brain gyrification, that is, the development of the highly folded
and convoluted cerebral cortex of adult humans. Their data indicate that
there are new patterns of folding in the human brain. The origins of these
new patterns cannot be found in other primates and did not evolve from
features already present in human ancestors. Armstrong and colleagues
conclude that 'the data support the thesis that human cortical proportions
evolved when the brain enlarged in size and that the process was not one of
neoteny' (p. 56).

   One specific example of the failure of neoteny and hypermorphosis to
account for the direction of human development is the ontogeny of cranial

growth related to language development. The human newborn cannot produce the speech sounds (phonemes) used by adult speakers of any language. Lieberman *et al.* (1972) and Laitman & Heimbuch (1982) believe that the shape of the basicranium is the reason for this. They argue that newborns possess a basicranium with a relatively large angle of flexion (the angle formed by the junction of the occipital and vomer bones). This angle influences the shape of the soft tissues of the vocal track, especially the pharynx, and determines the nature of the vocal sounds the newborn can produce. During growth the angle of flexion becomes more acute, as the skull assumes child, juvenile, adolescent, and, finally, adult proportions. As this growth process takes place a greater range of linguistically recognizable phonemes are produced. In this one aspect of growth of the skull and its functional correlates neither neoteny nor hypermorphosis are useful concepts. Non-human primates never possess the human-type of basicranial anatomy at any stage of their development, nor would they develop this anatomy if they prolonged any of their stages of growth. As a consequence, non-human primates cannot produce human-like phoneme sounds.

A graphic case against neoteny and hypermorphosis can be made by considering the ontogeny of human body proportions (see Figure 2.3). From fetus, to child, and to adult, human body proportions are so much altered that the mature morphology cannot be simply predicted from earlier stages of growth. **Allometry**, differential rates of growth of parts of the body relative to that of the body as a whole (Huxley, 1932), is the rule in primate, including human, development. Both positive and negative allometry take place in the ontogeny of human development (e.g., leg versus trunk growth, or head versus body growth). These allometric changes bring about functional differences between the adult and child in physical appearance and performance.

Brian Shea (1989) published the most cogent allometric analysis to date that rejects both neoteny or hypermorphosis as a 'grand unification theory' for all of human growth and evolution. Shea is not anti-neoteny or anti-hypermorphosis *per se*, and in fact he allows that the human brain and cranium may have evolved by neoteny, but he argues that the human face, jaws, and the rest of the body did not. In Shea's view, a variety of heterochronic processes are responsible for human evolution. The others may be hypermorphosis, acceleration (defined as an increase in the rate of growth or development), and hypomorphosis (defined as a delay in growth with no delay in the age at maturation). In Figure 4.3 are illustrated Shea's estimates for body size and shape as a consequence of neoteny and two types of hypermorphosis. None of these acting as a single process can produce the human adult size and shape from the human infant size and shape. The same holds true for acceleration and hypomorphosis. According to Shea, there are at least two reasons why no single type of heterochronic process can account for human evolution. The first reason is that, in agreement with Schultz, Shea points out that '. . . we [humans] have extended all of our life history periods, not merely the embryonic or juvenile ones . . .' (p. 84–5). We are not 'permanent children' or juvenilized apes, and while we take longer to reach sexual maturity than any primate we do not become sexually mature while in a juvenile stage. Rather, we humans have our own adolescent and adult stages of life. Furthermore, the adult stage has been prolonged beyond that found in any other primate species. The consequences of that adult stage prolongation, such as menopause in women, are not discussed by Shea, but are discussed later in this chapter.

The second reason Shea offers against a single heterochronic process is that humans have altered not only the length of each growth stage but also rates of growth from those found in other primates and possible ancestors. The mid-growth and adolescent growth spurts are, again, obvious examples of such rate changes. The heterochronic models of neoteny and hypermorphosis predict that a greater or lesser amount of time for growth produces the differences in size and shape between humans and chimpanzees. However, the empirical data gathered from growth and endocrine research with chimpanzees and humans, which was presented in the previous chapter, show that it is the sensitivity of specific skeletal parts to testosterone, determined by DNA and cellular activity, that results in the differences in limb size and shape between adult humans and chimpanzees. The time available for growth is largely irrelevant. As the hormones that regulate growth and development are, virtually, direct products of DNA activity, Shea proposes that the best place to look for evidence of the evolution of ontogeny is in the action of the endocrine system. In Shea's view, several independent genetic changes or adjustments must have occurred during human evolution to bring about all of these differences in size, shape, and rate of growth. According to Shea, differences in endocrine action between humans and other primates negate neoteny or hypermorphosis as unitary processes and instead argue for a multiprocess model for human evolution. Subsequent analysis of the evolution of the patterns of growth of the human pelvis (Berge, 1998) and of the human femur (Tardieu, 1998) generally confirms Shea's analysis and conclusions.

A closer inspection of human cognitive development also fails to support heterochrony. Parker's hypermorphosis hypothesis for the evolution of human cognitive capacities depends on prolongation of the developmental patterns of monkeys and apes in order to attain higher Piagetian stages.

Neoteny

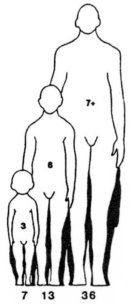

Time hypermorphosis

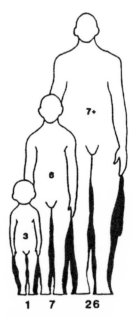

Rate hypermorphosis

Parker does not explain how a simple heterochronic change in developmental timing can account for all the details of human cognitive ontogeny from infancy to adulthood. Human development involves new cognitive competencies, such as those related to language and social intelligence (e.g., complex marriage and kinship systems), which have never been found in other primates. The stage theory of Piaget can be used to provide a descriptive and theoretical understanding of the development of human intelligence; however, that theory implies that human intelligence is due to the maturation of increasingly sophisticated and flexible cognitive processes, rather than due to the retention or extension of infantile or juvenile intellectual abilities. The neoteny and hypermorphosis arguments correctly assert that adult humans possess an intellectual plasticity and curiosity usually found only in the young of other species. Eccles (1979) has shown, however, that the adult human potentials for playfulness, creativity, and intellectual advancement are not derived via heterochrony. Eccles reviews much anatomical and physiological research of the human brain and nervous system and he shows that human intelligence arises from new competencies derived from a constant remodeling, restructuring, and maturation of the neurological architecture in the central nervous system. Humans possess anatomical regions of the brain that cannot be explained as developmental extensions of those in other primates, as the human regions are not found in any other primate. Human intelligence is, metaphorically, like the elephant's trunk – based on old parts but with new structures and functions.

Mathematical evidence against heterochrony

Clearly, the human pattern of growth from birth to maturity is qualitatively and quantitatively different from the pattern for other primates. The quantitative differences can be expressed in amounts, rates, and timing of

Figure 4.3 (left). Silhouettes of size and shape change during human growth. Numbers under silhouettes indicate age in years. Numbers above or on silhouettes indicate relative shape. Top left: actual size and shape change during normal human development. Top right: neoteny, note that at adult size shape 3 is still maintained. Bottom left: time hypermorphosis, the growth period is extended to 36 years yielding a peramorphic giant (size and shape of the descendant beyond that of the ancestor). Bottom right: rate hypermorphosis, growth ends at age 26 but proceeds at a faster rate producing another peramorphic giant. Note that in both cases the adult shape at 7 + is outside the range of normal development. From Shea (1989) with permission.

growth events, and are so reported in many standard textbooks of human growth (e.g., Tanner, 1962; Bogin, 1988b). These quantitative differences may also be expressed in terms of the type and number of mathematical functions that are needed to describe growth. The distance and velocity curves for most mammalian species can be estimated by a single function, such as a simple polynomial or exponential function. Even the monkeys and apes, with the addition of the juvenile stage, require no more than two such relatively simple functions (as shown in Chapter 3). The insertion of the mid-childhood and adolescent spurts into human ontogeny means that at least three mathematical functions are needed to adequately describe the shape of the velocity curve (Figure 4.4). Not only more, but also more complex functions are needed (Bock & Thissen, 1976; Karlberg, 1987). It is vitally important to stress here that all of this quantitative knowledge of the biology of human growth is well established and widely available. This information unequivocally negates neoteny, hypermorphosis, or any other single heterochronic process as the primary process of human evolution. Lamentably, the works on neoteny cited above – with the exception of Shea's work – make little or no reference to studies of the physical growth and development of living people.

The last word on heterochrony, so far at least, is the work of Godfrey & Sutherland (1996). These researchers do not argue for or against one or the other of the many heterochronic mechanisms that may have influenced human evolution. Instead, they develop a quantitative method for the fair testing of any or all heterochrony hypotheses. Their methodology is a major innovation in heterochrony research. They first show that Gould's clock model is based on three mathematically definable vectors, which they call AGE, SHAPE, and SIZE (Figure 4.5). In Figure 4.5(a) they illustrate these vectors and define them in mathematical terms. In Figure 4.5(b) they apply these mathematical vectors to Gould's clock model. Godfrey & Sutherland then show that linear vector distortions (Figure 4.5(c)), that is, changes in the AGE, SHAPE, or SIZE parameters may be exactly translated to equivalent movement of the clock's hands or age bar (Figure 4.5(d)). The translation of the clock model to linear vectors makes mathematical analysis of neoteny, hypermorphosis, and the other heterochronic processes for evolution relatively easy. Using vector analysis, the authors can predict the exact differences between ancestor and descendent species for the different types of heterochrony. Godfrey & Sutherland carry out several prediction analyses and find that many of the assertions made in the past about the role of heterochrony in human evolution are not correct. In the end, the authors find no support for the predictions of neoteny, sequential hypermorphosis or any currently published heterochronic models for

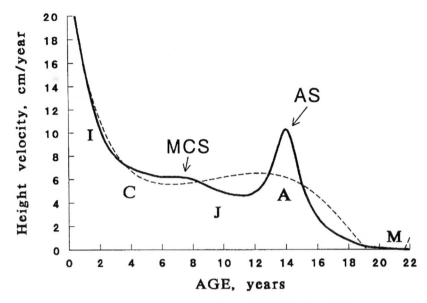

Figure 4.4. Idealized velocity curve of human growth for boys (solid line): I, infancy; C, childhood; J, juvenile; A, adolescence; M, mature adult. The dashed line is a sixth degree polynomial curve fit to the velocity curve data. The polynomial curve does not fit well to real growth data due to the pulses of the mid-childhood spurt (MCS) and the adolescent spurt (AS). The human velocity curve cannot be fit adequately by a single continuous mathematical function. At a minimum, three functions are required.

human evolution. Interested readers may consult the Godfrey & Sutherland article for the details of their analyses of specific heterochronic models.

The other side of the coin

In the tribulations and trials of heterochrony the jury is currently 'out', and a verdict on what type of heterochrony has shaped which aspects of human growth and development must await new research and testing. Forgotten by all parties in the litigation surrounding heterochrony is that there is another process by which evolution works. To requote Gould, 'Evolution occurs when ontogeny is altered in one of two ways': the first is '. . . when new characters are introduced at any stage of development with varying effects upon subsequent stages . . .' (the second is by heterochrony). Much of human evolution, especially the evolution of childhood and adolescence, the human capacity for symbolic language, and culture, are the result of the

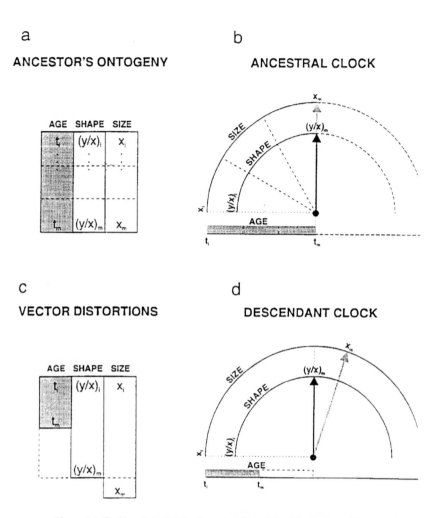

Figure 4.5. Godfrey & Sutherland's conversion of Gould's clock model to a series of mathematically definable vector distortions. (a) and (b): the pattern of growth of the ancestor in terms of age, shape, and size. In part 'a' the vector blocks are set at equal lengths, matching the concordance of the shape and size 'hands' and age bar of the clock in part (b). The quantitative value of each row or cell value in the vector blocks can be determined from the clock model by converting the proportional distances of the clock hands and the age bar to radians. For example, the dashed lines in parts (a) and (b) show the corresponding placements of vector cells to clock coordinates. In part (c) the vector blocks are distorted to match changes in size and age of maturation, relative to shape, of the descendant's clock, part (d). Reproduced with permission from Godfrey & Sutherland, 1996.

introduction of new life stages into the general pattern of primate growth and development.

As discussed in previous chapters, human growth and development from birth to reproductive maturity may be characterized by five stages: (1) infancy, (2) childhood, (3) juvenile, (4) adolescence, and (5) adulthood. It was explained in Chapter 2 how each stage may be defined by rate of growth, by characteristics of the dentition, by changes related to methods of feeding, by physical and mental competencies, and by maturation of the reproductive system and sexual behavior. The following discussion focuses on why and when the stages of childhood and adolescence were added to the human pattern of growth.

Human childhood

Human beings are not 'permanent children' but we do pass through a childhood stage of growth and development. The onset of the childhood stage is marked by a change in growth rate, from the rapid decline during infancy to a more steady period of gentle decline, the eruption of all of the deciduous teeth, weaning (the end of breast-feeding), and maturation of both new motor and cognitive skills. Though weaned, children are still dependent on older individuals for feeding and protection. The end of childhood is proclaimed by several biological events; eruption of the first permanent molar, the end to brain growth in weight, attainment of adult-style locomotion, and by a small increase in growth velocity, the mid-growth spurt. Childhood's end is also associated with new levels of cognitive maturity, the so-called '5–7-year-old shift'. Some of the milestones that mark the beginning and end of childhood were discussed in Chapter 2, but now these, and other events, are considered in terms of the evolution of human life history.

Weaning

Childhood begins after the infant is weaned. Weaning is defined here as the termination of lactation by the mother (other researchers may define weaning as the process of shifting from lactation to eating solid foods). In human societies the age at weaning varies greatly. Industrialized societies provide a poor indication of weaning age because bottle feeding and the manufacture of 'baby foods' allow either early termination of breast-feeding or no breast-feeding at all. Pre-industrialized human societies provide a better indication of the age at weaning, and hence the transition from infancy to childhood. Cited in Chapter 2 is one study of such a society that finds that the termination of breast-feeding occurs at a median age of 36 months (Dettwyler, 1995). Another review of the age at human weaning (Lee *et al.*, 1991) finds that in so-called 'food enhanced' societies, those where nutritional intake is good, weaning takes place as early as nine months of age. In 'food-limited' societies, where chronic undernutrition occurs, weaning takes place at 36 month median age. There are two fascinating corollaries of this comparison. The first is that in both the 'food enhanced' and the 'food limited' societies the mean weight of weaned infants is about the same, 9.0 kg and 9.2 kg respectively, or about 2.7 times birth weight (Lee *et al.* assume a mean birth weight of 3400 grams for full-term human beings). The second is that some solid foods are introduced into the diet when the infant achieves about 2.1 times birth weight. Lee and colleagues (Lee *et al.*, 1991; Bowman & Lee, 1995) compare the human data with data from 88 species of large-bodied mammals (32 non-human primates, 29 ungulates, 27 pinnipeds). They find that for all these species solid food is introduced, again, at about 2.1 times birth weight, but weaning takes place when the infant achieves between 3.2 to 4.9 times birth weight. For all primates the mean value is 4.6 times birth weight, with a range from 2.3 for the talapoin monkey (*Microcebus talapoin*) to 9.4 for the gorilla (*Gorilla gorilla*). The other great apes average at the following multiples of birth weight: *Pan troglodytes* (chimpanzees), 4.9; *P. paniscus* (bonobos), 6.1; *Pongo pygmaeus* (orang-utans), 6.4.

Humans are similar to other mammals in that we introduce solid foods at about 2.1 times birth weight. However, humans are unlike other mammals, even other species of primates, in that pre-industrial and traditional societies, including 'food limited' groups such as !Kung hunter-gatherers, wean at a relatively early stage of growth – before reaching 3.0 times birth weight. Even more startling is that human infants are weaned years before the first permanent tooth erupts. For all of the other primates, and virtually all other mammals, weaning is coincident with first molar eruption. Human mothers can wean their infants at this early stage of growth because the mother, or other people, will provide her child with specially prepared post-weaning foods. Children require specially prepared foods due to the immaturity of their dentition. The deciduous dentition, often called 'milk teeth', have thin enamel and shallow roots compared with the permanent dentition. Smith (1991a) and Smith *et al.* (1994) report that given this dental morphology, young mammals with only the deciduous dentition cannot process the adult-type diet. With eruption and occlusion of the first permanent teeth the infant mammal moves either to adulthood (most mammals have only two postnatal growth stages) or to the juvenile stage (social mammals). Smith reports that mammals with some permanent teeth are able to process the adult diet and are independent in terms of feeding. When human infancy ends, the deciduous dentition is still in place, thus the human child still requires a special diet and remains dependent on older individuals for food.

Children also require a special diet due to the small size of their digestive tracts relative to that of the adult. This need can be appreciated most acutely when it is not met. Behar (1977) studied the causes of growth retardation and undernutrition of children living in rural villages in Guatemala. He found that the children had access to sufficient food, but the traditional adult diet of corn and beans did not have the caloric density to meet their growth requirements. The result was chronic undernutrition, growth failure, and increased susceptibility to infectious disease for most of the children. Bailey *et al.* (1984) studied the growth of more than 1000 infants, children, and juveniles living in 29 villages in northern Thailand. The participants in the study lived in rural agricultural villages, with rice as the basic subsistence crop. The participants were between the ages of six months and five years old at the start of the study, and they were measured for height, weight, and several skinfolds about every six months for five years. It was found that, compared with local or international reference standards for growth, the rural Thai children and juveniles were delayed in growth for all the dimensions studied. Careful consideration of a number of factors that may influence growth, including disease or the lack of specific nutrients, such as protein, vitamin A, or iron, showed that the delays in growth were not due to any one of these, rather the delays were due to a deficiency in the total intake of calories. The most dramatic falloff in growth occurred at 18 months of age, which corresponds with the average age at weaning in these villages. Weaning foods were usually watered-down versions of adult foods. Although there were no food shortages in the villages, Bailey *et al.* reason that the small gastrointestinal tracts of the weaned infants and young children may not have been capable of digesting enough food to meet their caloric demands for maintenance and growth of the body. The work of Behar in Guatemala, Bailey *et al.* in Thailand, and many other similar studies (some of which are discussed in Chapter 5) show that without the use of appropriate weaning foods children will suffer calorie insufficiency leading to undernutrition, developmental delays, and growth retardation.

Another reason that children need a special high-energy diet is due to the rapid growth of their brain (Figure 2.7). In this regard, the research of Leonard & Robertson (1992) was discussed briefly in Chapter 2. They

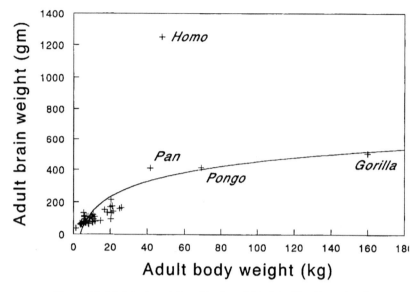

Figure 4.6. Adult body weight and brain weight plotted for 61 species of Cercopithcidae (Old World monkeys, apes, and people). The curve is logarithmic regression fit to the data for all species. Each part of the human brain enlarged during evolution, especially the size of cerebral cortex (data from Harvey *et al.*, 1986).

estimate that due to this rapid brain growth, 'A human child under the age of 5 years uses 40–85 percent of resting metabolism to maintain his/her brain [adults use 16–25 percent]. Therefore, the consequences of even a small caloric debt in a child are enormous given the ratio of energy distribution between brain and body (p. 191)'. In a related study, Leonard & Robertson (1994) also show that the size of the human brain relative to total body size necessitates an energy dense diet. At all stages of life after birth, human beings have brains that are significantly larger than expected given the human body size (Figure 4.6). These large, metabolically active brains demand a larger percentage of energy than any other primate. Aiello & Wheeler (1995) refine the relationship between human brain size, body size, and metabolic costs by noting that relative to total body size, adult human beings have an unexpectedly small gastrointestinal tract (gut) as well as an unexpectedly large brain (Figure 4.7). In contrast, the size of other organs, such as the liver, kidney and heart are about as big as expected. Aiello & Wheeler show that both brain tissue and gut tissue are 'expensive', meaning that both types of tissue have relatively high metabolic rates. Aiello & Wheeler present estimates of the percentage of total

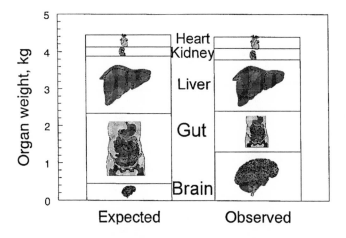

Figure 4.7. Expected and observed weight of several human organs. The histograms provide a comparison of the weight of the heart, kidney, liver, digestive system (gut), and brain. The observed size of the human heart, kidney, and liver are about equal to the expected size. The human gut is much smaller than expected, while the human brain is much larger than expected. The last three organs are shown pictorially in relative size. The expected size of the organs are based on the actual sizes of these organs in more than 20 species of monkeys and apes. The data are from Aiello & Wheeler (1995), where the details of data selection, sample sizes, and methods of analysis are described.

body basal metabolic rate for several tissues utilized by the typical 65 kg adult human male. For the brain the value is 16.1% and for the gut the value is 14.8%. Given these values, Aiello & Wheeler propose that during human evolution the gut decreased in size as a trade-off allowing, in part, for expansion of brain size. Had the gut remained as large as expected for a primate of human size the metabolic costs would have been too great to also support a large brain. The trade-off in size between brain and gut means that humans have a total metabolic rate that is about average for a placental mammal of our size.

Even with this trade-off the large and metabolically expensive human brain still requires a constant supply of energy. A smaller than expected gut size for humans means that less total food can be processed in a given amount of time. This presents humans with the problem of eating just the right kind of food to meet the nutritional demands of brain and the rest of the body. Both Leonard & Robertson (1994) and Aiello & Wheeler conclude that the way human beings satisfy nutritional demands is to consume a diet that is nutrient dense, especially in energy, and easy to digest. Human beings meet these dietary requirements in two ways. The first is by selection of appropriate foods. Leonard and Robertson compared the diet of five human foraging societies – !Kung, Ache, Hiwi, Inuit, and Pygmies – to 72 non-human primates species and found that diet quality, in part measured by energy density, of the human groups was almost twice that of other primates of the same body size. The human ability to include foods such as seeds, roots, and meat in the diet increases quality, as these are nutrient dense foods. Most other primate species rarely or never eat seeds or roots. Some non-human primates do eat meat, but meat comprises less than 10 percent of their diet. Human foragers studied in the twentieth century were found to obtain at least 30 percent of their dietary energy from animal foods. This dietary pattern is ancient in that it is estimated that Paleolithic foragers living in western Europe between 15 000 and 20 000 years ago consumed about 30 percent of total energy from animal protein (Bogin, 1997a). More ancient species of hominids seem to have eaten less animal protein, but the archaeological evidence indicates that they expended significant effort to acquire as much animal food as possible (ibid).

The second way humans meet nutritional requirements is by processing food to extract, concentrate, and enhance its nutritional content. Humans, and human ancestors, expend great effort in food preparation, using a variety of culturally learned methods (i.e., recipes) to cook, combine, flavor, and detoxify natural ingredients (ibid). Since at least the time of *Homo habilis* hominids have depended on technology to do this preparation (e.g., the hand-axes and fire used by *Homo erectus* or the food processors and microwave ovens of *Homo sapiens*). Both careful food selection and intensive food preparation are important for the survival of people of all ages, but they are particularly crucial for the survival of children. Children, of course, have a relatively larger disproportion between brain size and body/gut size than do adults (Figure 2.7) and children are dependent on older individuals for care and feeding. Added together, each of the constraints of childhood – an immature dentition, a small digestive system, a calorie-demanding brain that is both relatively large and growing rapidly, and feeding dependency – necessitate a special diet; one that must be procured, prepared, and provided by older members of the social group.

The passage from childhood and the mid-growth spurt

Important developments that allow children to progress to the juvenile stage of growth and development are the eruption of the first permanent molars and completion of growth of the brain (in weight). First molar eruption takes place, on average, between the ages of 5.5 and 6.5 years in most human populations (Jaswal, 1983; Smith, 1992). Functional occlusion occurs some weeks to months thereafter. Recent morphological and mathematical investigation shows that brain growth in weight is complete at a mean age of seven years (Cabana *et al.*, 1993). Thus, significant milestones of dental and brain maturation take place at about seven years of age. At this stage of development the child becomes much more capable of processing dentally an adult type diet (Smith, 1991a). Furthermore, nutrient requirements for the maintenance and the growth of both brain and body diminish to less than 50 percent of total energy needs.

The mid-growth spurt is another important milestone that signals the end of human childhood. This spurt is associated with an endocrine event called **adrenarche**, the progressive increase in the secretion of adrenal androgen hormones. Adrenal androgens produce the mid-growth spurt in height, a transient acceleration of bone maturation, the appearance of axillary and pubic hair, and seem to regulate the development of body fatness and fat distribution (Katz *et al.*, 1985; Parker, 1991). There is a little story that links the mid-growth spurt with neoteny. Louis Bolk (1926), the 'father' of the scientific hypothesis for human evolution via neoteny, speculated that for our early human ancestors sexual maturation took place at about six to eight years of age. The mid-growth spurt was first reported by Backman (1934) and following its discovery, several of Bolk's followers, without any additional supporting evidence, opined that the mid-growth spurt and adrenarche are vestiges of sexual maturation from our evolutionary past. We now understand that the mid-growth spurt is associated with the adrenarche, a maturation event of the adrenal gland, and not with gonadarche, the maturation of the gonads. Much research, from clinical medicine to anthropological fieldwork, shows that there is virtually no connection between adrenarche and gonadarche, as each are independently controlled events (Smail *et al.*, 1982; Weirman & Crowley, 1986; Worthman, 1986; Parker, 1991).

Bolk's idea about sexual maturation in the past may be wrong, but a connection between the mid-growth spurt and the evolution of the human pattern of growth is still a possibility. The mechanism controlling adrenarche is not understood as no known hormone appears to cause it. There are connections, however, between the production of adrenal hormones, growth, and maturation. Cutler *et al.* (1978) and Smail *et al.* (1982) measured the plasma concentration of the adrenal androgens dehydroepiandrosterone (DHA), dehydroepiandrosterone sulfate (DHAS), and delta^4-androstenodine (D^4) before and after sexual maturation in 14 species. These species include samples of rodents (rat, guinea pig, hamster), domestic animals (rabbit, dog, sheep, pig, goat, horse, cow), primates (macaques – including 76 *Macaca mulatta* and 80 *M. nemestrina* – a few baboons and 52 chimpanzees), and the chicken. Cutler and colleagues found that the plasma concentrations of DHA, DHAS, and D^4 were significantly higher in sexually mature primates species than in any of the other animals. However, the serum level of these adrenal androgens was not related to sexual maturation. Rhesus monkeys aged one to three years old, and not sexually mature, had the same high concentrations of all three adrenal androgens as older, sexually mature monkeys. The same was true for baboons. In contrast, chimpanzees seven years old or older had adrenal androgen concentrations that were, on average, 4.7 times greater than those for chimpanzees less than four years old. Cutler and colleagues concluded that among the animals examined so far, the chimpanzee and the human being are the only species that show adrenarche. Smail and colleagues generally confirmed the findings of Cutler *et al.*, but also measured adrenal function in sexually mature *M. nemestrina* monkeys. In the oldest group of these monkeys, ages six to nine years, there was a significant increase in the serum concentration of both DHA and DHAS. The authors state that this shows clearly that like chimpanzees and humans, this species of macaque has adrenarche. However, in the monkey adrenarche occurs *after* gonadarche, meaning that the adreanal changes and sexual maturation are completely independent.

Only human beings are known to have both adrenarche and a mid-growth spurt. The primate data reported by Cutler *et al.* (1978) and Smail *et al.* (1982) suggest a possible function for adrenarche. In those primate species with adrenarche, serum levels of adrenal androgens are relatively low after infancy and prior to adrenarche. Moreover, chimpanzees and humans have relatively slow growth and a long delay prior to the onset of sexual maturation. Given that one biological action of adrenal hormones is to speed up skeletal growth and maturation, perhaps the evolution of reduced adrenal androgen production prior to adrenarche may be explained as a mechanism that maintains slow epiphyseal maturation and skeletal growth in the face of the prolongation of the pre-pubertal stages of growth. In the human being, the delay is so protracted that it becomes possible to insert the childhood stage of development between the infancy and juvenile stages.

Synthesizing all of these data, it is possible to view the combination of adrenarche and the human mid-growth spurt as life history events marking the transition from the childhood to the juvenile stage of growth. In terms of physical growth, the effects of the adrenal androgens, to increase rate of skeletal growth, stimulate body hair growth, and regulate body fat distribution, are short-lived and quite small. Even so, these physical changes may be noticed by the child and his/her intimates, such as parents, and recognized as markers of developmental maturation. More to the point is the fact that while adrenarche may have only transient effects on physical development, there is a more permanent and important effect on cognitive function. Psychologists have long been interested in what is called the '5–7-year-old shift' in cognition (White, 1965; Rogoff *et al.*, 1975; Weisner, 1996). Weisner (1996, p. 295) states that 'The 5–7 shift involves changes in internal states and competencies of the maturing child – shifts in cognitive capacities, self concept, visual/perceptual abilities, or social abilities. The transition marks the emergence of increasing capabilities for strategic and controlled self-regulation, skills at inhibition, the ability to maintain attention and to focus on a complex problem, and planfullness and reflection'.

Using the terminology of Piaget, the 5–7 shift moves the child from the preoperational to concrete operational stage of cognition. This 5–7-year-old shift is found in all cultures so far investigated (Rogoff *et al.*, 1975). The shift has never been reported in any other primate species (Parker, 1996) and thus seems to be a human species phenomenon. Ethnographic and psychological research show that juvenile humans have the physical and cognitive abilities to provide much of their own food and to protect themselves from environmental hazards such as predation (children in foraging societies and rural agricultural societies are especially vulnerable to predation due to small body size) and disease (Weisner, 1987; Blurton-Jones, 1993). In addition to self-care and feeding, the post-shift juvenile becomes increasingly involved with domestic work and 'caretaking inter-actions with other children' (Weisner, 1996, p. 296). The association of adrenarche, the mid-growth spurt, and the 5–7 shift all seem to mark the progression of the child to the juvenile stage of development.

How and when did human childhood evolve?

The stages of the life cycle may be studied directly only for living species. However, there are lines of evidence that may be used to reconstruct the life cycle of extinct species. The fossil evidence of skeletons and teeth provide direct and tangible clues as to the life cycle of extinct species. Indirect evidence comes from the fields of comparative anatomy, comparative physiology, comparative ethology, and archaeology. We know from fossil evidence that one human characteristic, bipedalism, appears relatively early in hominid evolution – about four MYA. There are many hypotheses for the evolution of hominid bipedalism, but in the context of this discussion one of these is most important. Bipedalism 'allows individuals to walk long distances and carry objects' (Zihlman, 1997, p. 185). Zihlman explains that 'objects' include infants. There are significant connections between human infancy and bipedalism. During infancy, when the child is dependent on the mother for care and feeding, the rate of leg growth and the maturation of corresponding nerve cells in the motor-sensory cortex of the brain are slow relative to the growth of the head, the arms, and the nerve cells that control movement in the upper half of the body. By about two years of age, there is an acceleration in both the rate of leg growth and the maturation of the motor-sensory cortex region devoted to the legs (Tanner, 1978). At the time the child is weaned (about three years of age in many pre-industrial societies), independent locomotion takes on greater importance for the child and the rate of leg growth becomes faster than the rate of arm growth (Scammon & Calkins, 1929; Hansman, 1970).

Another method used to reconstruct life histories for extinct hominids is by analyzing teeth. Teeth are covered by enamel, the hardest substance in a mammal's body, which protects teeth from destruction. The jaw (mandible and maxilla) that support the teeth are composed of fairly dense and durable bone, and therefore both teeth and jaws (either in whole or part) are more likely to be preserved in the fossil record than any other body parts. This is fortuitous for the study of life history because the morphology and development of the dentition is highly conservative in evolution and the pattern of tooth development reveals a great deal about life history. The use of teeth to reconstruct the evolution of hominid life history is a very active area of research with a burgeoning literature that cannot be adequately reviewed in this book. Interested readers should consult Mann *et al.* (1990) and Winkler & Anemone (1996) for reviews, details of methodology, and alternative interpretations of the evidence. One example of this research is the work of Holly Smith. As discussed in the previous chapter, Smith (1991b) found a significant correlation between age of eruption of the first molar (M1) and adult brain size across a large number of mammalian species. She also found high correlations between age at M1 eruption and age at weaning ($r = 0.93$) and age at sexual maturity for both males ($r = 0.93$) and females ($r = 0.93$). Such high correlations mean that these life history events are linked to some more fundamental developmental process, and that a change in the timing of one of these events will probably result in a change in the timing of them all. We may conclude, then, that even though age at weaning and sexual maturity cannot be seen directly in the fossil record, they can be reconstructed based on the dentition.

Another example of the methods used to reconstruct the evolution of life history may be found in the work of Martin (1983, 1990) and Harvey *et al.*

(1986) on patterns of brain and body growth in apes, humans, and their ancestors. Martin showed that Old World monkeys and apes have a pattern of brain growth that is rapid before birth and relatively slower after birth. In contrast, humans have rapid brain growth both before and after birth (Figure 4.8). This difference may be appreciated by comparing ratios of brain weight divided by total body weight (in grams). At birth this ratio averages 0.09 for the great apes and 0.12 for human neonates. At adulthood the ratio averages 0.008 for the great apes and 0.028 for people. In other words, relative to body size human neonatal brain size is 1.33 times larger than the great apes, but by adulthood the difference is 3.5 times. The human–ape difference is not due to any single heterochronic process, that is, not the result of delay, prolongation, or acceleration of a basic ape-like pattern of growth. Rather it is due to new patterns of growth for the human species. The rate of human brain growth exceeds that of most other tissues of the body during the first few years after birth (Figure 2.7). Martin (1983, 1990) and Harvey *et al.* (1986) also show that human neonates have remarkably large brains (corrected for body size) compared with other primate species. Together, relatively large neonatal brain size and the high postnatal growth rate give adult humans the largest encephalization quotient (an allometric scaling of brain to body size) of all higher primates (Figure 4.6).

Martin (1983) hypothesizes that a 'human-like' pattern of brain and body growth becomes necessary once adult hominid brain size reaches about 850 cubic centimetres. This biological marker is based on an analysis of head size of fetuses and birth canal dimensions of their mothers across a wide range of social mammals, including cetaceans, extant primates and fossil hominids (Martin, 1983, pp. 40–1). Given the mean rate of postnatal brain growth for living apes, an $850 \, cm^3$ adult brain size may be achieved by all hominoids (living and extinct apes and humans) by lengthening the fetal stage of growth. At brain sizes above $850 \, cm^3$ the size of the pelvic inlet of the fossil hominids, and living people, does not allow for sufficient fetal growth. Thus, a period of rapid postnatal brain growth and slow body growth – the human pattern – is needed to reach adult brain size.

Martin's analysis is elegant and tenable. Nevertheless, the difference between ape and human brain growth is not only a matter of velocity; it is also a matter of life history stages. Brain growth for both apes and human beings ends at the start of the juvenile stage, which means that apes complete brain growth during infancy. Human beings, however, insert the childhood stage between the infant and juvenile stages. Childhood may provide the time and the continuation of parental investment necessary to grow the larger human brain. Following this line of reasoning, any fossil

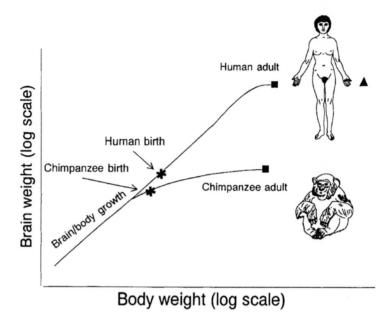

Figure 4.8. Growth curve for human brain and body compared with the chimpanzee. The length of the human fetal phase, in which brain and body grow at the same rate for both species, is extended for humans. Chimpanzee brain growth slows after birth, but humans maintain the high rate of brain growth during the postnatal phase. In contrast, the rate of human body growth slows after birth. If human brain/body growth rate were equal to the chimpanzee rate, then adult humans would weigh 454 kilograms and stand nearly 3.1 meters tall (indicated by the ▲ symbol). After Martin (1983).

human, or any of our fossil hominid ancestors, with an adult brain size above Martin's 'cerebral Rubicon' of 850 cm^3 may have included a child-hood stage of growth as part of its life history.

Given this background, Figure 4.9 is my own Schultz-inspired summary of the evolution of the human pattern of growth and development from birth to age 20 years (the evolution of adolescence is discussed later in this chapter). Figure 4.9 must be considered as 'a work in progress' for two reasons. The first is that only the data for the first and last species (*Pan* and *Homo sapiens*) are known with some certainty. The second is that this version of the figure supersedes earlier versions that were prepared without the advantage of more recent information about patterns of growth for fossil hominids (e.g., Bogin, 1994, 1995, 1997b; Bogin & Smith, 1996). Even with the latest information available, the patterns of growth of the fossil hominid species are tentative reconstructions, based on published analyses

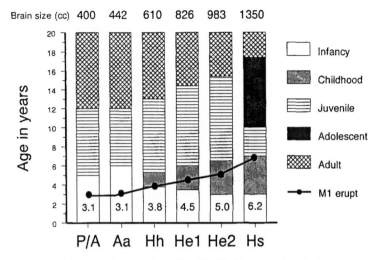

Figure 4.9. The evolution of hominid life history during the first 20 years of life. Abbreviated nomenclature as follows: P/A, Pan and *Australopithecus afarensis*; Aa, *Australopithecus africanus*; Hh, *Homo habilis*; He1, early *Homo erectus*; He2, late *Homo erectus*; Hs, *Homo sapiens*. Mean brain sizes are given at the top of each histogram. Mean age at eruption of the first permanent molar (M1) is graphed across the histograms, and given below the graph.

of skeletal and dental development of fossil specimens that died before reaching adulthood. Known ages for eruption of the M1 are given for *Pan* and *H. sapiens*. Estimated ages for M1 eruption in other species were calculated by Smith & Tompkins (1995). Age of eruption of M1 is an important life history event that correlates very highly with other life history events. Known or estimated adult brain sizes are given at the top of each bar; the estimates are averages based on reports in several textbooks of human evolution. Following Martin's analysis (1983, 1990), brain size is another crucial influence on life history evolution. One major message to take from the figure is that the prolongation of the total time for growth that plays such a prominent role in the hypotheses of neoteny and hypermorphosis is definitely a part of human evolution. However, I find that time prolongation is not sufficient to account for the insertion of the new stages of childhood and adolescence that are part of human growth.

Australopithecus afarensis appears in the fossil record about 3.9 MYA and is one of the oldest hominid fossil species. *A. afarensis* shares many anatomical features with non-hominid pongid (ape) species including an adult brain size of about 400 cm³ (Simons, 1989) and a pattern of dental development indistinguishable from extant apes (Bromage & Dean, 1985; Conroy & Vannier, 1991; Smith, 1991b). Therefore, the chimpanzee and *A.*

afarensis are depicted in Figure 4.9 as sharing the typical tripartite stages of postnatal growth of social mammals – infant, juvenile, adult (Pereira & Fairbanks, 1993). Following the definitions used throughout this book, infancy represents the period of feeding by lactation, the juvenile stage represents a period of feeding independence prior to sexual maturation, and the adult stage begins following puberty and sexual maturation. The duration of each stage and the age at which each stage ends are based on empirical data for the chimpanzee. A probable descendent of *A. afarensis* is the fossil species *A. africanus*, dating from about 3.0 MYA. To achieve the larger adult brain size of *A. africanus* (average of 442 cm³) may have required an addition to the length of the fetal and/or infancy periods. Figure 4.9 indicates an extension to infancy of one year.

The first permanent molar (M1) of the chimpanzee erupts at 3.1 years, but chimpanzees remain in infancy until about age five years. Until that age the young chimpanzee is dependent on its mother, and will not survive if the mother dies or is otherwise not able to provide care and feeding (Goodall, 1983; Nishida *et al.*, 1990). After erupting M1 the young chimpanzee may be able to eat adult-type foods, but still must learn how to find and process foods. Learning to successfully open fruits that are protected by hard shells and to extract insects from nests (such as ants and termites) requires more than one year of observation and imitation by the infant of the mother. For these reasons, chimpanzees extend infancy for more than one year past the eruption of M1. Based on brain size details of dental anatomy, the mean age of M1 eruption, and eruption of M2 and M3 for *A. afarensis* and *A. africanus*, is estimated to be identical to the chimpanzee. It is likely that these early hominids followed a pattern of growth and development very similar to chimpanzees. Behavioral capacities of these fossil hominids also seem to be similar to living chimpanzees. For these reasons, the early hominids also may have extended infancy for at least one year beyond the age of M1 eruption.

About 2.2 MYA fossils with several more human-like traits including larger cranial capacities and greater manual dexterity appear. Also dated to about this time are stone tools of the Oldowan tradition. Given the biological and cultural developments associated with these fossils they are considered by most paleontologists to be members of the genus *Homo* (designated as *H. habilis*, *H. rudolfensis*, or early *H. erectus* – referred to collectively here as *H. habilis*). The rapid expansion of adult brain size during the time of *H. habilis* (650 to 800 cm³) might have been achieved with further expansion of both the fetal and infancy periods, as Martin's 'cerebral Rubicon' was not surpassed. However, the insertion of a brief childhood stage into hominid life history may have occurred. Christine

Tardieu (1998) shows that *H. habilis* has a pattern of growth of the femur that is distinct from that of the australopithecines, but consistent with that of later hominids. The distinctive femur shape of the more recent hominids is due to the addition of a prolonged childhood stage of growth. *H. habilis*, then, may have had a short childhood stage of growth.

A childhood stage of growth for the earliest members of the genus *Homo* is also supported by a comparison of human and ape reproductive strategies. There are limits to the amount of delay possible between birth and sexual maturity, and between successful births, that any species can tolerate. The great apes are examples of this limit. Chimpanzee females in the wild reach menarche at 11 to 12 years of age and have their first births at an average age of 14 years (Goodall, 1983). The average period between successful births in the wild is 5.6 years, as infant chimpanzees are dependent on their mothers for about five years (Teleki *et al.*, 1976; Goodall, 1983; Nishida *et al.*, 1990). Actuarial data collected on wild-living animals indicate that between 35 percent (Goodall, 1983) and 38 percent (Nishida *et al.*, 1990) of all live-born chimpanzees survive to their mid-twenties. Although this is a significantly greater percentage of survival than for most other species of animals, the chimpanzee is at a reproductive threshold. Goodall (1983) reports that for the period 1965 to 1980 there were 51 births and 49 deaths in one community of wild chimpanzees at the Gombe Stream National Park, Tanzania. During a 10-year period at the Mahale Mountains National Park, Tanzania, Nishida *et al.* (1990) observed '. . . 74 births, 74 deaths, 14 immigrations and 13 emigrations . . .' in one community. Chimpanzee population size in these two communities is, by these data, effectively in equilibrium. Any additional delay in age of females at first birth or the time between successful births would likely result in a decline in population size. Galdikas & Wood (1990) present data for the orang-utan which shows that these apes are in a more precarious situation. Compared with the 5.6 years between successful births of chimpanzees, the orang-utan female waits up to 7.7 years, and orang-utan populations are in decline. Lovejoy (1981) calls the state of great ape reproduction a 'demographic dilemma' (p. 211).

The great apes, and fossil hominids such as *Australopithecus*, may have reached this demographic dilemma by extending the length of the infancy stage and forcing a demand on nursing to its limit (Figure 4.9). Early *Homo* may have overcome this reproductive limit by reducing the length of infancy and inserting childhood between the end of infancy and the juvenile period. Free from the demands of nursing and the physiological brake that frequent nursing places on ovulation (Ellison, 1990), mothers could reproduce soon after their infants became children. This certainly occurs among modern humans. An often cited example, the !Kung, are a traditional hunting and gathering society of southern Africa. A !Kung woman's age at her first birth averages 19 years and subsequent births follow about every 3.6 years, resulting in an average fertility rate of 4.7 children per woman (Short, 1976; Howell 1979). Women in another hunter-gather society, the Hadza (Blurton-Jones *et al.*, 1992), have even shorter intervals between successful births, stop nursing about one year earlier, and average 6.15 births per woman.

For these reasons, a brief childhood stage for *H. habilis* is indicated in Figure 4.9. This stage begins after the eruption of M1 and lasts for about one year. That year of childhood would still provide the time needed to learn about finding and processing adult-type foods. During this learning phase, *H. habilis* children would need to be supplied with special weaning foods. There is archaeological evidence for just such a scenario. *H. habilis* seems to have intensified its dependence on stone tools. There are both more stone tools, more carefully manufactured tools, and a greater diversity of stone tool types associated with *H. habilis* (Klein, 1989). There is considerable evidence that some of these tools were used to scavenge animal carcasses, especially to break open long bones and extract bone marrow (Potts, 1988). This behavior may be interpreted as a strategy to feed children. Such scavenging may have been needed to provide the essential amino acids, some of the minerals, and especially the fat (dense source of energy) that children require for growth of the brain and body (Leonard & Robertson, 1992).

Further brain size increase occurred during *H. erectus* times, which begin about 1.6 MYA. The earliest adult specimens have mean brain sizes of 826 cm^3, but many individual adults had brain sizes between 850 to 900 cm^3. This places *H. erectus* at or above Martin's 'cerebral Rubicon' and seems to justify insertion and/or expansion of the childhood period to provide the biological time needed for the rapid, human-like, pattern of brain growth. It should be noted from Figure 4.9 that the model of human evolution proposed here predicts that from the *Australopithecus* to the *H. erectus* stage the infancy period shrinks as the childhood stage expands. Perhaps by early *H. erectus* times the transition from infancy to childhood took place before M1 eruption. Of course, it is not possible to know if this was the case or to state the cause of such a life history change with any certainty. Maybe the evolution of ever larger brains led to a delay in M1 eruption, which in turn led to both the need for a childhood stage and the expansion of the childhood stage as brains continued to enlarge. Alternatively, a delay in dental maturation may have precipitated the need for childhood, and in turn the biocultural ecology of childhood and its effects on hominid social learning and behavior selected for ever larger brains. No matter what the cause of childhood may be, if an expansion of childhood led to a shrinking of the infancy stage, then *H. erectus* would have enjoyed a greater reproductive advantage than any previous hominid. This seems to be the case, as *H. erectus* populations certainly did increase in size and began to spread throughout Africa and into other regions of the world.

Later *H. erectus*, with average adult brain sizes of 983 cm^3, are depicted with further expansion of the childhood stage. In addition to bigger brains (some individuals had brains as large as 1100 cm^3), the archaeological record for later *H. erectus* shows increased complexity of technology (tools, fire, and shelter) and social organization (Klein, 1989). These techno-social advances, and the increased reliance on learning that occur with these advances, may well be correlates of changes in biology and behavior associated with further development of the childhood stage of life (Bogin & Smith, 1996). The evolutionary transition to archaic, and finally modern, *H. sapiens* expands the childhood stage to its current dimension. Note that M1 eruption becomes one of the events that coincides with the end of childhood. Perhaps no further extension of childhood beyond M1 eruption is possible, given the significant biological, cognitive, behavioral, and social changes that are also linked with dental maturation and the end of childhood. With the appearance of *H. sapiens* comes evidence for the full gamut of human cultural capacities and behaviors. The technological, social, and ideological requisites of culture necessitate a more intensive investment in learning than at any other grade of hominid evolution. The learning hypothesis for childhood, while not sufficient to account for its origins, certainly plays a significant role in the later stages of its evolution. The *H. sapiens* grade of evolution also sees the addition of an adolescent stage to post-natal development. The evolution of human adolescence is discussed later in this chapter.

Who benefits from childhood?

Brain sizes of extant and fossil apes and hominids provide some idea of when human life stages may have evolved, but do not explain why they evolved. Bonner (1965) shows that the presence of a stage, and its duration, in the life cycle relates to such basic adaptations as locomotion, reproductive rates, and food acquisition. To make sense out of the pattern of human growth one must look for the 'basic adaptations' that Bonner describes. The most basic of these adaptations are those that relate to evolutionary success. This is traditionally measured in terms of the number of offspring that survive and reproduce. Biological and behavioral traits do not evolve unless they confer upon their owners some degree of reproductive advantage, in terms of survivors a generation or more later. Three 'textbook' reasons for the evolution of human childhood were listed at the start of this chapter. These reasons emphasized the role of learning in human adaptation, and they are valid reasons inasmuch as learning does confer an adaptive advantage to pre-adult individuals. However, the 'textbook' explanations cannot account for the initial impetus for the insertion of childhood into human life history. A childhood stage of development is not necessary for the type of learning listed here. The prolonged infancy and juvenile period of the social carnivores (Bekoff & Byers, 1985) and apes (Bogin, 1994b) can serve that function. Rather childhood may be better viewed as a feeding and reproductive adaptation for the parents, as a strategy to elicit parental care after infancy, as a strategy to minimize the risks of starvation for the child, a means of shifting the care of offspring from the parents, especially the mother, to juveniles and older, post-reproductive, adults (i.e., grandmothers), and as a mechanism that allows for more precise 'tracking' of ecological conditions via developmental plasticity during the growing years.

Thus in addition to the three 'textbook' explanations given at the beginning of this chapter, I propose that there are at least five additional reasons for the evolution of childhood:

1.   *Childhood is a feeding and reproductive adaptation*

A childhood growth stage may have originally evolved as a means by which the mother, the father, and other kin could provide dependent offspring with food. This frees the mother from the demands of nursing and the inhibition of ovulation related to continuous nursing. This decreases the interbirth interval and increases reproductive fitness.

Consider the data shown in Figure 4.10, which depicts several hominoid developmental landmarks. In comparison with living apes, human beings experience developmental delays in eruption of the first permanent molar, age at menarche, and age at first birth. However, humans have a shorter infancy and shorter birth interval, which in apes and traditional human societies are virtually coincident. Dental development is an excellent marker for life history in the primates, and it was shown above that there is a very strong correlation between age at eruption of M1 and many life history events. In general, primate mothers wean their infants about the time M1 erupts. This makes sense, since the mother must nurse her current infant until it can process and consume an adult diet, and this requires at

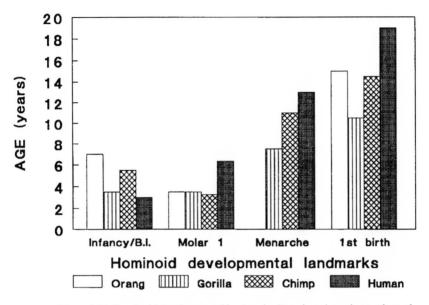

Figure 4.10. Hominoid developmental landmarks. Data based on observations of wild-living individuals, or for humans, healthy individuals from various cultures. Note that compared with apes, people experience developmental delays in eruption of the first permanent molar, age at menarche, and age at first birth. However, people have a shorter infancy and shorter birth interval, which in apes and traditional human societies are virtually coincident. The net result is that humans have the potential for greater lifetime fertility than any ape. Species abbreviations are: Orang, *Pongo pygmaeus*; Gorilla, *Gorilla gorilla*; Chimp, *Pan troglodytes*; Human, *Homo sapiens*. Developmental landmarks are: Infancy/B.I., period of dependency on mother for survival, usually coincident with mean age at weaning and/or a new birth (B.I. = birth interval); Molar 1, mean age at eruption of first permanent molar; Menarche, mean age at first estrus/menstrual bleeding; 1st birth, mean age of females at first offspring delivery. Sources: Bogin (1988b, 1994a), Galdikas & Wood (1990), Nishida *et al.* (1990), Smith (1992), Watts & Pusey (1993).

least some of the permanent dentition. As discussed in Chapter 2, several primate and social carnivore species provide food to their young during the post-weaning period. In these species the molar and premolar teeth may erupt before the post-weaning dependency period ends. These primates and carnivores are similar in that both are predatory (the primates hunt insects) and it takes the young time to learn and practice hunting skills. The effect of the post-weaning dependency on the mother is delayed reproduction. In a review of reproductive behavior of the carnivores Ewer (1973) finds that female lions, hyenas, the sea otter (a fish-hunter), and many bears wait two or more years between pregnancies. The same delay holds for the more carnivorous primates, such as marmosets, tamarins, and chimpanzees. From this we may conclude that reproduction in the social and predatory mammals occurs either at, or well after, the age at which the first permanent teeth erupt.

The human species is a striking exception to this relationship between permanent tooth eruption and birth interval. Women in traditional societies wait, on average, three years between births, not the six years expected on the basis of M1 eruption. The short birth interval gives humans a distinct advantage over other apes, as we can produce and rear two offspring through infancy in the time it takes chimpanzees or orang-utans to produce and rear one offspring. The basic point I am emphasizing here is that by reducing the length of the infancy stage of life (i.e., reducing the period of lactation) and by developing the special features of the human childhood stage, humans have the potential for greater lifetime fertility than any ape. Well-adapted species are those that survive for many generations, and they do this, in large measure, by being able to adjust rates of reproduction to maintain population sizes in equilibrium with environmental resources. When population size declines, for example, after an episode of disease or starvation, a well-adapted species can increase birth rates to restore the equilibrium. With a shortened birth interval, humans can make adjustments in birth rates more quickly than any ape. The human advantage is not that our species can have more offspring, for that might lead to overpopulation. Rather, the human advantage lies in our species' ability to regulate population size more efficiently. We accomplish this by prolonging growth and development, inserting new life cycle stages, and developing a larger, more complex brain. Each of these contribute to human biocultural adaptations to the environment. At the beginning of this chapter I quoted Allison Jolly with reference to human mental agility replacing the reproductive agility of other animals. Given the nature of human biocultural behavior it is perhaps more accurate to conclude that human beings have both mental and reproductive agility.

The evolution of the human childhood stage gave our species a reproductive advantage, but also introduced new liabilities. Children are not fed by nursing, but they are still dependent on older individuals for feeding and protection. Moreover, children must be given foods that are specially chosen and carefully prepared to meet their nutritional demands. The peoples of traditional societies solve the problem of child-care by spreading the responsibility among many individuals. A review of the literature I conducted with Holly Smith (Bogin & Smith, 1996) finds that in Hadza society (African hunters and gatherers) grandmothers and great-aunts are observed to supply a significant amount of food and care to children. In

Agta society (Philippine hunter-gatherers) women hunt large game animals but still retain primary responsibility for child care. They accomplish this by living in extended family groups – two or three brothers and sisters, their spouses, children and parents – and sharing the child care. Among the Maya of Guatemala (horticulturists and agriculturists), many people live together in extended family compounds. Women of all ages work together in food preparation, manufacture of clothing, and child care. In some societies, fathers provide significant child care, including the Agta and the Aka Pygmies, a hunting-gathering people of central Africa. Summarizing the data from many human societies, Lancaster & Lancaster (1983) call this type of child care and feeding 'the hominid adaptation', for no other primate or mammal does all of this.

A stimulus to release these parental behaviors towards children may be found in the very pattern of growth of the children themselves, in that:

2.   *The allometry of the growth of the human child releases nurturing and care-giving behaviors in older individuals*

The central nervous system, in particular the brain, follows a growth curve that is advanced over the curve for the body as a whole (Figure 2.7). The brain achieves adult size when body growth is only 40 percent complete, dental maturation is only 58 percent complete, and reproductive maturation is only 10 percent complete. The allometry of the growth of the human child maintains an infantile appearance (large cranium, small face and body, little sexual development), which stimulates nurturing and care-giving behaviors in older individuals. A series of ethological observations (Lorenz, 1971) and psychological experiments (Todd *et al.*, 1980; Alley, 1983) demonstrate that these growth patterns of body, face, and brain allow the human child to maintain a superficially infantile (i.e., 'cute') appearance longer than any other mammalian species (see Box 4.1). The infantile appearance of children facilitates parental investment by maintaining the potential for nurturing behavior of older individuals towards both infants and dependent children (Bogin, 1988b pp. 98–104; 1990; McCabe, 1988).

3.   *Children are relatively inexpensive to feed*

The relatively slow rate of body growth and small body size of children reduces competition with adults for food resources, because slow-growing, small children require less food than bigger individuals. A five-year-old child of average size (the 50th centile of the NCHS reference curves for

Box 4.1.   The evolutionary psychology of childhood

Reproductive success is the major force behind the evolution of all species. Part of the reproductive success of the human species is due to the intense investment and care that parents, and other individuals, lavish on infants and children. In the course of human evolution, at least since the appearance of the genus *Homo* in the last two million years, patterns of growth were shaped by natural selection to promote and enhance parental investment. One way this was accomplished was by stimulating what may be called the 'psychology of parenting'.

Lorenz (1971) stated that the physical characteristics of mammalian infants, including small body size, a relatively large head with little mandibular or nasal prognathism, relatively large round eyes in proportion to skull size, short thick extremities and clumsy movements, inhibit aggressive behavior by adults and encourage their caretaking and nurturing behaviors. Lorenz believed that these infantile features trigger 'innate releasing mechanisms' in adult mammals, including humans, for the protection and care of dependent young. Gould (1979) questions the innateness of the human response to infantile features. Such behavior may be '. . . learned from our immediate experience with babies and grafted upon an evolutionary predisposition for attaching ties of affection to certain learned signals' (p. 34). The important point is that whether innate or learned the resultant behavior is the same.

There seems to be a pan-human ability to perceive the five stages of human postnatal development and respond appropriately to each. An elegant series of experiments performed by Todd *et al.* (1980), show that human perceptions of body shape and growth status are consistent between individuals. When adult subjects (about 40 college students, all childless) were shown a series of profiles of human skull proportions, they could easily arrange them correctly into a hierarchy spanning infancy to adulthood. The subjects could also ascribe maturity ratings to skull profiles that were geometrically transformed to imitate the actual changes that occur during growth (Figure B4.1). This perception was selective because a variety of other types of geometrical transformations elicited no reports of growth or maturation. When the growth-like mathematical transformations were applied to profile drawings of the heads of birds and dogs, human subjects reported identical perceptions of growth and maturation, even though in reality the development of these animals does not

CARDIOIDAL STRAIN

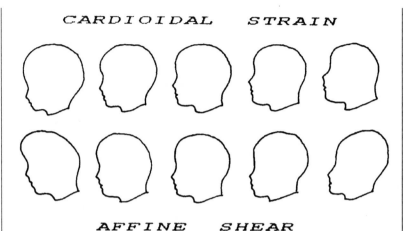

AFFINE SHEAR

Figure B4.1. Two of the mathematical transformations of human head shape used in the experiments of Todd *et al.* (1980). The middle profile in each row was drawn from the photograph of a ten-year-old boy. The transformations were applied to this profile of a real child. The cardioidal strain transformation is perceived by most adults as growth. The affine shear transformation is not perceived as growth.

follow the human pattern of skull shape change. Even more surprising is that subjects reported the perception of growth when the growth-like mathematical transformations were applied to front and side-view profiles of Volkswagen 'beetles', objects which do not grow.

In another series of experiments, Alley (1983) studied the association between human body shape and size, and the tendency by adults to protect and 'cuddle' other individuals. In the first experiment, subjects were shown two sets of drawings. One set was based upon two-dimensional diagrams depicting changes in human body proportion during growth. Alley's version of these diagrams are called 'shape-variant' drawings (Figure B4.2a). Alley's second set of figures were called 'size variant' drawings (Figure B4.2b). He used the middle-most, 'six-year-old' profile in the shape-variant series, to construct sets of figures that varied in height and width, but not in shape. Note that these figures have no facial features, or genitals. Perceptual differences between figures are due to body shape or size alone.

In the first experiment, the subjects were shown pairs of the shape-variant drawings (i.e., profiles of a newborn and a six-year-old, a two-year-old and a 12-year-old, etc.) or pairs of the size-variant drawings and asked to state which one of the pair they '. . . would feel most compelled to defend should you see them being beaten'. In another experiment they were asked about their feelings to 'hug or cuddle' the person depicted. The

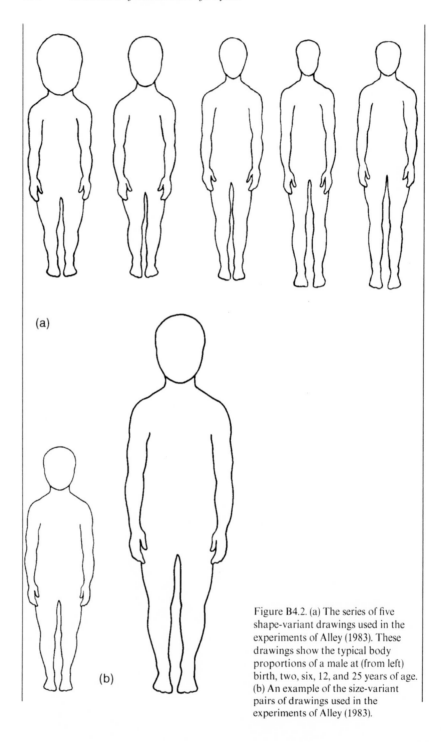

(a)

(b)

Figure B4.2. (a) The series of five shape-variant drawings used in the experiments of Alley (1983). These drawings show the typical body proportions of a male at (from left) birth, two, six, 12, and 25 years of age. (b) An example of the size-variant pairs of drawings used in the experiments of Alley (1983).

results of both experiments, summarized in Table B4.1, find a fairly strong reported willingness to defend 'newborns' and 'two-year-olds', and a moderate willingness to defend 'older' persons. The reported willingness to cuddle decreased with the 'age' of the drawings. Placed in the context of the ethological study of parental caregiving in mammals and birds, Alley believes that his results demonstrate a general tendency, to protect or cuddle others based on the perception of maturational status.

McCabe (1988) reviews the work of Alley and other similar studies. Taken together, these studies indicate that adults are more likely to protect or nuture individuals with 'neotenous' facial features. McCabe defines such features as having a relatively large ratio of cranium size to lower face size. McCabe also cites studies of the facial features of nursery-school-aged children under court protection for abuse compared with non-abused age-matched controls. The abused children had smaller ratios of the cranium/lower face – that is, they were less 'neotenous' or 'cute' – than the non-abused controls.

These psychological experiments and case control studies provide support for the arguments developed in this chapter for the evolution of human childhood. In particular, small body size and a superficially infantile appearance promote appropriate parental behavior by older individuals toward children.

Table B4.1. *Mean reported willingness to defend or cuddle persons of different body proportions (Alley, 1983)*

Age portrayed (years)	Defend	Cuddle
Newborn	7.7 (1.7)	3.4 (2.3)
2	7.1 (1.7)	3.4 (2.0)
6	5.8 (1.8)	3.2 (2.2)
12	5.0 (1.9)	3.0 (1.9)
25	4.3 (1.9)	2.7 (2.2)

Note: Standard deviations are given in parentheses.

growth) and activity, for example, requires 22.7 percent less dietary energy per day for maintenance and growth than a ten-year-old juvenile on the 50th growth centile (Ulijaszek & Strickland, 1993; Guthrie & Picciano, 1995). Thus, provisioning children, though time consuming, is not as onerous a task of investment as it would be, for instance, if both brain and body growth were both progressing at the same rapid rate. Moreover, in times of food scarcity children are protected from starvation by this unique pattern of brain and body growth.

Case (1978) proposed another advantage of small body size in relation to slow growth. He describes people as having 'low growth efficiency', by which he means that people have the slowest postnatal growth of virtually any mammal, but a high basal metabolic rate. For any given body size, human children require more energy for maintenance and growth than most mammals of the same body size. This is due, of course, to the relatively large brain of human children. Small body size during human infancy and childhood helps to reduce the total number of calories needed to maintain a big brain and high metabolic rate.

The task of child care becomes even less onerous because:

4.   *'Babysitting' is possible*

As children do not require nursing, any competent member of a social group can provide food and care for them. Early neurological maturity versus late sexual maturity allows juveniles and young adolescents to provide much of their own care and also provide care for children (Bogin, 1994a). Grandmothers and other post-reproductive women also provide much child care (Bogin & Smith, 1996). Again, this frees younger adults, especially the mother, for subsistence activity, adult social behaviors, and further child-bearing. This type of care-taking is rare in other primates, even for apes. Usually, an infant non-human primate must be cared for by its mother, or it will die. Adoptions of orphaned infants by females do occur in chimpanzee social groups, but only infants older than four years and able to forage for themselves survive more than a few weeks (Goodall, 1983). Goodall noted deterioration in the health and behavior of infant chimpanzees whose mothers had died. The behavioral changes include depression, listlessness, drop in play frequency, and whimpering. Health changes such as loss of weight were seen. Goodall reported that even those older infants who survived the death of their mother were affected by delays in physical growth and maturation.

It is well known that human infants and children also show physical and behavioral pathology after the death of one or both parents (Bowlby, 1969).

It seems, though, that the human infant can more easily make new attachments to other caretakers than the chimpanzee infant. The ability of a variety of human caretakers to attach to one or several human infants may also be an important factor. The psychological and social roots of this difference between human and non-human species in attachment behavior are not known. The flexibility in attachment behavior that hominid ancestors evolved allowed, in part, for the evolution of childhood and the reproductive efficiency of the human species.

One common pattern of child care in many traditional cultures is to have juveniles assume caretaking responsibilities for children. This occurs among two well-studied African hunting and gathering cultures, the !Kung and the Mbuti. Mothers carry their infant and still-nursing children (nursing a child to age four is common in these cultures) with them while foraging. Weaned children must stay 'home' at the base camp, as pre-adolescent children have neither the strength nor stamina to follow their parents while gathering or hunting (Draper, 1976; Konner, 1976; Turnbull, 1983a, b). At !Kung camps children of various ages play together within the camp boundaries while juveniles discharge many caretaking functions for younger children. The children seem to transfer their attachment from parents and other adults to the juveniles, behaving toward them with appropriate deference and obedience. The **age-graded play group** functions to transmit cultural behavior from older to younger generations and to facilitate the learning of adult parental behavior (Konner, 1976). Of course, the children and juveniles are never quite left on their own, as there is always one adult, or more, in camp at any time, but this person is not directly involved in child care. Rather, he or she is preparing food, tools or otherwise primarily engaged in adult activity. The Mbuti (nomadic hunters and gatherers of central African rain forests) have a similar child care arrangement. After weaning, toddlers enter the world of the *bopi*, the Mbuti term for a children's playground, but also a place of age-graded child care and cultural transmission. Between the ages of two or three to eight or nine, children and juveniles spend almost all of their day in the *bopi*. There they learn physical skills, cultural values and, even, sexual behavior. 'Little that children do in the *bopi* is not of full value in later adult life' (Turnbull, 1983b, p. 43–4).

The age-graded play group provides for both the caretaking and enculturation of the young, freeing the adults from these tasks so that they may provide food, shelter and other necessities for the young who may be at various stages of development. A woman may be pregnant, have a child weaned within the past year and have one or more older offspring simultaneously. Thus, adults may be able to increase their net reproductive output during a relatively short period of time. This benefit, and the selective advantage of a greater number of surviving offspring afforded by age-graded caretaking, may in part account for the evolution of the prolonged childhood and juvenile growth periods of hominids. The play group, in the protective environment of the home base or camp, provides the children with the freedom for play, exploration and experimentation, which Beck (1980) has shown encourages learning, socialization and even tool using.

A further important reason for the evolution of childhood is that:

5.   *Childhood allows for developmental plasticity*

Following the discussion by Stearns (1992, p. 62) and Mascie-Taylor & Bogin (1995, several chapters), the term plasticity means a potential for change in the phenotype of the individual caused by a change in the environment. The fitness of a given phenotype varies across an environment's range of variation. When phenotypes are fixed early in development, such as in mammals that mature sexually soon after weaning (e.g., rodents), environmental change is positively correlated with high mortality are. Social mammals (carnivores, elephants, primates) prolong the developmental period by adding a juvenile stage between infancy and adulthood. Adult phenotypes develop more slowly in these mammals. They experience a wider range of environmental variation, and the result is a better conformation between the individual and the environment. In Chapters 5 and 6 the nature of human plasticity in growth and adaptation are discussed in detail. The point here is that plasticity leads to increased evolutionary fitness, meaning that more offspring can survive to reproductive age. In mammalian species without a juvenile stage less than 10 percent of live-born offspring survive to reproductive age, while between 10 and 30 percent survive in the social mammals with a juvenile growth stage (Lancaster & Lancaster, 1983; Pereira & Fairbanks, 1993). The insertion of a childhood stage between infancy and the juvenile period in humans results in an additional four years of relatively slow physical growth and allows for behavioral experience that further enhances developmental plasticity. The combined result is increased fitness (reproductive success). Lancaster & Lancaster (1983) report that humans in traditional societies, such as hunters and gatherers and horticulturalists, rear about 50 per cent of their live-born offspring to adulthood.

Summary of childhood

These five themes of childhood – feeding, nurturing, low cost, babysitting, and plasticity – account for much of the evolution and pattern of growth of our species. Understanding these themes helps to resolve the paradox of human growth and evolution – lengthy development and low fertility. In reality, humans have greater reproductive economy than any other species since we raise a greater percentage of offspring to adulthood. These successfully reared young adults then begin their own reproduction and thus ensure some 'intimation of immortality' for their parents. In the center of it all is human childhood. For the child is indeed, to paraphrase Wordsworth, parent to the reproductively successful and well-adapted adult.

When and why did adolescence evolve?

The evolution of an adolescent stage is also depicted in Figure 4.9. Human adolescence is the stage of life when social, economic, and sexual maturation takes place. All three are needed for successful reproduction. The case was presented in the previous chapter for the special growth and development characteristics of human adolescence. Evidence for when and why human adolescence evolved are presented here.

The single most important feature defining human adolescence is the skeletal growth spurt that is experienced by virtually all boys and girls. There is no evidence for a human-like adolescent growth spurt in any living ape. There is no evidence for adolescence for any species of *Australopithecus*. There is some tentative evidence that early *Homo*, dating from 1.8 MYA, may have a derived pattern of growth that is leading toward the addition of an adolescent stage of development. This evidence is based on an analysis of shape change during growth of the femur (Tardieu, 1998). Modern humans have a highly diagnostic shape to the femur, a shape that is absent in fossils ascribed to *Australopithecus*, but present in fossils ascribed to *Homo habilis*, *Homo rudolfensis*, or early African *Homo erectus*. The human shape is produced by growth changes during both the prolonged childhood stage and the adolescent stage. The more human-like femur shape of the early *Homo* fossils could be due to the insertion of the childhood stage alone, or to the combination of childhood and adolescent stages. Due to the lack of fossils of appropriate age at death, the lack of dental and skeletal material from the same individuals, and the lack of sufficient skeletal material from other parts of the body it is not possible to draw any more definitive conclusions.

A remarkable fossil of early *Homo erectus* is both of the right age at death and complete enough to allow for an analysis of possible adolescent growth. The fossil specimen is catalogued formally by the name KMN-WT 15000, but is called informally the 'Turkana boy' as it was discovered along the western shores of Lake Turkana, Kenya, by Kamoya Kimeu in 1984 (Brown *et al.*, 1985; Walker & Leakey, 1993). This fossil is 1.6 million years old, making it an early variety of *Homo erectus*. The skeletal remains are almost complete, missing the hands and feet and a few other minor bones. It is the most complete known specimen of *H. erectus*. Holly Smith (1993) analyzed the skeleton and dentition of the Turkana fossil and ascertained that, indeed, it is most likely the remains of an immature male. The youth's deciduous upper canines were still in place at the time of death, and he died not long after erupting second permanent molars. These dental features place him firmly in the juvenile stage by comparison with any hominoid. The boy was 160 cm tall at the time of death, which makes him one of the tallest fossil youths or adults ever found.

Part of Smith's analysis focused on patterns of growth and development, especially the question 'Did early *H. erectus* have an adolescent growth spurt?'. Based on her analysis the answer to that question is no. Judged according to modern human standards, the Turkana boy's dental age of 11 years is in some conflict with his bone age (skeletal maturation) of 13 years and his stature age of 15 years. If the Turkana boy grew along a modern human trajectory, then dental, skeletal and stature ages should be about equivalent. By chimpanzee growth standards, however, the boy's dental and bone ages are in perfect agreement, both at seven years of age. As *Homo erectus* is no chimpanzee, the Turkana boy's true age at death was probably between seven and 11 years. What is clear is that the Turkana boy followed a pattern of growth that is neither that of a modern human nor that of a chimpanzee. Based on Smith's analysis the boy's large stature becomes more explicable. The reason for his relatively large stature-for-age is that the distinct human pattern of moderate to slow growth prior to puberty followed by an adolescent growth spurt had not yet evolved in early *Homo erectus*. Rather, the Turkana boy followed a more ape-like pattern of growth in stature making him appear to be tall in comparison with a modern human boy at the same age. At the time of puberty, the chimpanzee has usually achieved 88 percent of stature growth, while humans have achieved only 81 percent. Smith & Tompkins state that the human pattern of growth suppression up to puberty followed by a growth spurt after puberty had not evolved by early *H. erectus* times. 'Because of this, any early *H. erectus* youth would seem to us to be too large' (1995, p. 273).

Unfortunately there are no appropriate fossil materials of later *H. erectus* available to analyze for an adolescent growth spurt. There is one fossil of a Neanderthal in which the associated dental and skeletal remains needed to assess adolescent growth are preserved, and this fossil is being analyzed by Jennifer Thompson and colleagues. So far, they have published a description of the skull (Thompson & Bilsborough, 1997) and abstracts of preliminary descriptions of growth and development (Thompson, 1995; Thompson & Nelson, 1997). The specimen is a juvenile, most likely a male. It is called Le Moustier 1 and was found in 1908 in Western France. The specimen is dated at between 42 000 and 37 000 years BP (before present). Thompson uses information on crown and root formation of the molar teeth to estimate a dental age of 15.5 ± 1.25 years. Compared with modern human standards for length of the long bones of the skeleton, Thompson & Nelson estimate that Le Moustier 1 has a stature age of about 11 years, based on the length of his femur. The dental age of 15.5 years and the stature age of 11 years are in very poor agreement, and indicate that like the Turkana boy, Le Moustier 1 may not have followed a human pattern of adolescent growth. The dental age indicates that by human standards Le Moustier 1 was in late adolescence at the time of death, but the stature age indicates he was still a juvenile or had just entered adolescence. Quite unexpectedly, these differences in dental and skeletal maturity are exactly the opposite of those for the Turkana boy.

What is clear is that adolescent growth, at least the pattern of adolescent growth found in modern humans, does not seem to be present in either the Turkana boy or in the Le Moustier 1 fossil. In modern humans, certain diseases, prolonged undernutrition, and unusual individual variations in growth may produce the skeletal and dental features of the fossils. While it is possible that these two fossil specimens fall into one of these categories of unusual growth, the most parsimonious conclusion that one may draw from these findings is that the adolescent growth spurt in skeletal growth evolved recently. Quite likely this would be no earlier than the appearance of archaic *H. sapiens* in Africa at about 125 000 years ago. If Neanderthals are direct ancestors to modern humans, then the adolescent skeletal growth spurt may be less than 37 000 years old.

Why did the growth spurt evolve?

One often cited reason for the adolescent growth spurt is the prolonged time required to learn technology, social organization, language and other aspects of culture during the infant, child, and juvenile stages of growth. At the end of this period, so the argument goes, our ancestors were left with proportionately less time for procreation than most mammals, and therefore needed to attain adult size and sexual maturity quickly (Watts, 1985, 1990). But surely this cannot be the whole story. Consider first that there is no need to experience an adolescent growth spurt to reach adult height or fertility. Historical sources describe the *castrati*, male opera singers of the seventeenth and eighteenth centuries who were castrated as boys to preserve their soprano voices, as being unusually tall for men (Peschel & Peschel, 1987). Also, children who are born without gonads or have then removed surgically prior to puberty (due to diseases such as cancer) do not experience an adolescent growth spurt, but do reach their normal expected adult height (Prader, 1984). Of course, *castrati*, whether or not opera singers, do not become reproductively successful. There are, however, normal individuals, for the most part very late maturing boys and many girls, who have virtually no growth spurt. Nevertheless, these late maturing individuals do grow to be normal-sized adults, and they become fertile by their early twenties – not significantly later than individuals with a spurt.

Another problem with the 'lost time' argument for the adolescent growth spurt is that it does not explain the timing of the spurt. Girls experience the growth spurt before becoming fertile, but for boys the reverse is true. Why the difference? Sexual dimorphism in adult height is one obvious consequence of the timing difference, and hence a possible positive value for the growth spurt itself. Adult men are, on average, 12 to 13 cm taller than adult women (Eveleth & Tanner, 1976).

Dimorphism in stature is only one of a series of sex-based differences in development that take place during adolescence. The order in which several pubertal events occur in girls and boys is illustrated in Figure 4.11 in terms of time before and after peak height velocity (PHV) of the adolescent growth spurt. In both girls and boys puberty begins with changes in the activity of the hypothalamus and other parts of the central nervous system. These changes are labeled as 'CNS puberty' in the figure. Note that the CNS events begin at the same relative age in both girls and boys, that is, three years before PHV. This is also the time when growth rates change from decelerating to accelerating. In girls, the first outward sign of puberty is the development of the breast bud (B2) and wisps of pubic hair (PH2) (see Chapter 2 for an explanation of the system of staging the breast and pubic hair development of girls, and genital and pubic hair development in boys). This is followed, in order, by (1) a rise in serum levels of estradiol which leads to the laying down of fat on the hips, buttocks, and thighs; (2) the adolescent growth spurt; (3) further growth of the breast and body hair (B3 & PH3); (4) menarche; (5) completion of breast and body hair development (B5 & PH5); and (6) attainment of adult levels of ovulation frequency.

The path of pubertal development in boys starts with a rise in serum levels of luteinizing hormone (LH) and the enlargement of the testes and then penis (G2). This genital maturation begins, on average, only a few months after that of girls. However, the timing and order of other secondary sexual

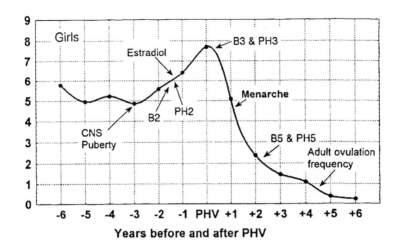

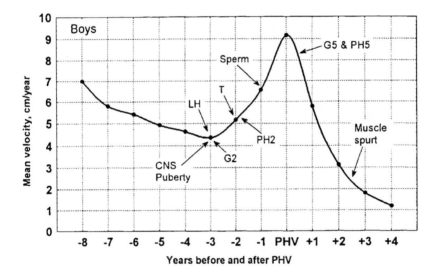

Figure 4.11. The ordering of several sexual maturation events for girls (top panel) and boys (bottom panel) during the adolescent growth spurt. The velocity curves are calculated using data derived from a sample of healthy, well-nourished girls and boys living in Guatemala. See text for an explanation of each labeled event.

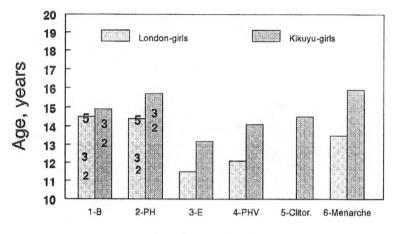

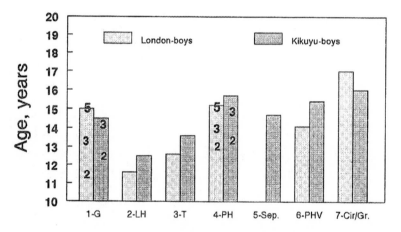

Figure 4.12. Comparison of biocultural events during adolescence in a sample of London and Kikuyu girls and boys. The data are abstracted from Worthman (1993), who provides references to the original studies. The events are presented in order of occurrence. The bars indicate the median age at onset of each event. Explanation of each event is given in the text. The abbreviations for the girls' events are: 1-B, breast stages; 2-PH, pubic hair stages; 3-E, first notable rise in serum estradiol concentration; 4-PHV, peak height velocity; 5-Clitor., clitoridectomy; 6-Menarche, age at menarche. Abbreviations for the boys are: 1-G, genital stages; 2-LH, first notable rise in serum luteinizing hormone concentration; 3-T, first notable rise in serum testosterone concentration; 4-PH, pubic hair stages; 5-Sep., separation to 'boys' house' for Kikuyu; 6-PHV, peak height velocity; 7-Cir/Gr., circumcision for Kikuyu, or graduation from secondary school for London boys.

characteristics is unlike that of girls. About a year after CNS puberty, there is: (1) a rise in serum testosterone levels (T) which is followed by the appearance of pubic hair (PH2); (2) about a year later motile spermatozoa may be detected in urine; (3) PHV follows after about another year, along with deepening of the voice, and continued growth of facial and body hair; (4) the adult stages of genital and pubic hair development follow the growth spurt (G5 & PH5); and (6) near the end of adolescence boys undergo a spurt in muscular development.

The sex-specific order of pubertal events tends not to vary between early and late maturers, between well-nourished girls and boys and those who suffered from severe malnutrition in early life, between rural and urban dwellers, or between European and African ethnic groups (Cameron *et al.*, 1988, 1990, 1993; Bogin *et al.*, 1992). In addition to these biological events there are behavioral and social events that also follow a predictable course during adolescence. Indeed, the biological and cultural events are usually tightly correlated. A comparison of the biocultural timing of adolescent events in two societies is given in Figure 4.12. Girls from a London, England sample and a Kikuyu (African) sample are compared in the upper panel and boys from the same two samples are compared in the lower panel (the data are reported by Worthman, 1993). The Kikuyu are a Bantu-speaking, agricultural society of the central highlands of Kenya. The London sample represent adolescents who are relatively well nourished and healthy. The Kikuyu sample represent adolescents who suffer from periodic food shortages and, perhaps, a higher incidence of infectious and parasitic diseases. The adolescent events for each sex are placed in chronological order and the bars indicate the median age at which each event occurs. These comparisons show two things. The first is, how differences in health and nutrition between human societies may influence the timing, but not the order, of adolescent events. The second is, how societies as diverse as urban Londoners and rural Kikuyu adjust the timing of some social events to the timing of the biological events of human adolescence.

For both London and Kikuyu girls, the first biocultural event is breast development (bars 1-B), and the second event is pubic hair development (bars 2-PH). The numbers on these bars indicate median age of entry to each stage of development (B2, B3, B5, etc.). The third event is a rise in serum estradiol concentration (bars 3-E). This hormonal event, and similar hormonal changes in boys, can only be detected by special tests, not by the adolescents or their parents. However, the rise of estradiol leads to biological and behavioral changes that are easily detectable in the form of fat deposits on hips, thighs, and buttocks and new levels of cognition (Piaget's formal operations stage). The fourth event is peak height velocity (bars 4-PHV). Note that for the Kikuyu, PHV occurs about two years later than for the London girls. For many Kikuyu girls the fifth biocultural event is clitoridectomy; about 40 percent of girls underwent this operation, which removes the tip of the clitoris, at the time of Worthman's research in 1979 and 1980. Clitoridectomy takes place just after PHV, at about breast stage 3, and just before menarche. The operation is timed so that it precedes the onset of sexual activity and marriage which follow menarche (Worthman, 1993). London girls may experience some adolescent rites of passage after PHV, but these are usually less well-defined and less traumatic than clitoridectomy. The sixth event is menarche, which is taken as a sign of impending sexual maturation in all cultures. In many cultures, menarche often precipitates intensified instruction about sexual behaviors and the practice of these behaviors (Schegel & Barry, 1991).

For both London and Kikuyu boys, the first two biocultural events are enlargement of the testes, or genital stage 2 (bars 1-G), and a rise in serum concentration of luteinizing hormone (bars 2-LH). In fact the order of the two events could be switched as it is the rise of LH that leads to testes enlargement. Here these two events are considered to be coterminous. The third event is a rise in the serum concentration of testosterone (bars 3-T), which precipitates a cascade of physical and behavioral changes. The fourth event is public hair development (bars 4-PH). For the Kikuyu, the fifth biocultural event is separation (bar 5-Sep.), which means that the adolescent boys leave their nuclear household and begin living in an age-graded adolescent male household. Worthman states that separation to the 'boys' house' is closely correlated with age at first emission, and that separation takes place at about the same age that girls undergo clitoridectomy. These events show that Kikuyu parents are able to recognize and respond to the sexual maturation of their adolescents. The sixth event is peak height velocity (bars 6-PHV). The seventh biocultural event for Kikuyu adolescents is circumcision (7-Cir), which is done to all young men and marks their entry into training for adulthood. Circumcision is timed to occur along with the spurt in muscle mass, which allows boys to perform physical labor at adult levels. London boys do not undergo a circumcision rite of passage, but within that same year they usually graduate from secondary school (7-Gr.). That event, which London girls also experience, is a rite of passage in most industrialized societies and often marks entry into the social world of adults.

More will be said about the biocultural significance of these adolescent events in the next few pages (see Chapter 8 for a discussion of biocultural models of human development). At this point it is important to focus on two general issues. The first is that the adolescent growth spurt is a biologically and socially significant event for both sexes. The second is that the order of adolescent events is different for each sex, for example the growth spurt occurs earlier in the sequence, as well as at an earlier age, in girls than in boys. Given this, the sexual dimorphism expressed in this sequence and timing of these events may be considered a species-specific characteristic. Evolutionary biologists usually find that species-specific traits evolve to enhance the survival and reproductive success. Thus, the human adolescent growth spurt must have its own intrinsic evolutionary value, and is not just a by-product of slow prepubertal development.

Why do girls have adolescence, or why wait so long to have a baby?

Differences between boys and girls in the timing, duration, and intensity of the adolescent growth spurt require that each sex be analyzed individually, if we are to understand how the separate paths of development each takes through adolescence promotes their reproductive success as adults. Moreover, the value of adolescence becomes apparent only when we consider human biology and behavior cross-culturally and historically. We need to include the role adolescence plays in contemporary Western, industrialized society, but thinking only of the present-day and one cultural milieu cannot reveal the evolutionary and biocultural nature of adolescence. Taking this approach from the female perspective, the human mother-to-be must acquire knowledge of (1) adult socio-sexual relations, (2) pregnancy, and (3) child care. The dramatic physical changes that girls experience during adolescence serve as efficient advertisements of their sexual and social maturation. So efficient, in fact, that they stimulate adults to include adolescent girls in their social circles and encourage the girls themselves to initiate adult social interactions, such as working with and learning from adult women, social and sexual bonding with adult men, and sexual intercourse. Psychologists call this the intensification of gender-related roles (Hill & Lynch, 1983). Ethnographic research shows that gender role intensification during adolescence is a universal feature of human cultures (Whiting & Edwards, 1988; Schlegel & Barry, 1991).

In human societies, juvenile girls are often expected to provide significant amounts of child care for their younger siblings. This stands in contrast to most other social mammals whose juveniles are often segregated from adults and infants. Whiting & Edwards (1988) surveyed information available in the *Human Relation Area Files*, a data base of 186 societies, concerning apprenticeships in child care. They find that 'the preferred age for child nurses in many of our sample communities is between 6 and 10 years' (p. 272). Whiting & Edwards believe that the best time for girls to learn child-care skills is during these juvenile years. Social and emotional bonds between the juvenile and her mother and other adults are shifting from ones of dependency to ones of interest in their family and the work of adult women. In a few years these juvenile girls enter adolescence '. . . and become less interested in the world of their family and more interested in sex, the future, and their own children' (p. 272). These findings show how the biocultural nature of gender role intensification is associated with human life history stages.

Schlegel & Barry (1991; see also Schlegel, 1995) used the *Human Relation Area Files* in their cross-cultural survey of adolescence. Their interest was more in social, emotional, and cultural role of adolescence than in biology. Nevertheless, they take a biocultural approach and find that both a biological and a social stage of life that may be called adolescence is found universally in human cultures. This is noteworthy only because some psychologists, educators, and historians have claimed that adolescence is an 'invention' of industrialized societies. Schlegel & Barry do not make a clear distinction between the juvenile and adolescent stages of life. Even so, they confirm the earlier work of Whiting & Edwards showing that during both the juvenile and adolescent periods both learning and social restructuring take place. They state that there is a 'cognitive and affective reorganization away from the behavioral modes of childhood and toward adult modes . . . The adolescent assumes greater autonomy, more peer relationships with same-sex adults, and an interest in sexual activities' (p. 8). The authors characterize the juvenile and adolescent stages as periods of 'unlearning and relearning'. The human capacity to makes these shifts in learned behavior, such as from childhood dependency to adolescent autonomy, attests to the biological and psychological plasticity of our species and negates the notion that adult humans are neotenous, that is, permanent children in any sense. These shifts do not support other models of human development via heterochrony, such as sequential hypermorphosis. Schlegel & Barry emphasize that with the onset of the juvenile stage, most societies segregate boys and girls socially and each sex experiences very different types of learning. Juvenile girls are included in the social and work world of the family and of older women, while boys are encouraged to form social bonds with other juvenile boys. By working closely with adult women, juvenile girls are assisted in the restructuring of their learning, and acquire first-hand experience in child care and many domestic tasks.

By the time these juvenile girls enter adolescence they have gained considerable knowledge of the needs of infants and children. To complete their restructuring and relearning for reproductive success as adult women, these girls need to gain knowledge of sexuality and reproduction. As adolescents they can do this efficiently because they look mature sexually, and are treated as such, several years before they actually become fertile. The adolescent growth spurt serves as one important signal of sexual maturation. Early in the spurt, before peak height velocity (PHV) is reached, girls develop pubic hair and fat deposits on breasts, buttocks and thighs. By the time of PHV, girls have achieved 91 percent of their adult height. Since the adolescent spurt and PHV of girls occurs about two years earlier than that of boys, the girls are, on average, taller than boys of the same age (Figures 2.5, 2.12; Tobias, 1970). In essence, all of these changes in body composition, in absolute stature, and in stature relative to boys help to make the girls look like women and to appear to be maturing sexually. About a year after peak height velocity, girls experience menarche, an unambiguous external signal of internal reproductive system development. However, most girls experience one to three years of anovulatory menstrual cycles following menarche, meaning that they cannot become pregnant. Two studies of girls and young women living in Switzerland and Finland examined the frequency of ovulation for 4.5 years following menarche (reviewed in Worthman, 1993). Ovulation frequency varied from zero to ten percent of menstrual cycles at six months post-menarche. The frequency increased to about 30 percent after 1.5 years, varied between 40 and 55 percent after 2.5 years, and leveled off at 60 to 65 percent after 4.5 years. Since the mature level of ovulatory frequency is about 65 percent of menstrual cycles, it appears that it takes about five years for healthy, well-nourished girls to achieve adult maturity for fertility. Adolescent girls, and the adults around them, may or may not be aware of this period of 'adolescent sterility'. Everyone in the social group is aware of the dramatic changes taking place in the adolescent girl, and these changes certainly stimulate both the girls, and adults around them, to participate in adult social, sexual, and economic behavior. For the post-menarche adolescent girl this participation provides the learning and experience she will need to be a successful woman and mother, and it is 'low risk' in terms of pregnancy for several years.

It is noteworthy that female chimpanzees and bonobos, like human girls, also experience up to three years of post-menarche infertility, so this time of life may be a shared hominoid trait. As with human adolescents, the post-menarchial but infertile chimpanzees and bonobos participate in a great deal of adult social and sexual behavior. Primate researchers observing these apes point out that this participation, without pregnancy, allows for practicing many key behaviors that are needed to rear an infant successfully (Goodall, 1983). Although ape and human females may share a year or more of adolescent sterility, apes reach sexually mature adulthood at about 12 years of age, much sooner than humans. This limits the learning and practice period for the apes.

Full reproductive maturation, in human women, is not achieved until about five years after menarche. The phrase 'full reproductive maturation' means the biological, social, and psychological maturation of the woman have reached the point where risks of pregnancy are near the minimum for both the mother and her offspring. Menarche occurs at a median age that varies from 12.1 to 13.5 years in healthy populations (the normal range in age at menarche is 8 to 17 years), which means that the average age at full sexual maturation occurs at age 17 or 18 years. Fertility may occur earlier, even as early as six months after menarche. Fertility, however, does not equal reproductive maturity. Becoming pregnant is only a part of the business of reproduction. Maintaining the pregnancy to term and raising offspring to adulthood are equally important. Girls under 17 years old have difficulty with both of these, since the risks for spontaneous abortions, complications of pregnancy, such as high blood pressure in the mother, and low birth weight babies are more than twice as high as those for women 20 to 24 years old. The likelihood of these risks declines, and the chance of successful pregnancy and birth increases, markedly after age 18.

Another feature of human growth, not found in the African apes, is that female fertility tracks the growth of the pelvis. Ellison (1982) and Worthman (1993) find that age at menarche is best predicted by **biiliac width**, the distance between the iliac crests of the pelvis. A median width of 24 cm is needed for menarche in American girls living in Berkeley, California, Kikuyu girls of East Africa, and Bundi girls of highland New Guinea. The pelvic width constant occurs at different ages in these three cultures, about 13 years in California versus 16 to 17 years in Kenya and New Guinea, due to chronic malnutrition and disease among the Kikuyu and the Bundi. Marquisa LaVelle (Moerman, 1982) also reports a special human relationship between growth in pelvic size and reproductive maturation. She finds that the crucial variable for successful first birth is size of the **pelvic inlet**, the bony opening of the birth canal. LaVelle measured pelvic X-rays from a sample of healthy, well-nourished American girls who achieved menarche between 12 and 13 years. These girls did not attain adult pelvic inlet size until 17 to 18 years. Quite unexpectedly, the adolescent growth spurt, which occurs before menarche, does not influence the size of the pelvis in the same way as the rest of the skeleton. Rather, the female pelvis has its own slow pattern of growth, which continues for several years even after adult stature is achieved. La Velle's research indicates that the one cause of the high risks of adolescent pregnancy is a small pelvic inlet. While there are several other risks, a study by Tague (1994) provides support to the effect of pelvic inlet size. Tague examined the relationship between pelvic size and age at death in three prehistoric Native American populations. The populations, called Indian Knoll, Pecos Pueblo, and Libben, are represented by sizable collections of skeletal material from people who died from infancy to old age. Only adult pelvies were studied by Tague. He finds that the female pelvis continues to grow and remodel in adulthood for a longer time than the male pelvis. The age at death for women (assessed by morphology of the pubic symphysis and the auricular surface of the ilium, and by dental wear) correlates with pelvic inlet size – women with smaller inlets died at younger ages. Tague states that complications of pregnancy and birth were a leading cause of mortality in prehistoric populations. The correlation between pelvic inlet size and age at death in his sample of adult women, all of child-bearing age, seems to show that inlet size is a major predictor of reproductive success.

Cross-cultural studies of reproductive behavior shows that human societies acknowledge (consciously or not) this special pattern of pelvic growth. The age at first marriage, and childbirth, clusters around 19 years for women from such diverse cultures as the Kikuyu of Kenya, Mayans of Guatemala, Copper Eskimo of Canada, and both the Colonial period and contemporary United States (Bogin, 1994a). Why the pelvis follows this unusual pattern of growth is not clearly understood. Perhaps another human attribute, bipedal walking, is a factor. Bipedalism is known to have changed the shape of the human pelvis from the basic ape-like shape. Apes have a 'cylindrical' shaped pelvis, but humans have a 'bowl-shaped' pelvis. The human shape is more efficient for bipedal locomotion but less efficient for reproduction because it restricts the size of the birth canal (Trevathan, 1987, 1996 details the relationship between the evolution of bipedalism and human birth). Whatever the cause, this special human pattern of pelvic growth helps explain why girls must wait for many years from the age at menarche to the age of full reproductive maturity.

That time of waiting provides the adolescent girls with many opportunities to practice and learn important adult behaviors that lead to increased reproductive fitness in later life. It seems that there was selection pressure in favor of female adolescence, because girls with the extra developmental time prior to reproduction were able to learn social, economic, and parenting skills that would help ensure greater reproductive success later in life. There is direct evidence for the reproductive value of human adolescence when the data for non-human primates is examined. The first-born infants of monkeys and apes are more likely to die than those of humans. Studies of yellow baboons (Altmann, 1980), toque macaques (Dittus, 1977), and chimpanzees (Teleki *et al.*, 1976) show that between 50 and 60 percent of first born offspring die in infancy. In hunter-gatherer human societies, such as the !Kung, about 44 percent of children die in infancy (Howell, 1979). Just for comparison, it may be noted that in the United States, in the year 1960, about 2.5 percent of all live, first-born children died before the age of one year (Vavra & Querec, 1973).

Studies of wild baboons by Jeanne Altmann (1980) show that while the infant mortality rate for the first-born is 50 percent, mortality for second-born drops to 38 percent, and for third and fourth-born reaches only 25 percent. The difference in infant survival is, in part, due to experience and knowledge gained by the mother with each subsequent birth. Such maternal information is usually mastered by human women during adolescence, which gives the women a reproductive edge. The initial human advantage may seem small, but it means that up to 21 more people than baboons or chimpanzees survive out of every 100 first-born – more than enough over the vast course of evolutionary time to make the evolution of human adolescence an overwhelmingly beneficial adaptation.

Why do boys have adolescence?

The most important difference in the pattern of adolescent development for girls and boys is that boys become fertile well before they assume adult size and the physical characteristics of men. Analysis of urine samples from healthy, well-nourished boys age 11 to 16 years old show that they begin producing sperm at a median age of 13.4 years (Muller *et al.*, 1989). Yet the cross-cultural evidence is that few boys successfully father children until they are into their third decade of life. Traditional Kikuyu men do not marry and become fathers until about age 25, although they start seminal emissions at about 14.5 years (Figure 4.12, age at separation) and become sexually active following their own circumcision rite at around age 16 to 18 years (Worthman, 1986, 1993). The National Center for Health Statistics of the United States reports that only 3.09 percent of all births are fathered by men under 20 years old, while another nationally representative longitudinal survey shows that only seven percent of young men then aged 20 to 27 years old fathered a child while they were teenagers (Marsiglio, 1987). Among the Ache, who were traditionally foragers of the forests of Paraguay, adolescent boys did not become net food producers until age 17 years, and they did not marry until about age 20 years (Hill & Kaplan, 1988). In the Central Canadian Arctic, Inuit people living as traditional hunters did not even consider that an adolescent boy was ready for marriage until he was 17 to 18 years old (Condon, 1990). Even then, the adolescent man had to provide bride service to his prospective in-laws for several years before he became a father. All this delay in fatherhood occurred despite the fact that there was considerable pressure to reproduce because of, 'the slim margin of survival in the pre-contact period . . .' (p. 270).

The explanation for the lag between sperm production and fatherhood is not likely to be a simple one of sperm performance, such as not having the endurance to swim to an egg cell in the woman's fallopian tubes. A more likely explanation is the fact that the average boy of 13.4 years is only beginning his adolescent growth spurt (Figure 2.5). Growth researchers have documented that in terms of physical appearance, physiological status, psychosocial development, and economic productivity the 13-year-old boy is still more of a juvenile than an adult. Anthropologists working in many diverse cultural settings report that few women, and more importantly from a cross-cultural perspective, few prospective in-laws, view the teenage boy as a biologically, economically, and socially viable husband and father.

The reason for the delay between sperm production and reproductive maturity may lie in the subtle psycho-physiological effects of testosterone, and other androgen hormones, that are released following gonadal maturation and during early adolescence – effects that may 'prime' boys to be receptive to their future roles as men. Studies on a cross-section of youths in Europe, North America, Japan, and Africa establish that as blood levels of testosterone begin to increase, but before the growth spurt reaches its peak, there is an increase in psychosexual activity. Nocturnal emissions begin (Laron *et al.*, 1980) and there is an increase in the frequency of masturbation. Sociosexual feelings, such as infatuations and 'dating', intensify (Higham, 1980; Petersen & Taylor, 1980): in cross-cultural perspective 'dating' refers to overt interest in forming social bonds with eligible sexual or marriage partners. Alternatively, it is possible that physical changes provoked by the endocrine changes provide a social stimulus toward adult behaviors (Halpern *et al.*, 1993). Whatever the case, early in adolescence sociosexual feelings intensify, including guilt, anxiety, pleasure, and pride. At the same time, boys become more interested in adult activities, adjust their attitude to parental figures, and think and act more independently. In short, they begin to behave like men.

However – and this is where the survival advantage may lie – they still look like boys. One might say that a healthy, well-nourished 13.5-year-old human male, at a median of 160 cm tall (5'2") 'pretends' to be more child-like than he really is. Because their adolescent growth spurt occurs late in sexual development boys appear to be juvenile-like for much longer than girls. During the adolescent years, boys are even shorter than girls of roughly the same chronological age, furthering an immature image (Figure 2.5; Tobias, 1970). Even more to the point is that the spurt in muscle mass of adolescent males does not occur until an average age of 17 years (Malina, 1986). At peak height velocity the typical boy has achieved 91 percent of his adult height, but only 72 percent of his adult lean body mass. Since most of the lean body mass is voluntary muscle tissue, adolescent boys cannot do the work of men. This is one important reason why the Kikuyu, the Inuit and many other cultures do not even think of younger adolescents as man-like. As Schlegel & Barry (1991) found in their cross-cultural survey, adolescent boys are usually encouraged to associate and 'play' with their age mates rather than associate with adult men. During these episodes of 'play' these juvenile-looking adolescent males can practice behaving like adult men before they are actually perceived as adults. The activities that take place in these adolescent male peer groups include the type of productive, economic, aggressive/militaristic, and sexual behaviors that older men perform. However, the sociosexual antics of adolescent boys are often considered to be more humorous than serious. Yet, they provide the experience to fine tune their sexual and social roles before either their lives, or those of their offspring, depend on them. For example, competition between men for women favors the older, more experienced man. As such competition may be fatal, the juvenile-like appearance of the immature, but hormonally primed, adolescent male may be life-saving, as well as educational.

Summary of adolescence

Adolescence became part of human life history because it conferred significant reproductive advantages to our species, in part, by allowing the adolescent to learn and practice adult economic, social, and sexual behaviors before reproducing. The basic argument for the evolution and value of human adolescence is this: girls best learn their adult social roles while they are infertile but perceived by adults as mature; whereas, boys best learn their adult social roles while they are sexually mature but not yet perceived as such by adults. Without the adolescent growth spurt, and the sex-specific timing of maturation events around the spurt, this unique style of social and cultural learning could not occur. Over the course of time and space, the styles of learning these behaviors have come to vary considerably cross-culturally. The evolution of human adolescence, therefore, has to be modeled in terms of both its biological and cultural ramifications.

Viewing human adolescence in this life history and biocultural perspective has significant implications. Some of these may pertain to medical treatment of growth disorders, for example, the effects of early or delayed puberty on the physical and psychological well-being of the adolescent. Other implications relate more to economic, social, and legal policies that impact the lives of adolescents. We will investigate these issues in more detail in Chapter 6, with reference to environmental influences on human growth and development.

The valuable grandmother, or could menopause evolve?

In addition to childhood and adolescence there is another unusual aspect of human life history: menopause. One generally accepted definition of menopause is, '. . . the sudden or gradual cessation of the menstrual cycle subsequent to the loss of ovarian function . . .' (Timiras, 1972, p. 531). The process of menopause is closely associated with the adult female post-reproductive stage of life, but menopause is distinct from the post-reproductive stage. Reproduction usually ends before menopause. In traditional societies, such as the !Kung (Howell, 1979), the Dogon of Mali (B. Strassman, personal communication), and the rural-living Maya of Guatemala (Ministerio de Salud Publica, 1989), women rarely give birth after age 40 years and almost never give birth after age 44. Menopause, however, occurs after age 45 in these three societies. In the United States, between the years 1960 and 1990, data for all births show that women 45 to 49 years gave birth to fewer than one out of every 1000 live born infants. In contrast, there were 16.1/1000 live births to women aged 40–4 years (National Center for Health Statistics, 1994). Similar patterns of birth are found for the Old Order Amish, a high fertility, non-contracepting population residing primarily in the states of Pennsylvania, Ohio, and Indiana. Amish women aged 45 to 49 years born before 1918 gave birth to an average of 13 infants per 1000 married women, while those women between 40 and 44 years of age gave birth to an average of 118 infants per 1000 married women (Ericksen *et al.*, 1979). Thus, even in the United States of 1960–90, with modern health care, good nutrition, and low levels of hard physical labor, and even among social groups attempting to maximize lifetime fertility, women rarely give birth after age 45 years. As for the !Kung, Dogon, and Maya, menopause occurs well after this fertility decline, at a mean age of 49 years for United States living women (Pavelka & Fedigan,

1991). After age 50, births are so rare that they are not reported in the data of the National Center for Health Statistics or for the Amish (but are sensationalized in the tabloids sold at supermarket check-outs).

I report these ages for the onset of human female post-reproductive life versus the ages for menopause for two reasons. The first is that some scholars incorrectly equate menopause with the beginning of the post-reproductive stage, so one must read the literature carefully to interpret in what sense the term 'menopause' is used. The second reason is that menopause, and a significant period of life after menopause, are claimed by some scholars to be uniquely human characteristics. Other scholars assert that menopause is a shared trait with other mammals.

In a review of menopause from a comparative primate and evolutionary perspective Pavelka & Fedigan (1991) find that menopause is a virtually universal human female characteristic and that menopause occurs at approximately the age of 50 years in all human populations. In contrast, Pavelka & Fedigan note that wild-living non-human primate females do not share the universality of human menopause, and human males have no comparable life history event. In a review of the data for all mammals Austad (1994, p. 255) finds that no wild-living species except, possibly, pilot whales '. . . are known to commonly exhibit reproductive cessation . . .'. Female primates studied in captivity, including langurs, baboons, rhesus macaques, pigtailed macaques, and chimpanzees, usually continue estrus cycling until death, although there are fertility declines with age (Fedigan & Pavelka, 1994). These declines are best interpreted as a normal part of aging. Fedigan & Pavelka's review of the literature finds that one captive bonobo over 40 years old (Gould *et al.*, 1981) and one captive pigtail macaque over 20 years old (Graham *et al.*, 1979) ceased estrus cycling. These two very old animals showed changes in hormonal profiles similar to human menopause and upon autopsy were found to have depleted all oocytes. Finally, Pavelka & Fedigan (1991) point out that in contrast to the senescent decline in fertility of other female primates, the human female reproductive system is abruptly 'shut down' well before other systems of the body, which usually experience a gradual decline toward senescence. Moreover, human women may live for decades after oocyte depletion (menopause), but other female primates die before or just after oocyte depletion (these events are correlated but not necessarily causal to each other).

Figure 4.13 illustrates the timing of the onset of the adult female post-reproductive stage and menopause in the context of the evolution of human life history. Again, as for Figure 4.9, the data for fossil hominids are speculative, extrapolated, in part, from evidence provided by extant chim-

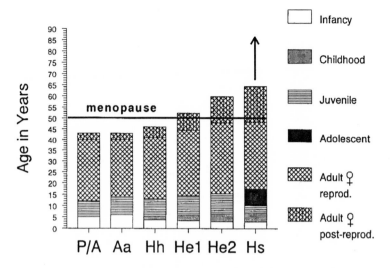

Figure 4.13. The evolution of human female life history emphasizing the post-reproductive stage. Life expectancy estimated by the formula of Smith (1991b). The arrow above the *H. sapiens* column represents Sacher's (1975) estimate of maximum longevity to 89 years. Increased human longevity extends the post-reproductive stage, not earlier stages of the life cycle. Abbreviations as in Figure 4.9.

panzees and human beings. Nishida *et al.* (1990) report that wild-living female chimpanzees give birth to their last offspring in their late thirties or early forties. They may then experience between 2.5 to 9.5 years (median of 3.9 years) of post-reproductive life, but most of these females continue estrus cycles until death. Based on these findings, a median age of 40 years for the onset of the chimpanzee post-reproductive stage is used for Figure 4.13. For human females, the data available from the industrialized nations, and a few traditional societies, provide mean ages of menopause from 48 to 51 years (Timiras, 1972; Pavelka & Fedigan, 1991), and as discussed above virtually no women over age 50 give birth. Accordingly, 50 years is used as a representative age for menopause and also the maximum age for onset of the human female post-reproductive stage. It is also possible to propose that 50 years is the effective upper limit for the age at menopause (oocyte depletion) of hominoids in general, based upon the human condition and the one known bonobo to experience menopause.

The estimates of life expectancy depicted in Figure 4.13 are based on regression formulae developed by Smith (1991b). The formulae predict life expectancy using data for body weight and brain weight. Smith's estimate for the chimpanzee (43 years) and for *H. sapiens* (66 years) accord well

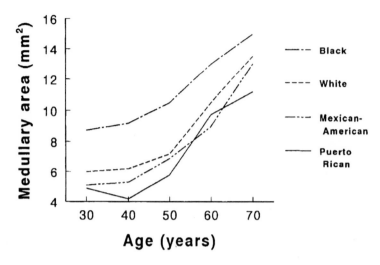

Figure 4.14. Age changes in medullary area for several nationally representative samples of populations of women residing in the United States. Medullary area increases after the third decade of life in all populations and the rate of bone loss increases after age 50, that is, at the time of menopause (data kindly provided by Professor S. M. Garn).

with data for wild chimpanzees (the maximum lifespan of captive chimpanzees is 50 years) and traditional human societies (e.g., Nishida *et al.*, 1990 for chimps; Neel & Weiss, 1975; Howell, 1979 for humans). The *H. sapiens* column also includes an extension of predicted life expectancy to 89 years. This estimate is based on the formula of Sacher (1975) for maximum longevity, and is being approached by populations of the most highly industrialized nations. Smith's formula is also used for predictions of life expectancy for the fossil species. The predictions are hypothetical, and based on the best available estimates of body weight and brain weight.

Age at onset of the post-reproductive stage for female fossil hominids is based on an extrapolation between the known mean ages for chimpanzees and humans as given above. A linear interpolation was used to calculate the ages for the fossils. A curvilinear fit, a step function, or some other discontinuous function may better represent the true nature of change in the age of onset of a post-reproductive life stage. Empirical research is needed to determine the best model. It is possible that empirical evidence for the evolution of the post-reproductive stage for women will be recoverable from the fossil record because of the biology of menopause. The hormonal changes associated with the menopause process have profound effects on bone mass and the histology of tubular bones. As defined by Garn (1970) there is a gain of bone mass and an increase in deposition on the endosteal surface of tubular bones during the 'steroid mediation phase' of life, e.g., during adolescence and reproductive adulthood. Moreover, the endosteal gain is greater in women than in men. By the fifth decade of life the apposition of endosteal bone stops and resorption begins. Data for women living in the United States of European, African, Mexican, and Puerto Rican origin are illustrated in Figure 4.14. The process is found in all populations so far investigated, although there are apparent differences in the absolute amount of bone remodeling.

Skeletal changes of this sort can be detected in both archeological samples, e.g., archaic/pre-contact Native American burials (Carlson *et al.*, 1976; Ruff, 1991), and paleontological collections (Ruff *et al.*, 1993; Trinkhaus *et al.*, 1994). Based upon the predictable relationship of these skeletal changes in the bone remodeling of women to menopause, a post-reproductive life history stage should be detectable in the fossil record.

Recovering these data may help settle the question of why human women have considerable longevity beyond menopause. Basically, there are two models for the evolution of menopause and a post-reproductive life stage. One model posits that a post-reproductive life stage could evolve if there are major risks to reproduction for an older female and if the old female can benefit her younger kin. The extraordinary duration of the human female post-reproductive life stage correlates with cross-cultural ethnographic research showing the crucial importance of grandparents as repositories of ecological and cultural information, and the value of grandmothers for child care (see above). Hamilton (1966) formalized the 'grandmother model' into a hypothesis based on his models of kin selection theory, but until recently the hypothesis was not tested scientifically. Nishida *et al.* (1990) demonstrate that a few wild chimpanzee females have a post-reproductive stage. They point out that the kin selection hypothesis does not correlate well with chimpanzee behavior: '. . . the evolutionary advantage of menopause [sic] in female chimpanzees is puzzling, since they rarely, if ever, care for younger relatives . . . aged females typically live a lonely life . . .' (p. 95). In an attempt to test the kin selection hypothesis with human data, Hill & Hurtado (1991) were unable to show that it would ever be advantageous to stop reproducing altogether. The authors used several hypothetical models that covered the range of reasonable estimates of maternal cost versus grandmother benefits. They also tested their predictions against objective ethnographic data derived from their work with the Ache, hunter-gatherers of South America. The Ache data show that offspring with grandmothers survive at somewhat higher rates than those without grandmothers, but the effect is not nearly enough to account for menopause. In a recent review of the Ache data and other cases derived from hunting–gathering and agricultural societies Austad (1994, p. 255) finds no evidence '. . . that humans can assist their descendants sufficiently to offset the evolutionary cost of ceasing reproduction'.

The second model for menopause may be termed the 'pleiotropy hypothesis'. In now classic works on the biology of senescence Medwar (1952) and Williams (1957) argued that aging is '. . . due to an accumulation of harmful age-specific genes . . . [or] . . . pleiotropic genes which have good effects early in life, but have bad effects later . . .' (Kirkwood & Holiday, 1986, p. 371). Kirkwood (1977), Charlesworth (1980) and others refined this hypothesis further in terms of a general theory of aging. Pavelka & Fedigan (1991) apply this line of reasoning to menopause. According to their application of the 'pleiotropy hypothesis', menopause is a secondary consequence of the female mammalian reproduction system. This system has a physiological limit of about 50 years because of limitations on egg supply or on the maintenance of healthy eggs. Female mammals produce their egg supply during prenatal development, but suspend the meiotic division of the eggs in anaphase. Approximately one million of these primary oocytes may be produced but most degenerate, so that in the case of female humans only about 400 are available for reproduction. During human adolescence and adulthood the remaining primary oocytes complete their maturation and are released in series during menstrual cycles. By about age 50 all of these eggs are depleted. If the woman lives beyond the age of depletion she will experience menopause. The physiological connection between oocyte depletion and the hormonal changes of menopause have yet to be elucidated. However, the pleiotropy hypothesis does account for the observation that few female mammals reproduce after 40 to 50 years of age, even though some species, such as humans, may live another 25 or 50 years.

Menopause and the post-reproductive life stage of women are, then, an inevitable consequences of the age-limited reproductive capacities of all female mammals. However, even if menopause is a pleiotropic consequence of mammalian reproduction, grandmotherhood may still be an important biological and sociocultural stage in the human female life cycle. The universality of human menopause makes it possible to develop biocultural models to support a combination of the pleiotropy and 'grandmother' hypotheses. Basically, if a 50-year age barrier exists to female fertility, then the only reproductive strategy open to women living past that age is to provide increasing amounts of aid to their children and their grandchildren. This strategy is compatible with Hamilton's kin selection hypothesis. The analyses cited above show that kin selection alone cannot account for the evolution of menopause or grandmotherhood. Holly Smith and I (Bogin & Smith, 1996) proposed a biocultural model for the evolution of grandmotherhood that combines the pleiotropy and kin selection hypotheses. In favor of our biocultural model is the ethnographic evidence cited above showing that significant numbers of women in virtually every society, traditional or industrial, live for many years after menopause. Moreover, the ethnographic evidence also shows that grandmothers and other post-reproductive women are beneficial to the survival of children in many human societies (Wolankski & Bogin, 1996). Old women control important cultural information and experience that are of value to the entire society. In past times, this may be especially the case during periods of food scarcity, epidemic disease, and other threats to the society that may occur infrequently during the lifetime of any individual.

Little comparative mammalian data on the value of grandmotherhood exists because the females of the wild-living species of primates, and other social mammals, only rarely survive to a post-reproductive stage. There are some exceptional species, for example, hyenas. Grandmother caretaking occurs in this species, including the nursing of grandoffspring. Indeed, when both are still fertile, mother and daughter hyenas take turns nursing each other's young (Mills, 1990). It is not known if this practice is wide-spread in other social carnivores. Nevertheless, the point is that when females do survive regularly past their reproductive stage of life, the basis for affiliative behaviors, including some grandmother interaction and care of young, exists in social mammals.

During hominid evolution, a post-reproductive life stage of significant duration, and menopause, became commonplace as life expectancy increased beyond 50 years. When this occurred is not known, but may be investigated as described above. The regular occurrence of a post-reproductive female hominid life stage would select for the females (and males?) of the species to develop biocultural strategies to take greatest advantage of this situation. Viewed in this context, human grandmotherhood may be added to human childhood and adolescence as distinctive stages of the human life cycle.

Conclusion

Perhaps the best summary of the importance of taking a life history perspective of human evolution was stated by Bonner (1993, p. 93): 'The great lesson that comes from thinking of organisms as life cycles is that it is the life cycle, not just the adult, that evolves. In particular, it is the building period of the life cycle – the period of development – that is altered over time by natural selection. It is obvious that the only way to change the characters of an adult is to change its development'. The stages of human postnatal life from birth to maturity – infancy, childhood, juvenile, and adolescent – shape the biology and behavior of adults and confer upon them greater reproductive success than any other mammalian species.

Human reproductive success is due to the biocultural adaptations of our species. These adaptations may have arisen as both a consequence of, and a response to, the evolution of the human life cycle. The stages of the life cycle and the growth patterns of the human body, the face, and the brain facilitate parental investment in offspring by releasing the potential for nurturing behavior of adults towards infants and older, but still physically dependent, children and socially dependent juveniles. Human culture, in large part, is a response to the need to nurture, protect, and teach these young people. The physical features of childhood and juvenile stages are lost during the time of the adolescent growth spurt. At the end of adolescence, boys and girls enter the social world of men and women. In physical features, interests, and behaviors these young adults are more similar to their parents than to their pre-adolescent selves of just a few years ago. Each new generation follows the cycle of reproduction, growth and maturation that was phylogenetically set in place millions of years ago, and continues to be expressed in the ontogenetic development of every human being born today.

5 Growth variation in living human populations

Population differences in stature, body weight, and other physical dimensions have been documented throughout recorded history. Ancient Egyptian sources mention groups of very short stature people living near the headwaters of the Nile River, possibly ancestors of central African 'pygmy' populations alive today (Hiernaux, 1974). Museum displays of medieval armor and fashion often provoke visitors to comment on how much bigger European people are today than in the past. Human biologists have recorded the variation in size that exists between living populations, and found that it is relatively easy to describe the differences in size, but much more difficult to explain why this variation exits. The causes of population differences in body size, including variation in amounts and rates of growth, are due to a wide range of hereditary and environmental factors. Examples of some of these factors were given in previous chapters. The present chapter begins with a review of some studies of population variation in growth and development and follows with a discussion of the interaction between hereditary and environmental causes of such variation. A consideration of the evolutionary value of population variation in body size completes this chapter.

Population differences in growth and development

The average height, weight, and weight/height (a simple measure of body proportion) for several human populations are given in Table 5.1. The data are listed in descending order according to the average height, an order followed by both the men and the women. Young adults in The Netherlands may be, on average, the tallest people in the world. Some of the shortest young Dutch men, those at the third percentile, have a height of 169.3 cm and, thus, are taller than the mean stature of Aymara men in Bolivia and Maya men in Guatemala. Young adults in the United States are, on average, shorter than the Dutch, but, relative to the other populations, are a 'tall' group of people. The Turkana, who are nomadic, animal herding pastoralists living in rural Kenya, are one of the tallest populations

225

Table 5.1. *Average height (cm), weight (kg) and weight/height, of young adult men and women in several populations*

Population	Age (years)	Height Men	Height Women	Weight Men	Weight Women	Weight/Height Men	Weight/Height Women
Netherlands, national sample, 1980 (medians)	20	182.0	168.3	70.8	58.6	0.39	0.35
United States, national sample, 1977 (medians)	20–1	177.4	163.2	71.9	57.2	0.41	0.35
Africa, Turkana pastoralists, 1970s (means)	20	174.3	161.6	49.8	47.4	0.29	0.29
Japan, Univ. of Tokyo students, 1995 (means)	*ca.* 20	171.6	159.1	63.3	50.7	0.37	0.32
Bolivia, Aymara Indians, 1970s (means)	20–9	162.0	149.0	58.1	52.4	0.36	0.35
Guatemala, Maya Indians, 1980s (means)	17–18	158.7	146.9	52.2	49.3	0.33	0.34
Africa, Efe Pygmy, 1980s (means)	19–29	144.9	136.1	43.3	40.6	0.30	0.30

Sources: Netherlands, Roede & van Wieringen (1985); United States, Hamill *et al.* (1977); Africa, Turkana, Little *et al.* (1983); Japan, T. Satake (pers. comm.); Guatemala, Bogin *et al.* (1992); Bolivia, Mueller *et al.* (1980); Africa, Efe, Dietz *et al.* (1989).

of Africa. The Tutsi of Rwanda are about two centimeters taller, on average, than the Turkana. It is a myth that Tutsi (sometimes called the Watutsi) are the tallest people in the world, and that they average more than 213 cm (7 feet) tall (Bogin, 1998). The sample of Japanese represent reasonably affluent university students. They are the tallest and heaviest, on average, of any group of young Japanese adults measured in this century, but they are considerably shorter and lighter, on average, than the Dutch or the Americans.

The Aymara of Bolivia and the Maya of Guatemala are native American peoples. Both groups are of very low socioeconomic status (SES). Both live in rural areas and many individuals suffer from mild-to-moderate malnutrition, along with repeated bouts of infections of the gastrointestinal and respiratory systems. Undernutrition and infectious disease are associated with growth retardation, and these are likely to be factors that account for the relative short stature of the Maya and Aymara. The African Efe pygmies may be, on average, the shortest people in the world, and their short stature appears to have a strong genetic component (Rimoin *et al.*, 1968; Merimee *et al.*, 1981; Hattori *et al.*, 1996). However, there is a wide range of variation in the stature of individual pygmies. Barnicot (1977) compared the distribution of male stature of the pygmies with the Tutsi, who have a mean male stature of 176.5 cm, and found that the tallest pygmy men were larger than the shortest Tutsi men. This analysis shows that average figures may be quite misleading for individuals within a population. Even so, the statistics for average size given in Table 4.1 indicate patterns of growth that are useful for descriptive purposes, and provide a starting point for the analysis of the causes of such variation.

The rank order of mean weights in Table 5.1 does not follow the same order as stature. On average, United States men are the heaviest and Efe women are the lightest of all the populations listed. The relatively tall Turkana men have a lower average weight than that of any other samples, save the Efe. Turkana women have a lower average weight than all samples, except the Efe and the Maya women. The ratios of weight for height show that the Turkana and the Efe have the lowest values, reflecting their linear body build. The Turkana, Efe, and many other sub-Saharan African peoples have arms and legs that are relatively long in proportion to their total stature (Eveleth & Tanner, 1976, 1990). The Turkana and the Efe are also absolutely lean, meaning that their bodies have less fat tissue than other human populations. Together, body proportions and body composition give these two groups a linear body build. The similarity in the proportion of height to weight between these African samples is striking, since the Turkana are, on average, 25.5 cm taller than the Efe pygmy sample.

Dutch and American men have the highest average ratios, meaning that there is, on average, relatively greater weight for height in these two populations than in the other samples of young men. Little *et al.* (1983) compared the Turkana with a United States reference population and found the greater weight for height of the Americans was due to both more fat and more lean tissue (e.g., muscle), but especially more fat. Since the year 1977, when the United States data were published, Americans have not changed in stature but have increased in mean weight. Today, Americans are the fattest population of all the industrialized countries (IBNMRR, 1995). Japanese university students, both men and, especially, women have lower average weight-for-height ratios than similarly aged Dutch or Americans. Why this is so is not known exactly. The Japanese sample represents a highly educated group, while the Dutch and American data are based on national samples. More highly educated people tend to be less fat than the population at large, at least in many industrialized nations.

Aymara and Maya men have weight/height ratios that are lower than for Japanese men, but the women have ratios that are higher than for Japanese women and virtually equal to the ratio for United States women. One factor influencing these ratios is that relative to the Dutch, Americans, and Japanese, the Aymara and Maya have short arms and legs in proportion to total stature (Eveleth & Tanner, 1976; Gurri & Dickinson, 1990). This means that the head and trunk of the body contribute disproportionately more to total weight. The Maya, and probably the Aymara as well, have less total body fat, on average, than Europeans or Americans, but they have more of it concentrated on their trunks (Bogin & MacVean, 1981b; Johnston *et al.*, 1984). This results in what is sometimes called a 'short and plump' physique, which elevates the weight-for-height ratio. In reality, adult Maya and Aymara develop this physique as a result of malnutrition and growth retardation in early life, and as adults they are absolutely shorter and lighter than populations of the industrialized nations.

Population differences in rate of growth

Rates of growth also vary considerably between human populations. Velocity of growth in height is presented in Figure 5.1 for four groups of boys: two samples from Guatemala, one of high SES Ladino boys and one of low SES Maya boys (Bogin *et al.*, 1992), a sample of low SES boys from The Gambia, West Africa (Billewicz & McGregor, 1982), and a sample of low SES boys from rural India (Satyanarayana *et al.*, 1989). The velocity curves were calculated from longitudinal measurements of height by fitting the measurements to the Preece–Baines model 1 function using algorithms developed by Brown (1983). This mathematical function estimates the mean-constant velocity (that is, the true average curve for longitudinal data) of growth from childhood to the attainment of adult height. Further detail about the Preece–Baines function is given in Box 5.1.

The Maya, the rural Indians (from the Hyderabad region of India), and the Gambians are described as suffering from poor living conditions, including high rates of disease and chronic undernutrition. In contrast, the high SES Ladino boys are generally healthy and well nourished. The Mayan, rural Indian, and rural Gambian boys have slower velocities of

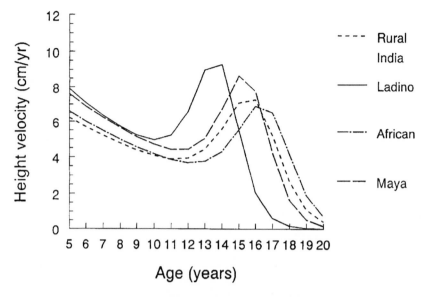

Figure 5.1. Mean-constant curves, estimated by the Preece–Baines model 1 function, for the velocity of growth in of Ladino and Maya boys from Guatemala, rural Indian boys, and rural Gambian boys.

growth during all stages of development. These three slower growing groups also have a longer period of childhood, juvenile, and adolescent growth; note that the Preece–Baines function estimates that growth continues beyond age 20 years. Despite this the Mayans, Indians, and Gambians end up significantly shorter than the high SES Ladinos (estimated mean adult heights are: high SES Ladino, 176.9 cm; Maya, 169.15 cm; rural Indians, 158.2 cm; Gambians, 170.8 cm).

Mean-constant velocity curves for high SES Ladina girls, Mayan girls, and rural Gambian girls are shown in Figure 5.2. Compared with Ladina girls, the Mayan and Gambian girls show a general pattern of slower growth and delayed maturation, although this is more pronounced for the Gambians. An exception to this general pattern is that Mayan girls grow more rapidly than the Ladina girls at age eight. Despite that anomaly, Mayan and Gambian girls are shorter at all ages, including adulthood, than Ladina girls (the estimated mean adult heights are: high SES Ladina, 162.95 cm; Maya, 151.8 cm; Gambian, 158.8 cm).

The difference in height between Mayans and Ladinos may be due, in part, to genetic determinants of amount of growth. However, it is not possible to assess that determination in these samples. Moreover, an explanation that relies heavily on the genetic limitation of Mayan growth is

Box 5.1.   Mathematical models of human growth

Preece & Baines (1978) derived a series of mathematical functions that describe the distance and velocity curves of growth in height from the ages of two years to maturity. The authors describe their method of derivation as, '. . . purely empirical and has made no pretense to true biological meaning' (p. 17). Marubini & Milani (1986) describe the Preece–Baines curves as being based on the assumption that the rate of growth is proportional to the difference between height at any age prior to maturity and height at maturity. The rate of growth is not a simple constant of proportionality, rather it is a function of age. This means, that at different ages the rate of growth may be relatively slow, or relatively fast, compared with other ages, which is the manner in which children actually grow. A set of differential equations was used to calculate the age function for the rate of growth, and the solution yields three functions, of which the following is preferred by Preece & Baines for application to growth data:

$$h = h_1 - 2(h_1 - h_c)/\{\exp[s_0(t - c)] + \exp[s_1(t - c)]\}$$

There are five parameters to be estimated in this model: (1) h is height at time t, (2) h_1 is final (adult) height, (3) s_0 and s_1 are rate constants, (4) c is a time constant, and (5) h_c is height at $t = c$. Although the model was derived empirically, Preece & Baines were able to correlate each of the five parameters of the model with 'biological' events that occur during growth. The rate constants s_0 and s_1 are highly correlated with the minimal prepubertal velocity of growth and the peak growth velocity during adolescence, respectively. Time c has a very high correlation to the age at peak height velocity during the adolescent growth spurt and h_c has a similarly strong association to height at peak height velocity. The relationship of these parameters to the velocity curve of growth are illustrated in Figure B5.1. By computation, other growth events may be estimated, such as age at the minimal prepubertal velocity (MPV), height at MPV, and amounts of growth between MPV, peak height velocity, and final height. The values of these estimates are useful when comparing the pattern of growth of one individual to another, or one population to another.

There are several other mathematical functions that attempt to describe growth. One is called the triple-logistic function (Bock & Thissen, 1976). The authors built on ideas proposed by Robertson (1908) and Burt (1937) that human growth could be described with the use of two or three

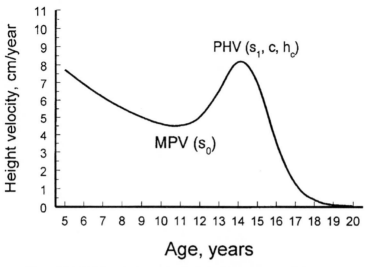

Figure B5.1. Height velocity curve of a boy with parameters of the Preece–Baines model indicated. MPV, minimal prepubertal growth velocity; PHV, peak height velocity during the adolescent spurt. Other terms as defined in text.

logistic curves. Since these curves have the same functional form they may be added together to produce a smooth model of human growth. Bock *et al.* (1973) developed a double-logistic function to describe growth in length. This model is the summation of two logistic curves; the first describes '. . . a component of prepubertal growth which continues in reduced degree until maturity, and the second term describes the contribution of the adolescent spurt' (pp. 64–5). This model fit the distance curve of growth with reasonable precision, but did not estimate well velocities of growth in height. Bock & Thissen (1976) refined the model by adding a third logistic term which, essentially, divided the prepubertal period into two components. The addition of the third term improved the fit of the model to both the distance and velocity curves of growth. The triple-logistic model has ten parameters to be estimated, and adult height must be known. In many cases, the analysis of longitudinal growth data for children does not include a measurement of final adult height and, often, there are ten or fewer measurements of growth. Thus, more mathematical parameters must be estimated than there are empirical data points, which is statistically undesirable from both a practical and theoretical standpoint.

When appropriate data are available, the triple-logistic model proves to be useful in describing distance and velocity curves of growth. The model is also of value since different growth periods, such as infancy, childhood–juvenile, and adolescence, may be described with precision. Velocities of growth during each period, the age of maximum velocity, and the contribution of each period of growth to final adult height may be estimated. The transition from one growth period to the next is smooth, and the contributions of each period may overlap, that is, the contribution of prepubertal growth is still active during the early phase of the adolescent growth period. Bock & Thissen, (1980) speculate that the genetic and endocrine determinants of development that characterize each period of growth operate in a similar fashion; making a smooth transition, and overlapping from one period to the next.

Several other mathematical models worth mentioning were developed by Jolicoeur *et al.*, 1992. Some of their models estimate growth in length from fertilization to adult height. These models are useful when prenatal growth in length needs to be studied. Other versions of their models estimate growth in height from birth to adulthood. The latter model, called the JPA-2 model, has only eight parameters. Jolicoeur and colleagues find that the JPA-2 model achieves a more accurate goodness-of-fit to real growth data than the ten-parameter triple-logistic model. As for the Preece–Baines function, the JPA-2 model is purely mathematical, with no pretext that its parameters have any biological meaning.

A discussion of mathematical models of growth with biological meaning is presented in Chapter 7.

not particularly useful, as shown by the enormous plasticity in growth of Maya refugees in the United States (described in the Introduction and re-evaluated later in this chapter). The low SES of the Mayans of Guatemala correlates with their chronic mild-to-moderate undernutrition, higher rates of disease, and generally unfavorable environment for growth. In this regard the Mayan pattern of growth in height is similar to that of other disadvantaged populations, such as the rural Indians and Gambians depicted in Figures 5.1 and 5.2. The result is shorter stature at all stages of growth for the Mayans, rural Indians, and Gambians compared with the high SES Ladinos. Because of the wide geographic, ethnic, and sociocultural differences between these groups, and despite possible genetic differences, it seems that it is the shared negative environment for growth that produces the similar pattern in amount and rate of growth in height in the Mayan, Indian, and Gambian samples.

Differences in growth between boys and girls

Population differences in growth may also be considered in relation to sex, in that we may treat boys and girls as belonging to separate biological and

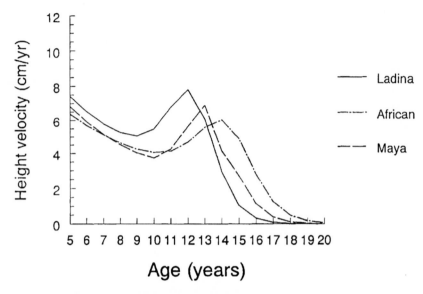

Figure 5.2. Mean-constant curves, estimated by the Preece–Baines model 1 function, for the velocity of growth of Ladino and Maya girls from Guatemala and rural Gambian girls.

social populations. In the analysis of Guatemalan boys and girls just described, Bogin and colleagues found that the overall environmental effect interacts with sex. The difference in height between Mayan and Ladino boys is established during childhood and remains fairly constant during juvenile, adolescent, and adult stages. In contrast, the difference in height between the Mayan and Ladina girls increases from childhood to adolescence and reaches its maximum at adulthood. The cause of this interaction is not known. However, the joint effect of two hypotheses about human growth may be considered to propose a reason for the interaction. One hypothesis is that growth during the childhood and juvenile stages is more sensitive to environmental factors and growth during adolescence is determined more by genetic factors. The discussion of the evolution of childhood and plasticity of growth in Chapter 4 lends theoretical support to this hypothesis and there are several empirical studies which directly support the hypothesis (Johnston *et al.*, 1976; Martorell *et al.*, 1977; Frisancho *et al.*, 1980). Of course, growth at all stages of development is controlled by the interaction of genetic and environmental factors. So, it is more accurate to state that this hypothesis apportions the relative weight of the contribution of genes and environment to this interaction differently during the pre-adolescent and adolescent stages of growth. The other hypothesis is that girls are better 'buffered' against environmental determinants of growth, especially negative influences such as undernutrition and disease, than boys. In the present context, 'buffering' means that the growth of girls is less likely to be knocked off-track by harmful environments. The theoretical and empirical evidence in favor of this hypothesis is reviewed by Stinson (1985). Suffice it to state here that the growth of boys, from prenatal life to adulthood, appears to be more 'sensitive' to environmental factors than the growth of girls. Why this is so is not known.

The first hypothesis predicts the observation that Mayan boys and girls are shorter than Ladino children during childhood. The environment for growth of the low SES Mayans may be so powerfully negative that despite the girls' 'buffering' (hypothesis 2) their growth deficit is about the same as that for the Mayan boys. Some evidence of that buffering does exist, in that the Mayan and Ladina girls do not differ in growth velocity at the start of the adolescent spurt, the so-called age at 'take-off'. 'Take-off' is that point at the end of the juvenile stage when growth velocity changes from decelerating to accelerating. In contrast to the girls, Mayan and Ladino boys do differ in growth velocity at 'take-off'. However, neither Mayan boys nor girls differ from Ladino boys or girls in the velocity of growth at PHV. That may be evidence of the relatively stronger contribution of a genetic determination of growth at that time. Compared with Ladino boys, the Mayan boys continue to show developmental delays as adolescence proceeds, for instance Maya boys have a later age at PHV. The Mayan boys' delayed maturation is probably due to the continuation of the negative environmental milieu in which they live. The developmental delay is not prolonged enough to compensate for the reduced rate of growth during childhood. Consequently, the height difference between Mayan and Ladino boys is maintained until adulthood.

In contrast to the Mayan boys, Mayan girls pass through adolescence at about the same rate as Ladina girls. The age at 'take-off' and PHV for the Mayan girls is a bit later than for the Ladinas, but the difference is not significant statistically. Perhaps this is due to the girls' genetic 'buffering' against the environment, combined with the relatively greater genetic determination of growth at adolescence. These two genetic factors working together may override the maturation delaying effects of the environment. Thus, Mayan girls proceed through adolescence more rapidly; that is, more in accordance with the genetically determined timing for development, than the Mayan boys. The Mayan girls' amount of growth is less than that of the Ladina girls during adolescence because the Mayans have a shorter time interval between 'take-off' and PHV. As a result, the difference in height between Mayan and Ladina girls increases to a greater extent than the difference in height between Mayan and Ladino boys.

Population variation in skeletal, dental, and sexual maturation

Other aspects of biological maturation may differ from one population to another. Masse & Hunt (1963) found that the dental and skeletal development, as measured by radiographs of tooth formation and of the appearance of ossification centers in the hand and wrist, of African blacks were, on average, more advanced at birth than European whites. However, by two to three years of age the Africans fell behind Europeans in these developmental measures. Jones & Dean (1956) and Garn & Bailey (1978) suggested that African blacks are 'genetically programmed' to develop more rapidly than European whites, however a postnatal delay in development occurs in the Africans, compared with the Europeans, due to an adverse nutritional, disease and socioeconomic environment for growth of the African children.

When infants and children of European ancestry (whites) and predominantly African ancestry (blacks) living in the United States are matched for socioeconomic status variables, such as mother's occupation and education, it is found that the black children are, on average, consistently advanced over white children in the formation and emergence of the permanent teeth, and advanced in radiological appearance of ossification centers of the skeleton and epiphysial union (Garn & Bailey, 1978). A tendency for precocity in dental and skeletal development of black over white children can also be observed during the prenatal period. These findings lend support to the hypothesis that a difference in the 'genetic programming' for rate of maturation exists between people of African and European descent.

A more recent review by Gillett (1998) of dental development finds that the average age at tooth emergence of populations in Africa is consistently advanced over that of populations in North America and Asia. The African populations live under generally less favorable environmental conditions than the other groups, and to the extent that poor living conditions tend to delay both growth and development, the findings are somewhat surprising. Especially puzzling is the fact that African-Americans are delayed relative to Zambians and Ugandans in the emergence of several teeth. There are some important factors that confound the relationship between genes, the environment, and age at tooth emergence. One is that tooth emergence is a brief and fleeting event. The criteria used to assess when, and by how much, a tooth has emerged through the gum can, and does, vary greatly from one study to the next. Data reviewed by Gillett come from eight different studies, spanning the years 1919 to 1995. These methodological variations can significantly alter the reported mean ages of emergence (Demirjian, 1986). A second problem is that the loss of a deciduous tooth can accelerate the emergence of the underlying permanent tooth. If children from poorer families lose more deciduous teeth, then they would also have earlier permanent tooth eruption. This may account for the differences between Zambians and Ugandans on the one hand and African-Americans on the other. Accordingly, it is difficult to make a clear case for the relative importance of genetic and environmental influences on dental development.

Rates of maturation of different populations may also be compared by examining the stages of development of the breast, the genitals, and pubic hair. Assessing the rate of maturation by using these sexual development events serves as a complement to skeletal and dental maturation, and may be used as an alternative test of the 'genetic programming' hypothesis. A survey of sexual maturation stages in girls between the ages of 3.0 and 12.99 years old in the United States was conducted by Herman–Giddens and colleagues (1997). The sample included 15 439 'white' girls and 1638 'African-American' girls, who were seen for well-child examinations at 65 pediatric clinics in 26 states and Puerto Rico. The authors of the study do not explain how the 'race'/ethnicity of each subject was assessed. The sample is not truly nationally representative, but it is the largest and most geographically diverse sample ever assembled for the assessment of sexual maturation in the United States. Clinicians participating in the study were trained in the Tanner Staging method and tested for accuracy before data were collected. The mean ages for the onset of pubic hair development (the PH2 stage) and breast development (the B2 stage) was 8.78 and 8.87 years, respectively, for the African-American girls and 10.51 and 9.96 years for the whites. Statistical analysis shows that the differences in mean ages between African-Americans and whites are significant. African-American girls were advanced over white girls in this sample at all ages. Even at age 3.0 years, when no sexual development would be expected, three percent of the African-American girls showed either breast (B2) and/or pubic hair (PH2) development, compared with only one percent of white girls.

This United States based study supports the genetic programming hypothesis, at least for the onset of sexual development. The African-American girls were also advanced over whites for B3 and PH3 stages, and for the mean age at menarche (12.16 years versus 12.88 years). Ideally, one would like to have comparative data for all five of the Tanner stages of sexual development to see if the differences between populations are consistent from the onset to the end of sexual maturation. The timing of all

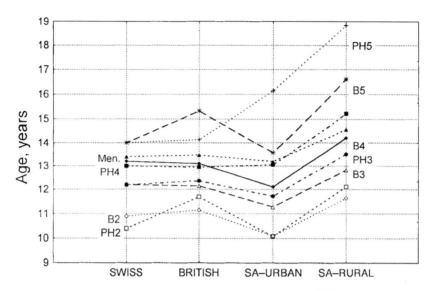

Figure 5.3. Pubertal stages for girls. The Swiss and British samples represent healthy, well-nourished European populations. The SA–Urban sample represents middle to high socioeconomic status South African blacks living in the Johannesburg area. The SA–Rural sample represents low SES South African blacks living in an impoverished rural area. The median age of attainment of the stages of breast development are indicated by B2 to B5. Median ages for the stages of pubic hair development are indicated by PH2 through PH5. The median age at menarche is indicated by 'Men'. Based on data published by Cameron *et al.* (1993).

five stages has been carefully studied for South African boys and girls by Noel Cameron and colleagues. In Chapter 2 their findings were discussed in relation to the sexual maturation of severely undernourished South African children. In a related study, Cameron *et al.* (1993) assessed the development of secondary sexual characteristics in 300 urban and 352 rural South African girls and boys. The sample ranged in age from 6.0 to 19.0 years and the ethnicity of both groups is described as 'black'. The rural children are of low SES, but of apparent good health as no child showed 'any overt sign of illness' (p. 585). The urban sample all attended private, fee-paying schools in Soweto and Johannesburg, indicating they come from middle- to upper-middle class families. The South Africans were compared with Swiss and British children, who represent healthy, well-nourished populations. The results are presented in Figure 5.3 for the girls and Figure 5.4 for the boys.

The South African urban girls show signs of sexual development, both

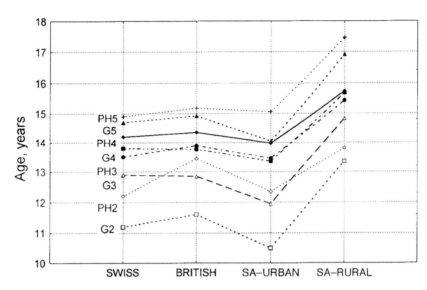

Figure 5.4. Pubertal stages for boys. The Swiss and British samples represent healthy, well-nourished European populations. The SA Urban sample represents middle to high socioeconomic status South African blacks living in the Johannesburg area. The SA Rural sample represents low SES South African blacks living in an impoverished rural area. The median age of attainment of the stages of genital development are indicated by G2 to G5. Median ages for the stages of pubic hair development are indicated by PH2 through PH5. Based on data published by Cameron *et al.* (1993).

PH2 and B2, at the youngest ages. The PH2 stage occurs 0.32 years earlier than for the Swiss girls. The first sign of sexual development for the British girls is the B2 stage, but even this occurs 1.06 years later than the B2 stage of the South African urban girls. Of these three better fed and healthy groups, the South African urban girls complete their sexual development at the latest age, reaching the PH5 stage 2.14 years after the Swiss girls. In fact, it take 6.1 years for the South African urban girls to proceed from PH2 to PH5, but it takes the Swiss and British girls only 3.6 years to complete all stages of sexual development.

   The data for boys shows the same basic pattern as for the girls. South African urban boys have the earliest onset of sexual maturation, with G2 occurring 0.7 years earlier than for the Swiss boys, 1.11 years earlier than for the British boys. All three groups reach the PH5 stage within 0.28 years of each other. The South African urban boys take 4.56 years to complete their sexual development, which is longer than the Swiss boys by 0.85 years, and longer than the British boys by 0.99 years. The rural South African boys begin puberty at the latest age and reach each stage of development at the latest age. Even so, these rural boys complete all the stages in 4.1 years, meaning that once started, their rate of sexual maturation is more rapid than the urban South African boys.

Given these data, the earlier onset of puberty of the South African girls and boys supports the 'genetic programming' hypothesis, but the later termination of adolescence of these same girls and boys negates the hypothesis. Turning to the data for the rural South Africans we see that for both the girls and the boys every stage of sexual maturation is delayed compared with the other three samples. The poorer living conditions of this rural sample seems to override any genetic predisposition to early sexual maturation. So, what is one to make of the 'genetic programming' hypothesis for earlier maturation in Africans, or African-Americans? The answer is, 'interesting', and possibly of importance in the clinical assessment of normal versus early or delayed sexual development. If we broaden our coverage to include samples of healthy boys and girls from other geographic regions the African-European comparisons become less interesting. Data compiled by Eveleth & Tanner (1990) indicate that Turkish girls begin sexual maturation, that is achieve the B2 stage, at a median age of 10.0 years, which is 0.09 years earlier than for urban South Africans. Girls in São Paulo, Brazil have an even earlier mean age for the onset of puberty, 9.7 years for the B2 stage. São Paulo boys have the earliest mean age for puberty onset of any population, with a G2 stage at 9.1 years. Brazilians have a diverse heritage, including African, European, Native American, and Asian genetic and cultural admixture. Viewed in this more global context, the African-European differences in rates of maturation are just a small part of the larger variability and plasticity in development that is found across many human populations.

Hereditary and environmental interactions as the cause of population variation

The factors responsible for individual and population variation in adolescent growth and development in all populations include heredity, nutrition, illness, socioeconomic status, and psychological well-being. The timing of menarche is, perhaps, the best-studied adolescent event known to be affected by these factors (no similarly well-marked and dramatic event occurs for boys). From a study of monozygotic (MZ) and dizygotic (DZ) twin girls, Tisserand-Perier (1953) and Fischbein (1977) showed that the difference in age at menarche was 2.2 months for MZ twins and 8.2 months for DZ twins.

MZ twins are genetically identical, while DZ twins share, on average, only half of their genes. Presumably, it is the genetic identity of the MZ twins that is responsible for their much greater concordance in menarchial age.

Nutrition, illness, and socioeconomic status are often linked together in human populations. For instance, poverty in the developing nations, and in the developed industrial nations, is almost always associated with high rates of undernutrition, high rates of infectious diseases, and illiteracy. Menarche is achieved at about 12.5 years of age in girls of the middle-class from many nations, and at 14 years of age or later in girls of the lower socioeconomic classes (see reviews by Johnston, 1974 and Eveleth & Tanner, 1976). The latest median age of menarche on record is 18 years of age for girls from the Bundi tribe of highland New Guinea (Malcolm, 1970). Malnutrition, heavy labor, infectious disease, and living at high altitude are some of the reasons for delayed maturation of this group. Since the 1970s, the living conditions of the Bundi have improved, especially in relation to nutrition and medical care, and the median age of menarche has declined (Worthman, 1993).

Finally, physical and psychological stress influence growth and development, including menarche and menstruation (Ruble & Brooks-Gunn, 1982). Malina *et al.* (1973) found that highly competitive female track athletes, who enter training before puberty, reach menarche later than girls in the general population. Frisch *et al.* (1980) found menarche is delayed in highly trained and competitive ballerinas. One explanation for the delay in menarche of these groups of girls is the stress of exercise. Several hormones, including progesterone, prolactin, and testosterone, are elevated by strenuous physical activity. These same hormones are known to delay the onset of menstrual cycles (Scott & Johnston, 1982). Psychological stress may also be a cause of delayed menarche. Prima ballerinas are extremely sensitive about their weight and body image, and compulsive behavior in relation to these and their dance is often reported (Warren, 1980). Post-menarchial ballerinas may stop menstruating before a performance, even before they begin intense training or dieting (Scott & Johnston, 1982). Adequate studies to determine the psychological component of menarche and menstrual regularity have not been carried out; however, the existing data are suggestive.

The old debate between the relative importance of genes versus the environment in human development is largely ignored by most researchers today. In reality, the biological development of the human being is always due to the interaction of both genes and the environment. It is erroneous to consider whether one or the other is more important; genes are inherited and '. . . everything else is developed' (Tanner, 1978, p. 117). The human **phenotype** is the outcome of this interaction. All of the measurable characteristics of the human body, of human behavior, and of the human mind are phenotypic traits. Although it is possible to describe a human **genotype** with great precision, such knowledge by itself provides very little information about how a human being will develop without also knowing about the environment with equal precision.

Body proportions

Comparisons of stature and body proportion between blacks and whites in the United States provide another example of gene–environment interactions and their affect on growth. Fulwood *et al.* (1981) published data from the first National Health and Nutrition Examination Survey (NHANES I) of the United States, which gathered anthropometric data on a nationally representative sample of blacks and whites aged 18 to 74 years. When the data are adjusted for differences between the two ethnic groups in income and education, urban or rural residence, and age, there is no significant difference in average height between black and white men. Nor is there a significant difference in average height between black and white women.

Although white and black adults in the United States have the same average stature, when education, income and other variables are controlled, the body proportions of the two groups are different. Krogman (1970) found that for the same height, blacks living in one American city had shorter trunks and longer extremities than whites, especially the lower leg and forearm. Hamill *et al.* (1973) found that this was also true for a national sample of black and white youths 12 to 17 years old. A genetic cause for the body proportion differences between blacks and whites seems likely, as the samples were matched for major environmental determinants of growth, and statures did not differ.

Differences in body proportion are known from other populations. Eveleth & Tanner (1976) surveyed studies of boys and girls of European (London), African (Ibadan), Asian (Hong Kong) and Australian Aborigine origin. In proportion to sitting height, the Australians had the longest legs followed, in order, by Africans, Europeans, and Asians. Expressed quantitatively, 'at a sitting height of 60 cm, for example, London boys have leg lengths averaging 43 cm, Ibadan boys 53 cm and Australian Aborigine boys 61 cm' (Eveleth & Tanner, 1976 p. 229). Over the past century, many researchers have assumed that the body proportion differences between geographic populations are explainable only in terms of a genetic model, though the mechanism is not known.

Recent work on the growth of Japanese children suggests that environmental factors also may be powerful determinants of body proportion.

Kondo & Eto (1975) found that between the years 1950 to 1970 the ratio of sitting height to leg length decreased for Japanese schoolchildren, meaning that the children became relatively longer-legged with time. Tanner *et al.* (1982) confirmed this finding by comparing both the rate of growth and the amount of growth for Japanese school children measured in 1957, 1967, and 1977. Each successive cohort of children grew faster, and grew larger, than the previous cohort. Between 1957 and 1977, sitting height showed practically no increase, while increased leg length accounted for almost all of the difference in height (4.3 cm for boys and 2.7 cm for girls). In 1977, adult Japanese had sitting height to leg length proportions similar to Northern Europeans, whereas 20 years earlier, the two populations were significantly different. The major influences on growth that changed during the past two decades are improvements in nutrition (especially greater intakes of both protein and energy), health care, and sanitation (Kimura, 1984; Takahashi, 1984).

Similar findings on the plasticity of body proportions are reported by several researchers working in Argentina, Poland and Mexico. A team of researchers in Buenos Aires measured 569 boys and girls, seven to 13 years old, attending several schools (Bolzan *et al.*, 1993). The sample was divided in groups according to age, sex, and occupational status of the father. Both boys and girls with fathers of lower occupational status (lower SES) were shorter, and especially shorter in leg length, than subjects of higher family SES. In Poland (Wolanski, 1979), improvements in living conditions in towns and villages, such as nutrition and health care, are associated with increases in leg length relative to stature. The leg-length-to-stature proportions of women living in Chiapas Mexico, all of low SES, differs according to the type of work the women do (Gurri & Dickinson, 1990). Essentially the same findings are reported for children and adults living in the Yucatan (Dickinson *et al.*, 1990; Murguia *et al.*, 1990). Another series of analyses finds that the body proportions of boys and girls, of both the Yucatan Maya and non-Maya populations, are influenced by family SES. Children from higher SES families are longer-legged than children from lower SES families (Wolanski *et al.*, 1993; Wolanski, 1995; Siniarska, 1995).

Size versus shape

After 1977, the average height of Japanese men and women continued to increase, but at a slower rate (Takaishi, 1995), and both leg length and sitting height seem to have increased at about the same rate, at least for young women (Hojo *et al.*, 1981). Since 1990 there is little evidence for further increase in stature. This means that the body proportions of Japanese and Northern Europeans remain similar, even though the two populations differ in mean stature (Table 5.1). It seems, then, that if there exists a genetic difference in growth between contemporary Japanese and some European populations it is more likely to be for total size, rather than for body proportions. However, it is difficult to establish a genetic difference when comparing populations that live on different continents, since important environmental variations that can affect growth are likely to exist.

Another approach to the study of population differences in growth and development is to compare children and adults of different national or geographic backgrounds living in the same, or very similar, environments. The differences found may indicate an hereditary determination of amount and/or rate of growth. Ashcroft & Lovell (1964) and Ashcroft *et al.* (1966) measured the heights and weights of four- to 17-year-old children and youths of European, African, Afro-European and Chinese background living in Kingston, Jamaica. All the children were from upper-middle to upper socioeconomic status and attended private fee-paying schools. There were no significant differences in height or weight between the European, African and Afro-European groups. However, the Chinese sample was significantly shorter and lighter than the other three groups at almost every age, suggesting a hereditary difference in amount or rate of growth between the Chinese children and the other samples of children.

More recent surveys of the growth of Chinese, Japanese, and other Asian children, adolescents, and adults find that they remain shorter and lighter, on average, than European and African populations living in the same cities or nations. However the differences in mean size have narrowed over the past 30 years (Eveleth & Tanner, 1990). This is too short a time for genetic change, and indicates that other factors must be influencing the size of members of these populations. In the United States, the only population group that has experienced a significant increase in mean stature since 1980 is the children of Asian immigrants (IBNMRR, 1995). The continued trend toward a narrowing of the gap in mean stature between Asians on the one hand, and European and Africans on the other hand in the United States, means that some environmental factors, as yet poorly understood, are the primary causes of population differences in growth.

Secular trends

The process that results in a change in the mean size or shape of a population from one generation to the next is known as the **secular trend** in growth. The word 'secular' has two meanings: (1) worldly, especially pertaining to the material, non-spiritual world, and (2) just once in an age, indicating a relatively long span of time. The process is aptly named because the factors influencing the secular trend are related to the material conditions of life and these conditions do act on human growth over long spans of time. In a review of secular trends taking place in the industrialized nations of the world, Hauspie *et al.* (1996) offer the following overview:

> Secular changes in body size and tempo of growth have occurred during the last century in almost all the industrialized countries and have been well documented . . . almost all of the secular increase in adult height is established during childhood . . . The secular increase in body length is due predominantly to an increase in leg length . . .; the legs are the fastest growing part of the body during early childhood when the impact of the environment is at its greatest . . . Secular changes in body dimensions has occurred simultaneously with secular change in the tempo of growth, as shown during the last 100 years by an advancement of age at menarche and at peak height velocity of about 3–4 months per decade in most European countries . . . The higher tempo of growth has resulted in adult height being reached at an earlier age: for males adult height is now reached at about 18 years but in the 1910s was only reached at about 26 years of age . . .   (p. 8).

Hauspie and colleagues review the evidence for these secular trends in 17 nations, including many European countries as well as Japan, Cuba, Brazil, North America, and Taiwan. Following World War II, the Japanese experienced the strongest secular trend so far recorded for an entire nation. In 1950 the mean height of Japanese young adult men was 160 cm, whereas in 1995 it is almost 172 cm. This is a rate of increase averaging 2.67 cm per decade over the entire 45 year period, but the rate of change was much faster in the first decade (about 4 cm) than in the last decade (about 1 cm). In contrast, the rate of secular increase in height in Sweden and Norway between 1952 and 1985 was only 0.3 cm per decade.

Age at menarche is commonly used to study the secular trend in tempo of growth. In Poland, age at menarche declined from 1955 to 1978 by about 4.15 months per decade for girls living in villages and towns (Wolanski, 1967, 1980; Hulanika & Waliszko, 1991). For city girls, the decline was 3.0 months per decade. Despite the greater rate of decline for village and town girls, the city girls have always had the earliest mean age at menarche. In 1955 the mean ages were: village – 14.3 years, town – 13.9 years, and city – 13.4 years. In 1978 these mean ages were 13.5, 13.1, and 12.9 years respectively. The Polish researchers attribute the differences between locales to the lower quality of nutrition and health care, and greater physical labor in towns and villages compared with cities. The overall decline in age at menarche in all locales attests to improvements in the quality of life in all three areas with time. Interestingly, from 1978 to 1988, which was a time of considerable political and economic turmoil in Poland, the age at menarche increased by an average of 1.7 months per decade. Such reversals in secular trends are common during times of environmental deterioration. These reversed trends are discussed in greater detail later in this chapter.

Secular trends take place not only within one country or region, but also when people move between places. As one example, it is well known that a positive secular trend (that is, increases in average size over time) is associated with migration from a low SES to a higher SES environment or *in situ* socioeconomic improvement (Garn, 1987; Bogin, 1988a). A few classic examples of the secular trend in the growth of migrant children include the work of Boas (1912, 1940) with European immigrants to the United States, the work of Shapiro (1939) with Japanese immigrants to Hawaii, the work of Goldstein (1943) and Lasker (1952) with Mexican immigrants to the United States, and the studies by Greulich (1976) of American-born children of Japanese descent (the effects of migration on human growth are reviewed in detail in the next chapter). Follow-up studies of these same populations show that, with time, the growth in height of each generation of the children of migrants continues to increase until it converges on that of the host population (Roche, 1979).

Sometimes the rate of the secular trend is much more rapid than these classic cases. The 'Maya in Disneyland' example, discussed in the Introduction, shows the power of the environment to bring about significant and rapid change in amount and rate of growth. In less than one generation, the Guatemalan Maya refugee children living in Indiantown, Florida and Los Angeles, California became 5.5 cm taller, on average, than their age-mates back in Guatemala (Figure I.1). The growth status of the Maya refugee in Florida was also compared with three other ethnic groups living in Indiantown (Bogin & Loucky, 1997). These groups are whites, blacks, and Mexican-Americans. The mean values by age and ethnicity for each of the growth variables are presented in Figures 5.5 (height and weight) and 5.6 (body composition). A statistical analysis shows that the Maya ethnic group is significantly shorter than each of the other ethnic groups of Indiantown. Mexican-Americans are shorter than blacks or whites, but there is no statistical difference between blacks and whites. Mean values of height for a national sample of United States children (Hamill *et al.*, 1977) are included in the comparisons for height. The white and black samples of

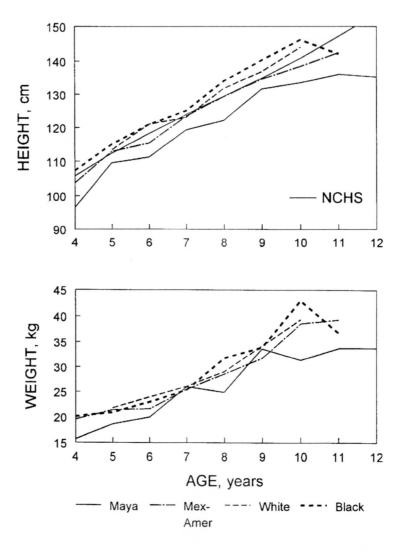

Figure 5.5. Mean height or weight of Indiantown children by ethnic group and age (from Bogin & Loucky, 1997).

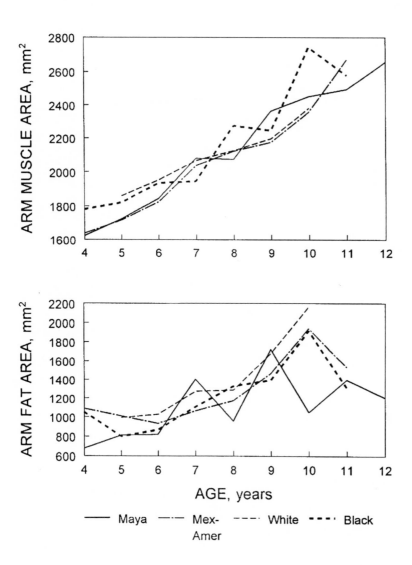

Figure 5.6. Mean arm muscle area or arm fat area of Indiantown children by ethnic group and age (from Bogin & Loucky, 1997).

Indiantown equal or exceed the mean height of the national sample. Mexican-Americans are at or below the mean height of the national sample. This is true for each of the other anthropometric variables as well.

The Maya, as a group, weigh significantly less than whites or blacks. There is no significant difference between Maya and Mexican-Americans, nor are there differences between the white and black ethnic groups. There are no ethnic differences in body composition measures, such as for arm fat area or for arm muscle area (Figure 5.6). This means that in terms of energy (fat) and protein (muscle) stores on the body, the Maya children appear to be generally healthy and well nourished. Why then, are they shorter than the other ethnic groups?

These results show that the plasticity of human phenotypes changes at different rates for different traits. In most studies that find a positive secular trend, the increase in mean height from generation to generation lags behind increases in weight and body composition. This happens because height reflects health and nutritional history, whereas weight and body composition reflect recent events. Indeed, as discussed in Chapter 2 with reference to the 'intergenerational effect hypothesis', a child's height is an historical record of both the individual and his or her parents. In the case of the Maya refugees, the effects of chronic undernutrition and disease suffered by the parents are still being expressed in the growth of their children. Conversely, children who are better nourished and healthier will give their own offspring a healthier prenatal start in life. Certainly, bigger mothers have longer, heavier babies who grow up to be taller children and adults (Garn *et al.*, 1984).

Another example of this phenomenon comes from Mexican immigrants to the United States who have become taller, on average, with each generation since the 1930s (Bogin, 1989). The most recent generation of USA-born Mexican-Americans (under 12 years old) have mean heights equal to NCHS references (Martorell *et al.*, 1984). The Mexican-American sample from Indiantown includes both immigrant and USA-born children and, as one would predict, are intermediate in stature between the black and white samples and the Maya sample.

Eight thousand years of secular trend

The study of recent Maya immigrants to the United States, and the study of changes in size of Mexican-Americans since the 1930s, shows that stature is a dynamic phenotypic trait, one that is very responsive to the quality of the environment for growth. To better understand the relationship between the stature of a population and its environment for growth,

Bogin & Keep (1998) surveyed the literature on the growth of Latin American populations. Several analyses of the height of children and adults were conducted, but here the focus is on the analysis of adult stature for Native Latin Americans. Native Latin American has two connotations. First, it means those people identified as the descendants of pre-Colombian forager, tribal, chiefdom, and state societies – such people are also referred to in the literature as American Indians or Amerindians. Secondly, Native Latin Americans may also be people of social groups that formed after European contact, but came to identify themselves culturally (by language, dress style, kinship organization, etc.) as Latin American. For example, these groups include both Amerindian societies that formed post-contact, as well as groups of rural *mestizos*, that is, people of mixed Spanish and Native American heritage. People of primarily European, Asian, and African descent living in the Americas are excluded from our analysis.

In total, 322 samples of adult height for men, representing 20 808 individual measurements, and 219 samples of adult height for women, representing 9651 individual measurements, examined between the years 1873 and 1989 were found in the literature. In all samples, adult height refers to the stature of people who are reported to be 18 years old or older at the time of measurement. In addition to these data for people measured in life, Bogin & Keep also assembled some estimates of stature for archaeological samples of pre-Conquest and early post-Conquest populations from the present-day Latin American region. These estimates are based on the measurement of skeletal remains of individuals of higher social status (burials from tombs) and lower social status (non-tomb burials). There are 29 samples for men and 27 samples for women, representing 1305 and 1158 individual measurements respectively. The pre-Conquest data are used to provide a deeper historical perspective on the dynamics of stature variation for Native Latin Americans.

The adult height data were analyzed by plotting the data points for each sample and then fitting a distance weighted least square regression to the data – this procedure fits a curve that, basically, passes through or near the mean height for each year with data. Separate regression equations were fit to the data for men and for women. The data for entire series of adult statures, and the regressions estimated by distance weighted least squares, are presented in Figure 5.7. For the archaeological samples (i.e., AD 1750 and earlier), these regressions were calculated to present an idea of trends in 'average' stature over time. Because intervals between archaeological data points are not equidistant, and since sample sizes are often small, and certainly not representative of all people alive at those times, it is not

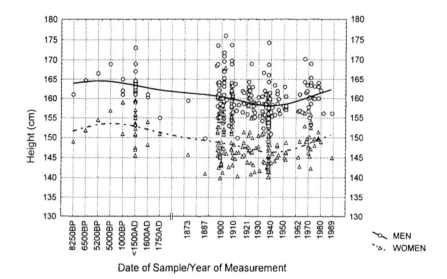

Date of Sample/Year of Measurement

Figure 5.7. Mean statures of Latin American men and women during the past 8250 years. Prior to 1873 the data are estimates of adult stature based on skeletal remains. From 1973 onward statures are based on measurements of the living. The fitted curves are trends in mean stature estimated by distance-weighted least-squares regression (see text for details). From Bogin & Keep (1998).

possible to perform formal statistical analyses. Nevertheless, a descriptive analysis reveals several important associations between estimated stature and the biological and sociocultural conditions for life.

The oldest data are for skeletal remains from the Vegas culture, a foraging people living along the southwest coast of Ecuador from 8250 to 6600 BP (the dates given in Figure 5.7 for the skeletal samples are always for the earlier point of a time range). These ancient foragers seem to have been taller than average, as the mean stature for the entire sample presented in Figure 5.7 is 159.2 cm for men and 147.6 cm for women. These foragers are known to have eaten a wide variety of foods, including abundant fish and shellfish (Ubelaker, 1994).

The next three archaeological samples are from the Paloma site of coastal Peru. This is a pre-ceramic period site with many indications that it was permanently settled from 6500 BP to 4500 BP (Benfer, 1984, 1990). The inhabitants of Paloma were horticulturists, producing a wide variety of garden foods, and also hunted and gathered wild animal and plant foods. The density of the population at Paloma was low to moderate relative to archaeological sites from later time periods. Socially, the Paloma people seem to have been organized into tribal-type political groups, with minimal social stratification. They were economically and politically autonomous from any other social groups in the region. Mean estimated stature increases from the earliest period (6500–5300 BP), to the middle period (5200–5000 BP), and finally to the latest period (5000–450 BP). Benfer (1990) points out that the increases in stature occur along with declines in skeletal and dental indicators of stress, such as bone loss or enamel hypoplasias. Benfer interprets these biological changes as evidence of increasing adaptation to sedentary life and improvement of the nutrition and health of the Paloma people. Indeed, by the latest period, the mean stature of Paloma men and women would be considered as tall, even by modern Latin American reference values.

The samples dated 1000 BP are from several coastal and montane sites in Ecuador, spanning the time from 1000 BP to AD 190. Archaeological remains indicate that the people at these sites subsisted mainly from intensive agriculture. Ubelaker states, 'By this time, agriculture was well established in both the highlands and the coast, a shift toward increased sedentism had occurred, and population densities were higher' (1994, p. 148). Several lines of evidence indicate that the reduction in stature of these samples, compared with the earlier pre-ceramic period Paloma samples, is the result of economic, social, and political changes associated with intensive agriculture. Agriculture may have produced a decrease in diet quality for the majority of people. An agricultural diet is usually restricted to a few intensively cultivated crops. Essential nutrient deficiencies are a common result of the restricted diet (Cohen & Armelagos, 1984). In addition to dietary restriction, the social and political control that is necessary to efficiently organize agricultural labor almost invariably leads to an increase in social stratification (i.e., workers versus ruling elites), and then to economic and political inequality (Cohen & Armelagos, 1984). Further exacerbating the plight of the lower social classes are the effects of warfare and military conquest, which were common in the Andean area at this time (Webster *et al.*, 1993). Together these nutritional and social factors may have brought about a decline in health for the lower social classes, a decline preserved in skeletons of shorter stature.

The next group of data (marked as < AD 1500) are for several pre-Conquest sites in Mexico, Guatemala, and Ecuador spanning the time from about 200 BC to the time of the European Conquest at about AD 1500. All are from complex, state level societies (e.g., Toltec, Aztec, Maya) with dense populations. It is noteworthy that by this late pre-Conquest period the variability in mean stature is almost as great as for the twentieth century. This variability is due to at least two factors; (1) the social status of the individuals within any sample, and (2) social, economic, and political changes over time between samples. Because these skeletons were recovered from state-level societies, there are marked social status differences between individuals within samples. High social status is indicated when individuals were buried in tombs. Lower social status is indicated when individuals were recovered from non-tomb graves. For all of the archaeological samples, the tomb burials are taller, on average, than the non-tomb burials.

The second source of variation in stature is due to changes in the conditions for growth over time. This effect is illustrated in Figure 5.8 for skeletons recovered from Tikal, a major Maya city-state and center of cultural life (agriculture, trade, religion, etc.) during the Classic Period (AD 250–900). During both the Early and Late Classic period, skeletons from tomb burials average greater stature than burials from small-sized homes or mid-sized homes (Late Classic period only). The size of homes is an indication of the wealth and social status of both the occupants and the burials (occupants of the home were usually interred under the floor of the house). Over time, the mean stature of both tomb and non-tomb burials declines by about 5.0 cm. The decline in stature occurs during a time of increasing population growth, increasing warfare between Maya city-states, increased investment in militarization (larger armies, weapons production, construction of fortifications, etc.), and declines in food production and public building (Webster *et al.*, 1993). The 'material and moral condition of that society' (Tanner, 1986) were directed away from the environmental factors that would promote growth and toward those factors that would inhibit growth (see Chapter 6 for a review of the negative effects of a war-time environment on human growth).

The samples dated at AD 1600 are from the Tipu site, Belize and post-Conquest cemeteries in Ecuador. Cohen *et al.* (1994) report that Tipu was a mission site, periodically visited by a Spanish priest, but otherwise entirely inhabited by Maya. The cemetery at Tipu was in use from 1567 to 1638. The Ecuadorian cemetery samples come from two historic churches in Quito. The remains date from 1500 to 1725 and include Indian, Mestizo, European, and some Africa slave remains (Ubelaker, 1994). Tipu males are slightly taller than the Ecuadorian males, but Ecuadorian females are noticeably taller than Tipu females. Given the mixture of ethnic and social groups represented by these samples, it may be best to note only that, on average, all of these skeletal samples are shorter than the majority of the pre-Conquest samples.

The trend for a decline in average stature continues with the samples dated at AD 1750, which are actually burials dating from 1750 to AD 1940

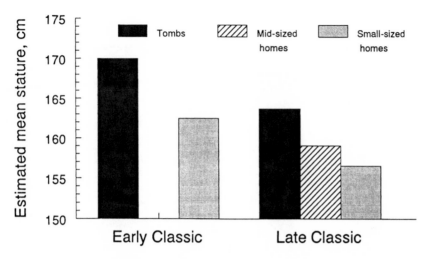

Figure 5.8. Mean stature of skeletons recovered from tombs, mid-sized houses, or
small-sized houses at Tikal during the Early Classic or Late Classic Periods
(redrawn from Haviland & Moholy-Nagy, 1992).

from the same two Ecuadorian cemeteries just discussed. Indeed, it is clear
that estimates of the statures of pre-Conquest Latin Americans (prior to
1500) are significantly greater than stature anytime after the Conquest
[pre-Conquest mean = 163.4 cm (SD = 3.4 cm) for men and 152.9 (3.8) cm
for women; AD 1600–AD 1989 mean = 159.5 (4.7) cm for men and 148.6
(4.8) cm for women].

Stature of the living

When the analysis is restricted to only the sample of people measured in
life (after 1973), it is possible to analyze upward and downward trends in
the regression curve for statistical significance. This is allowed because the
data are graphed by equidistant time intervals and are representative of
the larger native Latin American population. The overall change in mean
height from 1873 to 1989 is negligible and the linear regression coefficient
for this entire time period is not significantly different from zero. However, mean statures decrease between the years 1898 to 1939 and then
increase from 1940 to 1989 (the data from 1873 to 1897 are excluded as
there are only four samples). For men and women, the decline from 1898
to 1939 amounts to about 4.5 cm and 3.0 cm, respectively. From 1940 to
1980 the increase is about 5.0 cm for men and 4.0 cm for women. These
positive and negative secular trends are biologically significant. Separate

linear regression coefficients for these two time periods also show that these trends are statistically significant. The pattern of average change in stature is virtually identical for men and for women, and the difference in height between the sexes are almost constant. At the year 1900 the difference is 12.0 cm, at 1939 the difference is 11.5 cm, and at 1980 the sex difference is 12.5 cm.

What do secular trends mean?

There has been much discussion in the literature as to the meaning of secular trends. Various hypotheses and speculations have been offered to explain secular trends, such as notions that these trends reflect evolutionary tendencies toward larger size in most mammalian species, that the trends are due to break-up of genetically isolated (and hence inbred) human populations, and that the trends reflect changes in the quality of the environment in which people grow up (Wolanski, 1985). The trends in mean stature for the Latin American populations over the past 8000 years provide strong support for the 'quality of the environment' hypothesis.

Social, economic, and political determinants of secular trends

The post-Conquest decline in stature of Middle American adult men and women has been documented previously (Stewart, 1949; Newman, 1962; Malina, 1990). Newman (1962, pp. 242–3) reported that for the Valley of Mexico the decrease in stature is '. . . 4.0 cm over the 4000 year time span going back to the archaic period' while in highland Guatemala there is '. . . a 4–7 cm stature decrease to the modern Cakchiquel Maya now living in the area'. Bogin & Keep's research confirms this negative trend using larger sample sizes and data through the 1980s. There are fewer estimates of pre-Conquest adult stature for South America, but the available data indicate that from maximum mean statures prior to AD 1500 there was a decline of more than 10 cm for men and slightly less than 10 cm for women by AD 1950.

The post-Conquest period was a time of enormous change in the social, economic, and biological environments of Latin America. Eltis (1982, p. 473) writes, 'At once decimated by European disease, subjected to intense European cultural pressures, and largely insulated from the nutritional benefits of the industrialization process, the Indians were denied access to the land which had previously saved them from malnourishment'. Cook & Borah (1979) add that European imposition of forced labor, including slavery, placed additional biological and social demands on the Indians. To the extent that changes in mean stature for human populations reflect the 'material and moral condition of that society' (Tanner, 1986), the decline in adult stature after the year AD 1500 is probably due to the effects of the Conquest on native Latin Americans. Corroborating evidence for this interpretation comes from the pre-Conquest archaeological data. These skeletal remains show that socioeconomic disparity within Native American societies, such as the Classic period Maya, are associated with a decline in stature for the lower social classes. The highland Ecuador samples of AD 1000 also show that stature declines are associated with a restricted diet, increased social stratification, and, possibly, military subjugation by emerging state-level societies. The social and economic history of each Latin American nation following the Conquest is somewhat different. However, the general tendency was to have a political system of local dictatorship, with economic exploitation by European and North American nations. The health and nutrition of Amerindians and rural *mestizos* suffered under this system. These conditions remained in place up through the first half of the twentieth century in much of Latin America. The world-wide economic depression of the 1930s intensified these already deleterious conditions for the biological, economic, and social well-being of Native Latin Americans. The negative trend in stature until 1939 may be a consequence of these environmental conditions.

The positive trend in stature from 1940 to 1989 is associated with the world-wide economic recovery sparked by World War II. Latin America benefited from this recovery and did not suffer the ravages of the war in Europe, Asia, and the Pacific. Post-war economic growth continued, especially with investment from the United States. That investment had both monetary and political (i.e., cold war, anti-Communism) goals. Regardless of the motivation, the foreign investment expanded the economy and helped to increase the rate of urbanization in Latin America. The positive trend for stature may be an outcome of these changes in the standard of living as both a rising economy and urbanization are associated with increased amounts of growth (Bogin, 1988a).

Negative secular trends

The study of secular changes in body size and body composition has been, and remains, a very active area of research. The most useful of these studies go beyond the description of secular changes and attempt to explain the causes of these trends in growth. An especially useful approach to secular trend research is the discovery and analysis of **negative secular trends**, that is examples of decreases in body size from generation to generation. The analysis of stature change over the past 8000 years in Latin America is one example of such an approach. The landmark publication in this area is a paper by Phillip Tobias (1985) in which he presents a world-wide review of growth data for stature from the twentieth century for succeeding generations of people living under deteriorating conditions of low SES and/or under political repression. For each of the cases Tobias examines he finds either no secular change in stature or negative secular change. His own data from South Africa (Tobias, 1975; Price *et al.*, 1987) demonstrate a clear decline in mean stature for blacks from the late nineteenth century to the present day. These stature declines are linked to the deterioration of the social, economic, and political environment for blacks both prior to and during the apartheid era in South Africa.

In follow-up studies, it has been shown that between the years 1880 and 1970, South African whites had an increase in mean height of 4.5 mm/ decade and South African blacks had a mean increase of 2.4 mm/decade (Henneberg & van den Berg, 1990). The whites are predominantly of Dutch ancestry and quite unexpectedly the increase in stature of the South African whites was significantly less than that for the Dutch living in the Netherlands – 15 mm/decade – and measured in the same years. This finding is unexpected because the purpose of the apartheid policies was to guarantee political and social domination by the country's white minority over the non-white population. One expectation of apartheid was that it would ensure that South African whites would live and grow up under socioeconomic conditions equal to, or superior to, those of the industrialized nations of Europe and North America. However, the superior secular increase in stature of the Dutch in the Netherlands shows that the policy failed to do this.

Henneberg & van den Berg interpret these findings to indicate that factors other than a general socioeconomic improvement in the environment are responsible for secular increases in stature. Unfortunately they do not know what these other factors might be, and are '. . . in no position to present or advance a fully constructed alternate hypothesis' (p. 464). My own explanation for this failure is that the deterioration of living conditions for the black population caused by the apartheid policies could not be confined to that one ethnic group. Blacks are the majority population of South Africa. Of the almost 39 million inhabitants in 1991, 75.2 percent of the population were classified as black Africans, 13.6 percent were whites, 8.6 percent were known as coloureds, and 2.6 percent were Asians. When that many people live under poverty, the economic and social development of the country as a whole is likely to be arrested, and even the privileged social classes will be affected. The most meaningful single statistic in this regard is the infant mortality rate, which as explained in Chapter 2 is strongly associated with overall socioeconomic status. In 1985, the infant mortality rate for South African blacks was 68 per 1000 live births, and for South African whites it was 13 per 1000 (Cameron, 1997). By comparison, the infant mortality rate in 1987 for the Netherlands, for all ethnic groups, was only 7 per 1000.

An even clearer case of the negative secular trend comes from Guatemala during the period from 1974 to 1983, a time of intense civil war and political repression. Economic decline and political unrest due to the war is associated with a significant decline in the mean stature of cross-sectional samples of 10- and 11-year-old boys and girls (Bogin & Keep, 1998). Boys and girls from families from very high, moderate, and very low socioeconomic status were measured in each sample and all three SES groups show the decline in stature. A general deterioration of the quality of life in Guatemala, especially the quality of nutrition and health of the entire Guatemalan population, seems to be the cause of the negative secular trend. Even the very wealthy were not spared as the environmental decline affected municipal water supply systems and led to the outbreak of cholera and other epidemic diseases.

An economic crisis in Venezuela has also resulted in a negative secular trend in growth. The crisis began in 1983 and by the 1992 had only worsened. The World Bank blames the crisis on economic and political mismanagement by the Venezuelan national government (World Bank web page on Venezuela, April, 1997). From 1992 to 1995 the annual rate of inflation averaged 333 percent. Civil unrest, including riots which killed more than 300 people, was widespread. The gross domestic product (GDP, an indicator of economic activity within a country) fell precipitously and was a negative 0.4 percent in 1993 and negative 1.0 percent in 1996. There were two military coups during the crisis and several changes of government. National surveys of the growth of Venezuelan children find that positive secular trends in height and weight were evident in the decades before the start of the economic crisis. By 1987, however, negative shifts in growth were evident in infants and children under four years old (Lopez-Blanco, 1995). The negative secular trend is due in large part to a decline in the availability of food, which by 1990 had dropped to a mean of 2171 kilocalories per person per day. That value is below the estimated per capita average required to sustain the population (Lopez-Blanco, 1995). These examples from South Africa, Guatemala and Venezuela provide further evidence that the growth of human populations is a sensitive indicator of the quality of the social, economic, and political environment.

Population differences in body composition

Population variation in body composition has been a fascinating, sometimes contentious, but always important focus of research. The importance of this work is due to the association between body composition and disease. Of special significance is the relationship of both the amount of fat on the body, and the placement of this fat at specific sites, to risks for cardiovascular disease, diabetes, and some cancers. These diseases are major causes of death in the industrialized nations.

In the United States, black children and youths have, on average, less total **subcutaneous fat** (the fat layer just under the skin) than white children and youths. Black children and youths also tend to have relatively less subcutaneous fat on their extremities compared with the whites (Piscopo, 1962; Malina, 1966; Johnston *et al.*, 1974; Harsha *et al.*, 1980). Robson *et al.*, (1971) measured the triceps and subscapular skinfolds of children of African (black) descent, one month to 11 years old, living on the Island of Dominica. Only healthy, well-nourished children were included. The data were compared with similar measurements taken from a sample of English children of the same ages. The black children of Dominica were, on average, leaner than the English children, and this difference was entirely due to the Dominicans having significantly smaller triceps skinfolds. There were no significant differences between the populations in the mean subscapular skinfold thickness. The results of these studies suggest that during childhood and adolescence, blacks have less total body fat, and a different anatomical distribution of subcutaneous fat, than whites.

Working in the African nation of Cameroon, Ama & Ambassa (1997) found similar differences between white men of French and Italian ancestry and black men of Cameroon ancestry. The mean age of both groups was 24 years, and both groups were matched for height and weight. There were only 13 men in each group, but all were non-competitive swimmers. Swimmers are known to be self-selected to have more total body fat than other athletes (Malina *et al.*, 1982b). The extra fat aids in buoyancy and swimming efficiency. The white men in this study were found to have more body fat than the black men. Moreover, the whites have this fat concentrated more on the extremities of the body relative to the trunk. Due to both factors, the authors conclude that whites have better buoyancy than blacks.

Similar average differences in fatness and fat distribution have been found when other samples of African black children and American or European white children were compared (Eveleth & Tanner, 1976, 1990). Based on these studies, some researchers conclude that a genetic difference in mean fatness and typical fat distribution may exist between children of African and European origin. Of course, many individual children do not follow the mean tendency for body composition of their natal population. Moreover, Mueller (1982) showed that genetics, or so-called 'racial' differences, explain relatively little of the mean differences in body composition between blacks and whites living in the United States. In this study, Mueller used multivariate statistical analysis to mathematically identify factors that influence fatness and fat distribution. A factor Mueller called 'ethnicity' (i.e., black or white) was found to be statistically significant, but accounted for only about two percent of the variance in fatness and five percent of the variance in fat distribution between the two samples. Sex, age, and unspecified factors, of both genetic and environmental origin, accounted for most of the variance in fatness and fat distribution.

Two studies conducted in Guatemala lend support to an environmental determination of body composition. In the first study, Johnston *et al.* (1975) examined samples of children of European ancestry and Guatemalan Ladino ancestry living in Guatemala. Both groups of children were of high socioeconomic class, attending the same private school, so in several ways they were exposed to a common environment. As measured by skinfolds, there were no significant differences in fatness between the ethnic groups at either the triceps or subscapular skinfold sites. Since the triceps site correlates highly with other measures of extremity fatness and the subscapular site correlates highly with other measures of trunk fatness, it may be inferred that there is little ethnic difference in fat distribution in this case. Johnston *et al.* suggested that the common pattern of fatness and fat distribution of the Europeans and Guatemalan Ladinos might be due to their living under similar environmental conditions.

In the second study, Tim Sullivan and I (Bogin & Sullivan, 1986) found that the environment is the major determinant of the average amount of fatness and the fat distribution of children. We compared the fatness and fat distribution of four groups of children, age seven to 13 years, living in Guatemala. The groups (and sample sizes) were: Guatemalan Ladinos of high socioeconomic status (SES) (320 children), Europeans of high SES (164 children), Guatemalan Ladinos of low SES (340 children), and Guatemalan Indians of very low SES (669 children). Triceps and subscapular skinfolds were measured for each child. Previous research with these same groups had shown that SES was significantly associated with nutri-

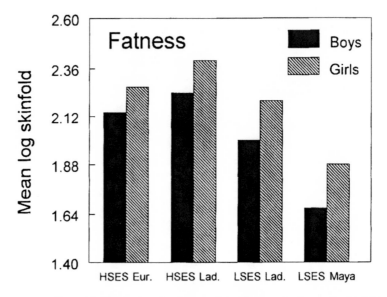

Figure 5.9. Mean values for the sum of the triceps and subscapular skinfolds of children living in Guatemala. The raw data were log transformed to normalize the distribution of the values. Larger log values indicate greater fatness. Sample abbreviations are: HSES Eur. – high SES Europeans; HSES Lad. – high SES Ladinos; LSES Lad. – low SES Ladinos, LSES Maya – very low SES Maya (from Bogin & Sullivan, 1986).

tional status, as reflected by skinfolds and other measures of body composition (Bogin & MacVean, 1981a, 1984). So, as expected, and as shown in Figure 5.9, the high SES Ladinos and Europeans had, as a group, larger skinfolds than the low SES Ladinos or very low SES Indians.

It was hypothesized that socioeconomic status would also be associated with fat distribution. Trunk fat may be more physiologically important than extremity fat; for example, trunk fat may serve to protect the internal organs, and is associated with reproductive development in women. Accordingly, if total body fatness is reduced in children from lower socioeconomic status populations, there should be a relatively greater reduction of extremity fat and a relatively greater retention of trunk fat. As shown in Figure 5.10, Tim Sullivan and I found that high SES Ladinos and high SES Europeans in Guatemala had a similar distribution of fat between the triceps and subscapular skinfold sites. The low SES Ladinos had significantly less fat at the triceps site than the two high SES groups, and the very low SES Indians had significantly less fat at the triceps site than any of the three other groups. Thus, as SES decreased fat distribution became more centripetal, that is,

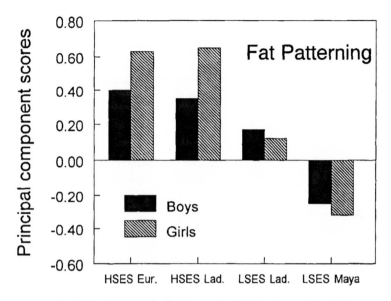

Figure 5.10. Relative distribution of subcutaneous fat, the fat patterning, at the triceps and subscapular skinfold sites for children living in Guatemala. Fat distribution is expressed as principal component scores; larger scores indicate relatively greater triceps fatness, smaller scores indicate relatively greater subscapular fatness. Sample abbreviations are: HSES Eur. – high SES Europeans; HSES Lad. – high SES Ladinos; LSES Lad. – low SES Ladinos; LSES Maya – very low SES Maya (from Bogin & Sullivan, 1986).

relative amounts of arm fat decreased and relative amounts of trunk fat increased. In accordance with the hypothesis, the relative fat distribution changed even as the absolute amount of subcutaneous fat decreased at both arm and trunk sites from high, to low, to very low SES. It seems that fat distribution in these samples of children was determined, at least partly, by socioeconomic status, which is a proxy measure for nutritional adequacy and other environmental variables. No evidence for an ethnic effect could be demonstrated for these samples of children.

Adult differences in body composition have been studied in several populations. Johnson *et al.* (1981), using data from a national sample of the United States, found that the average black–white differences in fatness and fat distribution persisted into adulthood for males, from one to 74 years of age. That is, at all ages and at all levels of fatness, black males were leaner, and especially so on the extremities, than white males. Black females were found to be leaner, on average, than white females from the ages of one to 24 years, but from age 25 to 74 years, black females had larger mean values for the triceps and subscapular skinfolds than white females. This suggests that as adults, black women are, on average, fatter than white women, and that the relative anatomical distribution of fat of the black women changed after childhood.

A study by Malina *et al.* (1982b), compared athletes participating in the Montreal Olympic games of 1976 for skinfold measures of subcutaneous fatness and fat distribution. The sample included 264 white male and 133 white female athletes, 38 black male and ten black female athletes, and seven Asian male and four Asian female athletes, representing 46 countries, 20 major sports, and 68 different Olympic events. The median age of the male athletes was 21 years and of the female athletes was 23 years. It was found that the black athletes were significantly less fat than the white or Asian athletes. The total variance in fatness was statistically partitioned into the following factors (and percentage of variance): sex (31 percent), sport (19 percent), ethnicity (3 percent), and age (3 percent). The other 44 percent of the variance in fatness is not explained by these factors and must be due to other (univestigated) causes. In terms of fat distribution, white athletes had significantly more fat located on their extremities than Asian athletes, who had more fat located on the trunk of their bodies. The fat distribution of the black athletes was intermediate between that of the white and Asian athletes. The factors, and percent of the variance, associated with fat distribution (as contrasted with the amount of fat) were: sex (35 percent), age (7 percent), ethnicity (2 percent), sport (2 percent), and residual (unspecified) factors are associated with the remaining 54 percent of the variance.

Malina *et al.* point out that the patterns of fatness and fat distribution of the highly trained Olympic athletes follow the same pattern as that of the general population. In the case of sexual differences, with sex accounting for more than 30 percent of the variance in fatness and fat distribution, there seems to be a strong genetic determination. Between populations, however, with the percentage of variance due to ethnicity being only two to three percent, genetic determinants of body composition appear to be relatively weak. Perhaps uninvestigated factors related to the many environmental conditions for growth and development, and their interactions with the genome, which differed between the athletes, account for the unexplained variance in fatness and fat distribution.

The significance of population variation

From this brief review of population variation in growth, it is possible to conclude that the differences between human groups in the average values for size, body proportions, and body composition are due to an interplay between genetic and environmental determinants of growth and development. These variations in the morphology of the human species have their own intrinsic fascination and scholarly appeal for study, but there are also important practical reasons for the analysis of these variations in growth. Cities and nations are becoming increasingly composed of people from many different geographic and ethnic origins. To monitor the health and welfare of the children of these diverse groups of people, clinicians and public health workers need to know about the normal range of amounts and rates of growth of children from different physical and cultural environments.

Furthermore, reference data for height, weight, body composition, and skeletal and dental development are used to assess health status, nutritional status, obesity, progress during treatment for disease, and relative risks for acquiring several acute and chronic diseases. For accuracy and reliability in their work, health care professionals may need population-specific reference data that reflect both the hereditary and environmental determinants of growth and development for the people they serve.

Adaptive value of body size in human populations

Implicit in the foregoing discussion of factors that influence population variation in growth is the fact that such variation occurs within the limits of biologically possible human phenotypes. Some of the evolutionary and ecological influences on the pattern of growth and development of the human phenotype were described in Chapters 3 and 4. In addition to research on the evolution of the pattern of human growth, there has been considerable interest in the evolution of human body size. One popular, but incorrect, notion is that the average height and skeletal mass of modern humans is greater than that of any of our ancestors. Based on several studies of more than 200 individual skeletons of early to late Pleistocene age (1.8 MYA to 10 000 BP), it seems that our ancestors, from *Homo erectus* to modern *H. sapiens*, were on average about 10 percent taller and 30 percent heavier than living humans (Mathers & Henneberg, 1995; Ruff *et al.*, 1997). While some researchers argue for a genetic explanation for the recent decline in body size, Ruff *et al.* (1993) explain that the difference is to be found in the way of life of ancient and modern humans, and is not due to any genetic change. Our ancestors were required to do more heavy labor, associated with a hunting and gathering way of life, and this imposed more mechanical loading (i.e., physical stress) on the skeleton. The increased mechanical stress seems to have occurred from an early age, and the skeleton responded by growing larger and more massive during the years of development. In related research, Styne & McHenry (1993) find that skeletal evidence '. . . from recent prehistory and the last 2000 years also reveal adult height in many groups to be equal to modern humans of the same region' (p. 3). Confirmation of this was found in the analysis of 8000 years of secular trend in Latin America presented in this chapter. The oldest skeletal material, of foragers from Ecuador, had a mean stature equal to that of Latin Americans alive in 1989. It seems, therefore, that the range of normal body size found in contemporary populations reflects the human, and pre-human, condition of the past 1.8 million years.

As human beings alive today, we retain the sizes and shapes that were best suited for the way of life of our ancestors. As past conditions for life may not prevail today, the size of living peoples may be unsuited for contemporary conditions. In this case, further adjustments in body size and shape may be evolving. The advent of agriculture and permanent human settlements during the past 10 000 years has significantly changed ways of life. Much archaeological and recent research shows that agriculture led to a decline in nutritional status, an increase in nutrient deficiency diseases and infectious diseases, and an exponential increase in world-wide population size (reviewed by Bogin, 1997a). Each of these changes contributes to a decline in the quality of life, which is often reflected in reduced growth in height. This may explain why our Pleistocene ancestors were larger, on average, than humans are today. However, the question still remains as to whether the reduction in body mass represents an evolutionary change in the genetic sense.

In 1980, an economist, David Seckler, proposed that the small size of children and adults living in poverty may be a genetic adaptation, acquired from generations of malnutrition. Seckler's basic argument is that smaller bodies need less total food to survive than larger bodies. For example, the Maya of Guatemala have suffered from undernutrition, heavy workloads, and disease for the 450 years since the conquest. Guatemala Maya are also a relatively short-stature population. Did they accommodate to poor living conditions by evolving a smaller and less energy-demanding body? According to recent data the answer is no. As shown in this chapter, the children of Mayan immigrants growing up in the United States average 5.5 cm greater in stature than Mayan children of the same age living in Guatemala. Clearly, if Mayan short stature were a genetic adaptation, fashioned over centuries, such a biologically and statistically significant difference in height could not occur in less than one generation.

It is relatively easy to demonstrate that Seckler's 'genetic hypothesis' is incorrect. Such a notion had been rejected earlier by Stini (1975), who reviewed several studies of the body size of adults from malnourished communities, including his own research conducted in Colombia, South America. Stini found that in Colombia, and in other countries, adults from undernourished communities were, on average, shorter and had less muscle mass than better nourished groups. According to Stini, muscle mass is a direct indicator of metabolic activity and correlates with the body's requirements for energy and protein. Thus, adults with reduced stature and muscle mass need to eat less total food than people with more skeletal and muscle tissue. This, Stini argued, was a beneficial accommodation to undernutrition, but not a genetic adaptation. Stini believed that the size reduction associated with malnutrition took place during the prenatal and early postnatal growth of the individual. It was a developmental change, adjusting the growth rate and size of an individual to his environment, and not the result of genetic selection for small body size. Stini argued for developmental plasticity, since '. . . genetic adaptation alone would frequently result in a stereotypic and potentially maladaptive and rigid response' (p. 35).

Seckler reacted to his critics by publishing a second paper in which he rejects his earlier genetic explanation in favor of what he calls the 'homeostatic theory of growth'. By this he means that '. . . the single genetic potential growth curve of the older view is replaced by the concept of an *array* of potential curves in several anthropometric dimensions . . . in a word, with the concept of a potential *growth space*' (1982, p. 129, author's italics). In Seckler's view, any path taken within this 'potential growth space' results in a normal healthy individual, that is, an individual without functional physical or cognitive impairment. Only in the extreme cases of under- or overnutrition does pathological growth failure or growth excess with functional impairment result. Seckler concludes '. . . that most of the people in the "mild to moderate" category of malnutrition are "small but healthy" people and should be considered "normal" in relation to their environment' (1982, p. 130). In this case, the 'small but healthy' category would include the Maya of Guatemala, South African blacks living under apartheid, and most other people living in poverty. According to Seckler, this means that nutrition interventions, while well-intended and likely to make children grow bigger, will not result in improved health or functional capacity.

At first glance this appears to be a most reasonable notion. Indeed, the concept of a 'homeostatic theory of growth' correlates well with the process of developmental plasticity. However, only in the sense that homeostasis or plasticity during development results in a change in morphology is Seckler's idea correct. The assertion that growth retardation associated with undernutrition is either a genetic adaptation or a homeostatic response without functional consequence is an abuse of the concept of biological adaptation. The consensus of research with undernourished peoples of the poor nations shows that the consequences of childhood undernutrition are: (1) reduced adult body size, (2) impaired work capacity throughout life, (3) delays and permanent deficits in cognitive development, and (4) impaired school performance (Pelto & Pelto, 1989). Spurr (1983) reviews the world-wide research for physical performance and finds that '. . . malnutrition is accompanied by a reduced PWC [physical work capacity] . . . and the degree of depression is related to the severity of the depressed nutritional status and to the loss of muscle mass' (1983, p. 21). In another review of research on the cognitive consequences of undernutrition, Pollitt & Lewis find that mild to moderate undernutrition '. . . affects aptitudes and abilities of pre-school children, and determine[s] in part the degree of success the child will have later within the school' (1980, p. 34). Furthermore, when such malnutrition '. . . is part of an economically impoverished environment, the probabilities are very high that the cognitive competencies of the child will be adversely affected' (ibid). Research in Guatemala with Mayan and *ladino* children confirm the general findings reported by Spurr and Pollitt (Freeman *et al.*, 1980; Bogin, 1988a, pp. 148–59; Martorell, 1989). Guatemalan children of low socioeconomic status lack adequate total food intake and suffer from high incidence of infectious disease, infant and child mortality, and cognitive delays and deficits. These are all indicators of the poor biological adaptation of Mayan populations living in poverty. Amounts and rates of growth are usually reduced as an adjustment to the nutritional and health constraints of poverty, but smaller size does not overcome poverty. Indeed, an environment of poverty in early life usually results in diminished opportunities for educational, economic and socio-political advancement in later life. This situation recycles poverty into future generations, causing further malnutrition and poor growth (Garn *et al.*, 1984).

The forgoing discussion of Seckler's views are presented here because they seem reasonable and, consequently, are popular both with the public and with policy-makers who might prefer to believe that undernourished children are 'small but healthy' rather than stunted and suffering. The popular view is comforting and cheaper – no economic aid is required. Nonetheless, reasons for short stature of people living under poverty are more likely to be associated with the social, economic, and political environments that cause poverty rather than with 'genetic adaptations', or 'homeostatic adaptations'. There is another, more hopeful, reason for discussing the effects of poverty on human development. Poverty is associated with just a few of the harmful stressors to which human beings are subjected. Even in the face of fairly severe undernutrition and high disease loads people are able to survive and strive to make improvements for future generations. An important component of survival is due to the biological plasticity of the human phenotype (Lasker, 1969). Human plasticity allows the individual to adjust to a very wide range of stressful environmental conditions, and gives the human species an adaptive advantage not found in those species obligated to develop according to a rigid, and predetermined, genetic plan. Plasticity also means that when environmental conditions improve, individuals can recover quickly and return to a more optimal size and shape.

6 Environmental factors influencing growth

The previous chapter described population variation in growth, and discussed the causes of this variation in terms of both genetic and environmental factors. To better understand the contribution of genes, the environment, and their interaction to growth, it is useful to artificially separate these categories and analyze each in turn. The present chapter extends the discussion of environmental factors influencing human growth and development to five comprehensive, and well-documented, categories: nutrition, altitude, climate, migration and modernization, and socioeconomic status.

Nutrients and food

In the growing human being, the multiplication of cells or their enlargement in size depends upon an adequate supply of nutrients. Nutritional biochemists have determined that the are 50 **essential nutrients** required for growth, maintenance, and repair of the body. Essential nutrients are those substances which the body needs, but cannot manufacture. These substances are divided into six classes: protein, carbohydrate, fat, vitamins, minerals, and water. Table 6.1 lists the essential nutrients in these categories. One way that nutrients are shown to be essential is via experiments with non-human animals. A young rat, pig, or monkey is fed a diet that includes all the known nutrients except the one being tested. If the animal gets sick, stops growing, loses weight, or dies it usually means that the missing nutrient is essential for that animal. Such experiments do not prove that the same nutrient is needed for people. Some controlled experiments were done in the twentieth century with humans, such as with prisoners and with residents of villages in underdeveloped nations. Since about 1980 these experiments have been considered unethical to conduct. Certain medical conditions deprive people of nutrients, and social, economic, and political conditions of life also deprive people of food and nutrients. By using these 'experiments of nature', and past research, it is possible to prove the necessity of the essential nutrients.

268

Table 6.1. *Essential nutrients of the human diet (after Guthrie & Picciano, 1995)*

Carbohydrate	*Micronutrient elements (continued)*
Glucose	Copper
	Cobalt
Fat or Lipid	Molybdenum
Linoleic acid	Iodine
Linolenic acid	Chronium
	Vanadium
Protein	Tin
Amino acids	Nickel
Leucine	Silicon
Isoleucine	Boron
Lysine	Arsenic
Methionine	Fluorine
Phenylalanine	
Threonine	**Vitamins**
Tryptophan	*Fat-soluble*
Valine	A (retinol)
Histidine	D (cholecalciferol)
Nonessential amino nitrogen	E (tocopherol)
	K
Mineral	
Macronutrient elements	*Water-soluble*
Calcium	Thiamin
Phosphorus	Riboflavin
Sodium	Niacin
Potassium	Biotin
Sulfur	Folic acid
Chlorine	Vitamin B_6 (pyridoxine)
Magnesium	Vitamin B_{12} (cobalamin)
Micronutrient elements	Pantothenic acid
Iron	Vitamin C (ascorbic acid)
Selenium	
Zinc	**Water**
Manganese	

Growth and nutrition are, therefore, closely correlated. However, people do not usually eat the essential nutrients directly as pure chemicals, rather we eat food. This was certainly true for all of our animal ancestors throughout evolutionary history. Human foods come from five of the six Kingdoms of living organisms: plants, animals, fungi (e.g., mushrooms), protists (e.g., species of algae referred to as 'seaweed') and eubacteria (e.g., bacteria used in fermented foods). The sixth Kingdom, archaebacteria, are not eaten directly, but are essential in the diet of other species that people do eat. Herbivores, for example, have archaebacteria in their guts to digest plant cellulose.

Adequacy of the total quantity of food consumed is a major determinant of growth. This is so because, in part, nutrients may be widely distributed across many different types of food. An adequate diet in terms of food quantity is important because of the energy (kilocalories) that food provides, and different kinds of food can substitute for each other to produce energy. During the years from birth to adulthood, the human body requires energy for several process, which may be summarized by the following formula:

$$\text{Energy required} = \text{Growth} + \text{Maintenance} + \text{Repair} + \text{Work}$$

where maintenance means the energy used in basal metabolism, repair means the energy used to restore cells, tissues, or systems following disease or damage, and work means the energy used in voluntary activity. After these requirements are met any energy that remains may be used for growth.

In populations where food shortages are present growth delays occur, and children are shorter and lighter than in populations with adequate or overabundant supplies of food. The famines of the World Wars I and II retarded the growth of children and adolescents exposed to them (Wolff, 1935; Howe & Schiller, 1952; Markowitz, 1955; Kimura & Kitano, 1959). For example, in a review of Japanese studies Kimura (1984) found that compared with pre-war levels, the mean stature of age-matched children decreased between 1939 and 1949. Average heights returned to pre-war levels in 1953 for girls, but not until 1956 for boys. Kimura also found that the growth of children who were between the ages of birth and 12 years during the war was affected more than the growth of older children (showing, once again, how pre-adolescent growth is more sensitive to the environment than adolescent growth). Finally, Kimura noted that the post-war recovery in height and weight '. . . occurred most rapidly in large cities, followed by small cities, and then rural mountain villages' (p. 200). Since improvements in diet also followed the same path, it is likely that the growth recovery was, in large part, due to better nutrition.

Populations that depend on subsistence agriculture for their food may face periodic food shortages due to variation in rainfall, temperature, crop diseases and pests, and inadequate food storage. Billewicz & McGregor (1982) analyzed the growth of children and adults living in two Gambian (West Africa) villages. Longitudinal measurements of height and weight, along with extensive medical histories, were recorded in these villages from 1951. The authors describe agricultural practices as 'primitive', meaning that the villagers grow food by traditional horticultural methods. Vegetable foods are the main source of subsistence, though meat is eaten on religious festival days. The agricultural cycle is determined by climate. There is a dry season and a rainy season, the latter lasting from late May to the end of October, with August and September being the wettest months. Food supplies are lowest from August to November and, typically, adults lose 2.5 kg in body weight during the rainy season. The mean weight of 157 men, over 25 years old, was 59.8, 56.8, and 58.8 kg in March 1966, November 1966, and March 1967. For a sample of 201 women the corresponding mean weights were 52.6, 50.4, and 53.1 kg. Children grow significantly faster in height and weight during the dry season than during the rainy season. For boys and girls aged five to nine years old, the dry season increase in height averages 6.1 cm and the rainy season increase averages 4.2 cm. Food shortages during the rainy season occur simultaneously with an increase in the incidence of malaria, intestinal parasites, and childhood gastroenteritis. These diseases may decrease the food intake, or intestinal absorption, of affected individuals, increase protein and energy expenditures to combat the disease, or decrease the nutrient value of foods consumed (due to parasite competition, diarrhea, etc.). The combination of these insults results in severe undernutrition during the rainy season, directly reflected in the poor growth of children and the weight losses of adults.

Billewicz & McGregor found that the growth deficits associated with rainy season undernutrition were not overcome during the dry season, and that over the year the people suffered from a net shortage of calories. As a result, the Gambian children grew less at every age compared with a reference sample of British children. For instance, at three months of age Gambian children were 1.0 cm shorter than British children, but by three years of age the difference averaged 6.9 cm, and this difference was maintained to adulthood. The adolescent growth spurt of Gambian boys is delayed compared with better-nourished boys (see Figure 5.1), peak height velocity takes place at 16.3 years for the boys compared with about 14 years for British or high SES Guatemalan boys. The intensity of the spurt is reduced, peak height velocity is 6.9 cm/yr compared with 8.2 cm/yr for British children and 9.3 cm/yr for high SES Guatemalan boys. The Gambian boys have a longer total growth period, which makes up some of the difference in height, but the net result of a lifetime of undernutrition is a significant reduction in height and weight compared with better nourished populations.

Populations that have not experienced the acute starvation of wartime famine, or seasonal food shortages associated with subsistence agriculture, may still show a strong relationship between malnutrition and growth. Findings of the Ten-State Nutrition Survey, conducted in the United States from 1968 to 1970, were reviewed by Garn *et al.* (1974), Garn & Clark (1975), and Lowe *et al.* (1975). Anthropometric, dietary, and biochemical indicators of nutritional status were collected from 40 847 individuals of white, black, and Hispanic background. Most of the sample were low income families, with the lowest incomes disproportionately represented. Children from lower income families were shorter and lighter than children from higher income families, and this was true across all levels of income and within each ethnic group. Children from lower-income families consumed significantly less total food, and thus received fewer calories from all sources, than children from higher-income families. Relative percentages of nutrients were about equal in the diets of all economic and ethnic groups, except for vitamin C, which was more abundant in the diet, and in the blood serum, of higher income families. The differences between income groups in stature and weight were significantly correlated with the difference in caloric consumption. Garn & Clark (1975) pointed out that the poor in the United States were not suffering from acute undernutrition, but the moderate, chronic undernutrition they did experience resulted in a cumulative growth deficit.

A similar relationship between family income, total food consumption, and growth of children and juveniles is still apparent in the most recent surveys conducted in the United States (IBNMRR, 1995). Curiously, this relationship changes during the adolescent and adult stages of growth, in that lower SES groups become heavier and fatter, on average, than the higher SES groups. Gordon-Larsen *et al.* (1997) conducted a study of secular changes in growth of economically disadvantaged African-American juveniles and adolescents (ages 11 to 15 years) in the urban center of Philadelphia, Pennsylvania. They find that these youths are significantly heavier and fatter, but not significantly taller, than the reference standards for the United States. Moreover, the most recent cohort of youths, measured in the 1990s, are heavier and fatter than cohorts measured in the 1960s and 1970s. The authors of this study do not attempt to explain why there is an inverse relationship between body weight, fatness and SES in the United States. Whatever the reasons, they have to relate to the energy formula given above, and are mostly probably to be found in levels of caloric intake that exceed the body's need for growth, maintenance, repair, and work.

Biocultural studies of growth and nutrition

The past 20 years has seen an increased emphasis on research that attempts to understand the causes of poor growth and poor health due to inadequate nutrition. Some of the best studies were carried out in the developing nations of the world. This is due, in part, to the fact that the social, economic, and political conditions of life in these countries between the poor and the wealthy are more clearly delineated than in the industrialized nations.

Carol Jenkins (1981) studied the growth and nutritional status of 750 children belonging to four ethnic groups living in Belize. The groups were Creole, Garifuna, Mestizo, and Maya. Creole and Garifuna are of Afro-Caribbean descent, although it is popularly believed that Creoles have a significant amount of European admixture. Both groups speak English and follow Black Carib cultural practices. Mestizos are of mixed Spanish-Indian descent, speak Spanish and follow Latin American cultural practices. The Maya are descended from Native American peoples, they speak languages of their local culture (Mopam and Kekchi) and follow traditional indigenous cultural practices. Within Belizian society, Creoles and Mestizos have higher socioeconomic status, based on parental education and occupation, than the Garifuna or Maya, although all groups are of lower socioeconomic status compared with white Europeans living in the country.

Jenkins found that Maya and Garifuna children were generally smaller and more frequently malnourished than Mestizo or Creole children. Poor growth and nutritional status were associated with the frequency and severity of diarrhea, a later age at introduction to solid food, and a larger number of children in the household. Episodes of diarrhea probably reflected the exposure of children to infectious disease and contaminated food and water. Age of introduction to solid food reflected cultural differences in infant feeding practices. Number of children is a measure of family size, and, since larger families have more 'mouths to feed', there is usually a negative correlation between family size and the growth and nutritional status of children. Jenkins concluded that ethnic variation in the growth of Belizean children reflects '. . . differential access to medical, sanitational, and dietary resources . . .' (p. 177) of the groups studied. Access is mediated by cultural practices and economic resources, and of these, family economics was the more important; undernutrition was nearly always associated with poverty.

Bailey *et al.* (1984) studied the growth of more than 1000 children living in 29 villages in northern Thailand. The children lived in rural agricultural villages, with rice as the basic subsistence crop. The villages had schools and received some type of health care, but none had electricity and only one was located near a major highway. The children were between the ages of six months and five years old at the start of the study, and they were measured about every six months for five years. Children born during the study were also measured. This study design provided cross-sectional and longitudinal data on recumbent length (at all ages), weight, head circumference, skinfolds, hand–wrist radiographs, nutritional biochemistry from blood samples, and parasite infestations. Disease histories were collected on each child, every 15 days, for three years of the study. Some of the children were given nutritional supplements (amino acids, thiamin, vitamin A, and iron) to assess the effect of supplementation on growth.

It was found that, compared with local or international reference standards for growth, these rural Thai children were delayed in growth of all the dimensions studied. For instance, compared with a sample of healthy middle-class children from Bangkok, the capital city of Thailand, the rural village children averaged about 4.7 cm shorter, 1.3 kg lighter, and 1.2 cm smaller in head circumference over the first 36 months of life. From ages six to 18 months, the rural Thai children were between the fifth and tenth percentiles of length and weight of a national sample of United States children. The growth of Thai children older than 18 months fell to below the third percentile of the American data. When nine-year-old children were compared with local and international reference data, skeletal maturation, as assessed from the hand–wrist radiographs, was delayed up to 34 months in girls and up to 13 months in boys.

Disease histories, parasite infestations, and mortality during childhood were not significantly associated with growth in this sample. Children receiving the nutritional supplement did not grow significantly larger or faster than did children on the unsupplemented diet. Bailey *et al.* concluded that the delays in growth were not due to disease or the lack of specific nutrients, such as protein, vitamin A, or iron; rather the delays were due to a deficiency in the total intake of calories. The falloff in growth at 18 months of age corresponded with the average age at weaning in these villages. Weaning foods were usually watered-down versions of adult foods. Although there were no food shortages in the villages, the small gastrointestinal tracts of the weaned infants and young children may not have been capable of digesting enough food to meet their caloric demands for maintenance and growth of the body.

Behar (1977) found that the same problem was faced by young children in rural villages in Guatemala. Young children had access to sufficient food, but the traditional diet of corn and beans did not have the caloric density to meet their growth requirements. Waterlow & Payne (1975) reviewed several studies that weighed and analyzed the food intakes of one-to two-year-old children not receiving any breast milk. Results of the analysis for protein and energy (calories from all sources) content of the food is presented in Figure 6.1, in which the percentage of the recommen-

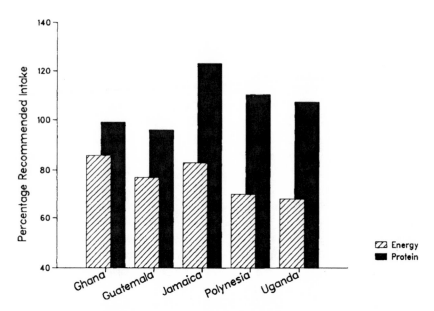

Figure 6.1. Percentage of United Nations (Food and Agricultural Organization and World Health Organization) recommended intakes of protein and energy of weaned children, one to two years old, in five developing nations. From Waterlow & Payne (1975) and Johnston (1980).

ded daily allowance, according to United Nations guidelines, are compared. Protein intakes are close to or exceed 100 percent of the recommended allowance in every country, but energy intakes are below recommendations in every case. These data support the conclusions of Bailey *et al.* (1984) and Behar (1977) that calorie insufficiency, due to feeding foods inappropriate for the digestive capabilities of young children, leads to undernutrition and growth delay.

This research indicates that cultural behavior related to infant and child feeding may determine differences between populations in nutritional status and growth. One specific example is a study by Jenkins *et al.* (1984) of the feeding practices and growth of infants and children of the Amele culture of Papua New Guinea. Amele parents classify several stages of child growth according to developmental landmarks, and ascribe to each stage appropriate foods. Newborns may only be fed breast milk or warm fruit and vegetable juices. Amele mothers believe that the quality of breast milk improves as the infant grows older, so breast-feeding is often prolonged up to four or five years. Infants who can sit alone may be given some mashed foods mixed with liquids, but not roasted foods, or those mixed with coconut milk, as these foods are too dry or slimy and may induce coughs. By the time a child can run and play with other children he may request and be given all types of food.

Jenkins *et al.* grouped 22 infants, for whom there were longitudinal growth data from birth to 24 months, according to the Amele developmental stages and calculated the percentage of children gaining, maintaining, or losing weight relative to international standards of weight for age. During the first growth stage, roughly from birth to six months, 77 percent of the infants gained weight relative to the standard. Thus, breast milk seemed adequate for the majority of these infants. During the next stage, from about six to 12 months, 82 percent of the infants lost weight on the, predominantly, breast milk dependent diet; the fruit and vegetable juice supplements were of little caloric or other nutritional value, since most of these supplements were more than 90 percent water. Breast milk and these liquid supplements did not meet the growth needs of the infants. The next two growth stages cover the period from roughly 12 to 24 months. As children began to sit, crawl, stand, and walk, progressively more solid foods of plant and animal origin were added to the diet. The percentage of children losing weight relative to the standard fell to 28 percent in the third stage and about 15 percent in the fourth stage. Although Amele children age 24 to 36 months appeared to grow adequately, they achieved heights and weights at only the fifth to tenth percentile of the international standard. Thus they were relatively small as children, and measurements of adults show that they remain small.

An experimental study of nutrition and growth

It is estimated that about 200 million of the world's infants and children under age five years are undernourished, and that of these approximately 12 million die each year (Bellamy, 1996). Most of the deaths may be attributable to specific infectious diseases, such as measles, but it is well known that it is the combination of undernutrition and infection that is the real cause of death. This combined effect is called a **synergistic action**, because neither undernutrition nor infection acting alone would result in so many deaths (Scrimshaw, 1968). The synergism of hunger and disease leaves its mark on those who survive past childhood, and usually that effect is growth retardation in either height, weight, or both, reduced intellectual capacity, and reduced work capacity (Adair & Guilkey, 1997). Each of these deficits impairs the individual, reduces the chances that the individual can lead a productive and satisfying life, and increases the chance that hunger and disease will be recycled into the next generation.

It seems obvious that one way to break the cycle of hunger, disease, and despair is to provide more food to infants and children. An experimental study, started in the 1960s, has been evaluating the effects of food supplementation on infants, children and adolescents. This is the INCAP 'Four Village Study' that was described briefly in Chapter 1. The original study ran from 1969–77 and involved the supplementation of pregnant women and all infants and children with one of two drinks, *atole* or *fresco*. Two villages received the experimental supplement called *atole*, and two villages received a type of placebo supplement called *fresco*. The *atole* drink provided a total of 163 kilocalories (kcal) per cup (180 milliliters) from 11.5 g of protein and 27.8 g of carbohydrate. The *fresco* provided 59 kcal per cup from 15.3 g of carbohydrate. After 1971, both supplements contained equal amounts of several vitamins, and both were given to pregnant women, infants, and children twice a day at feeding stations in the villages. The participants could consume all they wished, and intake was carefully monitored. The infants and children participating in this original INCAP study were followed until they were seven years old. Results of this study show that those people receiving the *atole* drink had higher birth weights, lower infant mortality, and improved growth for infants under three years of age (Martorell, 1995).

A follow-up study was conducted from 1988–9 to assess the original participants, who were then 11–27 years old. The reason for the follow-up was to see if nutritional supplementation up to age seven years had long-term effects. The answer is yes, there are several long-term effects of the *atole* drink, including greater stature and fat-free mass (the weight of the skeleton, muscles and organs of the body), especially for girls and women. *Atole* users also showed improved work capacity in the males and higher scores on tests of intellectual performance in both males and females. The *atole* group did not experience a faster rate of maturation, as measured by either skeletal development or age at menarche, than the *fresco* group (Martorell, 1995). From this study it is very clear that improved nutrition in early life has real benefits to human development at all stages of the life cycle from fetal life to adulthood.

The milk hypothesis

To what extent can a specific food make an impact on human growth and development? There has been considerable interest in this question and of all the foods studied so far, milk has generated the most important results. Takahashi (1984) made a strong case in favor of what may be called the milk hypothesis, which posits that increased consumption of milk by infants, children, and juveniles is directly related to greater average height of a population. Takahashi found an association between changes in dietary practices, especially milk consumption, and growth in Japan. Rice has been, and still is, the dietary staple of Japan. Until 1950, fish and shellfish were the major sources of animal protein, although most dietary protein was of plant origin, from soybean products. Post-war changes in Japan, including greater contact with Western cultures and economic development, altered the traditional diet. These diet changes began in the late 1950s, but became pronounced in the mid-1960s. From 1966 to 1976, rice consumption decreased from about 350 to 225 grams per person per day. During the same time, meat consumption rose from about 35 to 60 grams per person per day and milk consumption rose from about 55 to 100 grams per person per day. The height of school boys, aged six to 17 years, rose by an average of 4.1 cm between 1930 and 1960, a period of relatively great social and economic change, but rose by an average of 5.3 cm between 1960 and 1975. Takahashi attributes almost all of the increase in the 1960 to 1975 period to changes in diet, especially the increased consumption of milk. Other factors, such as lower rates of childhood disease and reduced family size, may also have contributed to the height increase, but the result is that between 1960 and 1975 the average height of 17-year-old Japanese boys increased from about 163 cm to 168 cm, a difference which is remarkably close to the 6.4 cm difference in average adult height between chronically undernourished Gambians of West Africa and the much better nourished population of Great Britain (Billewicz & McGregor, 1982).

To further support the 'milk hypothesis', Takahashi (1984) reviewed data on the growth of pastoralists, living in traditional, non-Western cultures, whose diet is based on animal milk, and agriculturalists, whose diet is usually devoid of milk and milk products. The pastoralists of Central Asia (peoples of the Gobi, Takola Makan, and Kavil Deserts) and the pastoralists of East Africa (the Masai, Samburu, and Datoga) were found to be taller than their rice- or grain-growing counterparts. In a related study, the growth of the Turkana, a pastoral people of Kenya, East Africa, was evaluated by Little *et al.*, (1983). Their findings support Takahashi's 'milk hypothesis'. The basic staple of the Turkana diet is milk, supplemented with blood and meat from their animals, and grains and sugar, which were acquired through trade. Little *et al.* found that the Turkana are significantly taller than their agricultural neighbors with whom they trade. As may be seen in Table 5.1, at age 20 the Turkana are relatively tall, but are shorter than Americans. Little *et al.* found that the Turkana continued growing until age 23 years, at which time they are as tall, on average, as Americans. The mean weight of the Turkana is significantly below that of the American blacks and whites at all ages. Little *et al.* suggested that the Turkana have high protein, but low calorie intakes that '. . . contribute to an adequate lean tissue disposition but limits the storage of adipose tissue' (p. 826). These findings indicate that the energy supplied by the milk, and other foods of animal origin, is not the factor that results in the increased growth in height. Rather, the 'height factor' is likely to be some other nutrient, or combination of nutrients, found in milk.

Takahashi was not the first to find an association between milk consumption and increased growth in height. Orr (1928) and Leighton & Clark (1929) gave schoolchildren, in several cities in Scotland, an extra pint of milk per day for seven months. Some children received whole milk and some skimmed milk. Both groups increased faster in height and weight than two control groups of children of the same ages, one group given no supplement and the other given a supplement of biscuits, equaling the milk in total calories. Since the biscuit supplement had no effect on growth, it was concluded that some factor in milk, either whole or skimmed, accelerated growth. Several other milk supplementation studies show similar results, including studies in the United States (Spies *et al.*, 1959) and New Guinea (Lampl *et al.*, 1978). In a reanalysis of the Scottish studies of the 1920s, Celia Petty (1989) found that the milk supplemented groups included '. . . a disproportionate number of children who were stunted (i.e., whose height for age was below the third centile) . . .' (p. 106) compared with either the group receiving the biscuit supplement or the unsupplemented group. Petty believes that the reason why the milk groups grew significantly faster than the non-milk groups may have been due to their catch-up growth following a more severe period of undernutrition.

Further support for the milk hypothesis comes from a study of heights of conscripts into the Bavarian army in the nineteenth century (Baten, 1998). The Kingdom of Bavaria was an independent state in the early nineteenth century; today it is a region of Germany. All young men of the Kingdom were measured for height as part of mandatory military conscription. Baten analyzed conscription lists for about 15 000 men born between 1815 to 1849, who were 21 years old at the time of measurement. He analyzed variation in height in relation to real wages, food production, urban–rural location, and estimates of disease impact. Average heights were greatest in the dairy herding regions, and milk consumption was the single most important variable associated with stature variation. Milk consumption was very high in the dairy regions because it was not possible to transport fresh milk to markets in other regions. Cheese was not an important trade good either at that time. Instead, dairy farmers produced clarified butter (which is essentially milk fat, devoid of protein and calcium) for trade to other regions. Conscripts from grain- or potato-producing regions, and from weaving districts where workers had money to purchase food, were significantly shorter than conscripts from the milk regions. So again, it seems that some factor in milk is the cause of increased height.

It is known today, of course, that milk contains several essential nutrients, including protein (amino acids), calcium, and vitamin D_3. The calcium and vitamin D_3 are essential for normal bone formation and growth in height. Some recent research indicates that calcium is the leading nutrient responsible for the milk effect. In one United States study a group of lactose-intolerant children and juveniles (ages six to 12 years), all identified as 'white' and including six boys and 13 girls with a mean age of 9.6 years, were measured for height, weight, and bone mineral content (Stallings *et al.*, 1994). These lactose-intolerant children and juveniles could not drink milk. Their dietary intakes were assessed by using both 24-hour recalls of foods eaten and a six-day food record. Their mean intake of calcium was 583 mg/day, which is only 61 percent of the appropriate age- and sex-specific recommended daily allowance (RDA). All other nutrients analyzed were consumed at rates above 100 percent of the RDA, except for total energy (calories) which were consumed at 89 percent of the RDA. The average height of these subjects is 134.2 cm, which is the 39th percentile of the reference charts for the United States (the 50th percentile is considered 'average'). Their mean weight of 32.1 kg places them at the 53rd percentile of the reference, and indicates that their growth in height is more affected than their growth in weight.

Similar results are reported in a study of 18 infants (1.0 to 3.5 years old) in Finland with clinically proven cow's milk allergy (Tiainen *et al.*, 1995). These infants were compared with healthy infants and it was found that both groups had similar intakes of total energy, but the allergic infants consumed less protein. The allergic infants were supplemented with calcium, but despite this they were 0.8 standard deviation units shorter than the healthy group. This is a relatively large difference in body length and suggests that the lack of milk in the diet has negative consequences on skeletal growth. In another study, 84 seven-year-olds living in Hong Kong were given either a calcium supplement of 300 mg/day or a placebo tablet over an 18 month 'double-blind' trial (Lee *et al.*, 1995). The Chinese typically consume little or no milk after infancy, and the pre-trial calcium intakes of the subjects were low, about 570 mg/day. The addition of the supplement, therefore, increased daily calcium intake to about 870 mg/day. After the 18 months the supplemented group had significantly greater increases in bone mineral density, which is a sign of positive bone growth, but there were no differences in the amount of growth in height during the trial. In another 'double-blind' study, 149 healthy Swiss girls, mean age 7.9 years, were provided a supplement of 850 mg/day or a placebo for 12 months (BonJour *et al.*, 1997). The supplemented group achieved both greater increases in bone mineral density and in height compared with the placebo group. The greatest differences in bone growth were found between the supplemented group and those girls in the placebo group with a natural calcium consumption below 880 mg/day. This finding may explain the lack of a significant effect on height growth in the Hong Kong study. In that study, even the supplemented group consumed, on average, less than 880 mg/day of calcium. That amount may be enough to influence bone density but not enough to promote additional growth in bone length.

Other nutrients

By definition, the lack of specific essential nutrients, such as those supplied by milk, can delay growth and may be the cause of some population differences in size. Iron deficiency is a serious problem in many parts of the world, affecting about two billion people (Ryan, 1997b). Iron deficiency causes anemia, and about 51 percent of children and infants under five years old are anemic. Iron-deficiency anemia may lead to delays in both physical growth and psychomotor development, reduced resistance to infectious disease, and increases in gastrointestinal disorders (Ryan, 1997b).

Iodine deficiency is another widespread nutrient problem, afflicting about 1.5 billion people world-wide. Iodine deficiency disease causes both physical and mental growth retardation in infants and children. Greene (1973) studied the growth of Ecuadorian Indians living in an area with chronic iodine deficiency. Men without clinical signs of iodine deficiency (goiter, cretinism, etc.) averaged 155.7 cm in height. Male deaf-mute 'cretins' averaged 146.2 cm. The difference is statistically significant, and a similar difference existed between normal and affected women. The height of normal men and women in the study area, and two other areas in the 'low iodine' zone of Ecuador, were compared with the heights of Quechua Indians from Peru. The Ecuadorian men averaged 155.5 cm and the Ecuadorian women averaged 144.8 cm. The Peruvian men and women averaged 160.0 cm and 148.0 cm in height. All the individuals were of Indian ethnicity, of low socioeconomic status, and lived at high altitude. Greene argued that the major environmental difference between the Ecuador and Peru samples, accounting for the difference in stature, was the low iodine availability in Ecuador.

In some places in the world, deficiencies of other nutrients, such as zinc or vitamin A, are serious problems. Although the deficiency of any essential nutrient will result in growth retardation, these kinds of nutritional problems are rare in comparison with mild-to-moderate energy deficiency, which is ubiquitous in the lower socioeconomic status populations of the developing nations of Africa, Asia and Latin America and also present among the disadvantaged peoples of the developed nations. The biological response of children to calorie undernutrition, in addition to the retardation of growth in height and weight, is to delay the rate of maturation towards adulthood. In principle, the delay in rate of maturation may allow for some 'catch-up' in size, since the malnourished individuals may grow for a longer amount of time. Bailey *et al.* (1984) reported that delays in skeletal maturation do allow undernourished rural Thai children to grow for, on average, at least a year longer than United States children. However, the delays in length that the Thai experienced over their lives meant that, as adults, they reached only the fifth percentile for height of the United States reference population. Mild-to-moderately undernourished children in rural Peru (Frisancho *et al.*, 1970) and Guatemala City (Bogin & MacVean, 1983) were found to be shorter and delayed in skeletal maturation compared with better-nourished children. In both cases, the relative difference in stature between the malnourished and better-nourished samples was greater than the relative delay in skeletal maturation. This suggests that, like the Thai children, the undernourished Peruvians and Guatemalans would also be of shorter stature as adults.

A similar pattern of growth was found for rural Indian boys by Satyanarayana *et al.* (1980), whose growth velocity data were illustrated in Figure 5.1. Boys who had been malnourished and delayed in growth at age five years entered adolescence about two years later than better-nourished boys. Despite their longer period of growth, the malnourished boys were significantly shorter than the better-nourished boys at age 18. At age five the two groups differed by 16.5 cm in height (105.0 cm versus 88.5 cm), and at age 18 the groups differed by 15.5 cm in height (164.5 versus 149.0 cm). Though the malnourished boys were likely to continue growing for about two years longer (and adding perhaps 5 cm in stature) than the better-nourished group, the significant difference in height would persist into adulthood. From studies such as those just described it is clear that patterns of growth established before the age of five years may often continue unabated to adulthood. It is crucial, therefore, to intervene at the earliest possible age – before birth when possible – in order to overcome the synergistic effects of undernutrition and disease on growth.

Altitude

More than 25 million people live in high altitude regions of the world, that is at altitudes of 3000 meters above sea level or higher (Baker, 1977). The effects of high altitude on the growth of populations living in the Andes mountains (South America), the high plains of Ethiopia (Africa), and the Himalaya Mountains and Tibetan Plateau (Asia) have been studied most intensively. High altitude environments impose a number of stresses on people, including hypoxia, high solar radiation, cold, low humidity, high winds, and rough terrain with severe limitations of agricultural productivity of the land. Of these, cold temperatures and **hypoxia**, the lack of sufficient oxygen delivery to the tissues of the body, has been considered to be the most important determinants of growth at high altitude. Cold requires a higher rate of basal metabolic rate and robs the body of energy that could be used for growth. Hypoxia may be the most severe stress, since it cannot be overcome by any cultural or behavioral adaptation available to native high altitude peoples. The cause of hypoxia is the low partial pressure of oxygen in the atmosphere above 3000 meters. The partial pressure of oxygen in the air and in the alveoli of the lung, the site of oxygen exchange between the lung and the red blood cells, determines, in part, the saturation of hemoglobin with oxygen. Red blood cells carry the oxygen-saturated hemoglobin from the lung to the tissues of the body where the oxygen is used to maintain normal metabolic activity. At sea level, the hemoglobin of the red blood cell is about 97 percent saturated with oxygen as it leaves the lung, but at 3000 meters the arterial hemoglobin is only about 90 percent saturated with oxygen. The decrease of oxygen saturation is sufficient to disrupt cellular metabolism (Luft, 1972), and may delay cell growth.

In humans, prenatal growth retardation at high altitude, as evidenced by low birth weight, was shown by Lichty *et al.* (1957) for infants born in Leadville, Colorado (USA), altitude 3000 meters, and by Hass *et al.* (1980) for infants born in La Paz,Bolivia, altitude 3,600 meters, compared with lowland births from the same populations. Haas *et al.*, controlled for the effects of maternal nutritional status and smoking, gestation length, ethnic background (Indian or Mestizo), and length of residence at high altitude. When statistically adjusted for these confounding variables the mean birthweight of low-altitude infants was 3415 grams and that for high altitude infants was 3133 grams, which is a statistically significant difference. The mean crown–heel length of the low altitude sample was 49.6 cm and that for the high altitude infants was 49.0 cm, also a significant difference. Thus, prenatal growth of the skeleton, as well as soft tissues, is reduced at high altitude. This study, and more recent research, indicate that problems with the delivery of oxygen to the fetus are part of the reason for the growth delay.

Growth after birth is also delayed at high altitude, and it was once almost dogma that the cause was largely due to hypoxia. In recent years, however, more evidence has accumulated in favor of undernutrition as the major cause of growth delay. An overview of some key studies is in order to show how and why this shift in reasoning occurred. One classic study, by Frisancho & Baker (1970), found that Peruvian Indian children living above 3000 meters are shorter and lighter, on average, than lowland Peruvian children of the same age. Frisancho & Baker (1970) characterized the high-altitude children as having a slow rate of growth, a prolonged growth period lasting to age 22, and a late and poorly defined growth spurt. Similar findings were reported by Beall et al. (1977) and Mueller et al. (1978, 1980). The Andean studies, carried out in Peru, Bolivia, and Chile, also found that chest dimensions, relative to stature, are greater in high-altitude children and adults than in lowlanders. Mueller et al. (1980) summarized these studies as indicating that hypoxia may induce growth of the oxygen transport system, chest size being one aspect of this system, even as it retards growth in height.

In contrast, Clegg et al. (1972) found that high-altitude-living Ethiopians were taller and heavier than ethnically similar people living at low altitude in Ethiopia. In this case, the lowland population suffered from malaria and intestinal parasites that may have compromised their growth. Altitude, meaning hypoxia, was shown to be of secondary importance in this early study. Similar findings were reported by Frisancho et al. (1975). In this case, chronically undernourished Peruvian Indian children living in the slums of Lima, Peru (altitude approximately 500 meters) were shorter and lighter than better-nourished Indian children at high altitude. Gupta & Basu (1981) found that Sherpas living at high altitude (3500–4500 meters) in Nepal were taller, heavier, and fatter than Sherpa migrants to lower altitude (1000–1500 meters) areas of India. As in the Ethiopian case, the lowlanders suffered from malaria and intestinal parasites, diseases that were relatively rare in the highlands. In all these studies, the insults of undernutrition and disease in low altitude environments may have a greater negative effect than the hypoxia of altitude on growth and development.

Perhaps the clearest evaluation of the effect of hypoxia on growth comes from the work of Stinson (1982). She measured 323 children, between the ages of eight and 14 years, attending a private school in La Paz, Bolivia. The children, mostly of Spanish or Spanish-Indian ancestry, were of middle to upper socioeconomic class, healthy, and well-nourished. The children had resided at high altitude for various amounts of time. Stinson found a negative correlation between stature and length of residence in La Paz, that is the shortest children had lived at high altitude the longest time. The altitude effect was small, however, with about a 2.5 cm difference in height between children with the longest and shortest residence at high altitude. Stinson's findings were generally repeated in research by Greska (1986), who studied both European and Amerindian children and juveniles who were born and raised in the city of La Paz, Bolivia.

So, there is a modest effect on growth that may be attributed to hypoxia, but this effect does not explain the very short stature of rural Andeans. The people of the rural town of Nuñoa, Peru (altitude 4000 meters) are among the shortest of all Andean populations. Nuñoa was the site of much of the high altitude research by Frisancho, Baker, and their colleagues cited above. From 1983 to 1985, new research in Nuñoa collected dietary and anthropometric data. The authors of this study (Leonard *et al.*, 1990), found that '. . . nutritional factors have played a significant role in shaping the statural growth . . . of Nuñoans' (p. 613). Most of the population live in poverty, with very low income and little access to land on which they can produce food or raise animals. These people eat a diet that is low in total energy relative to the needs of children and juveniles, and also very limited in terms of food diversity. Low diversity of food eaten often results in an inadequate intake of one or more specific nutrients. Moreover, there are serious food shortages for the lower income families of Nuñoa in the pre-harvest season. The result of all these dietary and nutritional stresses is that the average height of Nuñoans between the ages of two to 22 years is less than the fifth percentile of United States reference values. Leonard *et al.* extend these findings to several other rural high altitude populations in the Andes, and concluded that poor nutrition, not hypoxia or cold, accounts for the very short stature of all these Andean groups.

Research on high altitude populations in the former Soviet Union also comes to the conclusion that undernutrition, especially a deficit of protein and some vitamins, is the principle cause of growth retardation (Miklashevskaya *et al.*, 1973). However, a more recent study of Bolivian children living at high and low altitude (Post *et al.*, 1997) finds that poor nutrition and high energy expenditure are both factors that cause the short stature of low SES groups. The dietary intake and level of daily physical activity of low SES boys and girls, 10 to 12 years old, residing at high altitude (approximately 4000 meters) and low altitude (approximately 400 meters) samples were measured over a four years period. There was no difference in dietary intake between the high and low altitude groups, and both had inadequate intakes of total food to meet total energy requirements. Both groups also had relatively high levels of physical activity, which exacerbated their low energy intakes. The result is that both the high and the low altitude groups had statures at or below the tenth percentile of United States reference values.

Climate: effects on body size and shape

Heat, cold, and relative humidity are associated with variation in the size, proportions, and composition of the human body. Roberts (1953) found a significant negative correlation between body weight and mean annual temperature, that is higher body weight in colder regions, in a world-wide sample of populations. He noted also that the ratio of sitting height to total height decreased as temperature increased, meaning that people in warmer regions have relatively longer legs and arms. When Roberts separated the world sample into geographic groups he found these same temperature-dependent relationships within European, African, Asian, and Amerindian samples. Taken together, these results show that in colder climates people tend to be heavier with relatively larger trunks and shorter legs, while in hotter climates people tend to be lighter and relatively longer legged.

Newman (1953) found similar results for a large sample of Amerindian populations of North America, but within the more restricted geographic range of these populations the climate effect is much smaller than that found by Roberts. Schreider (1964a) used height and weight measurements from samples of people living at all latitudes to show that body surface area increases, on average, from colder to hotter climates. Froment & Hiernaux (1984) surveyed sub-Saharan African populations and found that body length measurements, such as stature, correlate positively with mean temperature of the hottest month and negatively with humidity of the driest month. That is, Africans living in hot–dry areas tend to be taller than Africans living in hot–wet areas. The findings of all of these studies have been confirmed and critically discussed in a comprehensive survey of both living humans and fossil hominids published by Ruff (1994).

The effects of climate on human variation in body size and shape conform to ecological 'rules' of mammalian biological adaptation to the thermal environment. In hot environments, excess body heat produced by mammalian metabolism and voluntary muscular activity must be dissipated to the environment to avoid hyperthermic stress. Such loss may occur by radiation (direct transfer of infra-red energy from the body to a cooler object), conduction (heat exchange by direct physical contact between the body and a cooler object), convection (heat exchange between the body and a cooler object via an intermediary medium, e.g. air flow), or evaporation (conversion of water, e.g. perspiration, to vapor using body heat). Relatively low body weight, or body volume, and relatively large body surface area, produced by having legs and arms relatively long in proportion to the size of the trunk of the body, assist in heat loss. Low body volume decreases the amount of metabolizing tissue, and also decreases the distance required for the radiation of heat from the internal organs and muscles to the surface of the body. Large body surface area increases the potential for convection, conduction, and evaporation. In cold environments, a relatively large body volume and small surface area (i.e., relatively short extremities in proportion to trunk size) is the body type best suited for heat retention. Body fatness, especially the thickness of the subcutaneous fat layer, may also increase in cold environments. Adipose tissue is relatively inert metabolically, due to poor vascularization, and acts as an insulating barrier against heat loss by radiation. In hot environments, a thin subcutaneous layer of fat helps minimize heat retention.

There are several studies that test the validity of these ecological 'rules' in specific human populations. Hiernaux (1974) believed that the small stature and lean body build of the Mbuti pygmies, and other pygmy peoples of Africa, is adapted to the tropical rain forest. In terms of climate, the most severe stress of the rain forest is the very high humidity of the rainy season which, combined with moderately high temperatures, limits heat loss from the body by evaporation. Radiation and convection must be maximized in this situation, and the small body mass of the pygmy peoples maximizes the avenue for heat loss by radiation. Populations inhabiting the Amazon rain forest also are, generally, of short stature and have small body mass. In some cases, such as the Maku people of Brazil and Colombia studied by Milton (1983), caloric undernutrition may be the cause of small body size. However, in a survey of other Amazonian cultures, Milton did not find evidence for undernutrition. Rather, the evidence favored the thermoregulatory value of small body mass in tropical rain forest adaptation.

Crognier (1981) surveyed 85 European, North African, and Middle Eastern populations and compared eight climate variables with 14 anthropometric measurements. Crognier hypothesized that mean annual low temperature would be the strongest influence on the body size and body proportions of the populations studied. The results showed this was true for most cranial measurements, such as head length and head breadth, but the post-cranial measurements had their strongest correlation with heat and dryness. Crognier explains these results in terms of the effective cultural adaptations to cold, such as fire and clothing, which have been in use since human groups first occupied the temperate latitudes. These means of combating cold may have relaxed selection pressures on biological adaptations to low temperature. In contrast, cultural adaptations to heat are, perhaps, less biologically effective than cultural adaptations to cold. This, Crognier believes, may explain why the populations surveyed showed the association between heat and post-cranial anthropometry. Crognier suggests a biological explanation for the correlation between cold and cranial measurements. He argues that the brachycephalic (rounder) head shape found in the populations studied is better suited for cold than doliocephalic (longer) head shape, since a sphere affords maximum volume for heat retention and minimum surface area for heat loss.

This 'ecological' argument for head shape, and other similar arguments for a biological function for head shape, is not very compelling. The same cultural means of protecting the body from cold could be used to protect the head, and are used by Eskimos who are about as doliocephalic as Africans. Furthermore, other cultural practices, such as normative sleeping positions for infants and children can alter head shape during growth. Swaddling practices that force infants to sleep lying on the back produce rounder skulls; sleeping with head turned to one side produces longer skulls. Boas (1912) showed that migration from southern Europe to New York, and other northern American cities, changed the shape of the migrants' skulls from the brachycephalic shape of the parents to the doliocephalic shape of their children in one generation. Given these caveats, there is little support for an adaptive or evolutionary explanation for head shape in any human population.

Climate: effects on seasonal rates of growth

Climate variation can also influence growth rates during the year. Buffon (1777) published the first data suggesting the existence of seasonal variation in the growth rates of healthy children. Studying the growth records of the son of Count Montbiellard, Buffon noted that most of the boy's height increase during the year took place in the spring and summer months. Scammon (1927) converted the old French metric units used by Buffon into modern metric values. These data are used to construct the curve of growth for Montbiellard's son shown in Figure 6.2. The boy was born on April 11, 1759 and his father measured his length or height at six month intervals, with a few exceptions. The boy's growth from April to

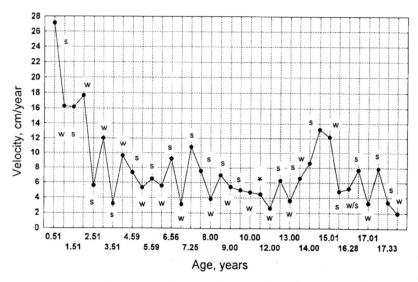

Figure 6.2. Seasonal variation in the velocity of growth of Montbiellard's son from age six months to 18 years. S indicates the 'summer' measurement, which was taken in October, meaning that it followed six months of growth over the spring and summer months. W indicates the 'winter' measurement, which was taken in April, meaning that it followed six months of growth over the fall and winter months. The * indicates an interval of one year between successive measurements, which precludes a seasonal effect. The W/S symbol indicates a time period from October to July, which includes some of the summer months.

October took place during the seasons of spring and summer in France, while the period of growth from October to April covered the seasons of fall and winter. Until about age four years, the boy grew faster during the fall and winter, but after that age grew faster during the spring and summer. During his adolescent height spurt, from ages 13 to 15.6 years, the seasonal effect seems to be obliterated.

Since Buffon's time, dozens of investigations of seasonal growth, using larger sample sizes, have generally confirmed that at temperate latitudes, healthy well-nourished children grow more quickly in height during the spring and summer than they do during the fall and winter (Bogin, 1977, 1978; Satake, 1994). Studies of seasonal variation in weight gain and loss find that for healthy individuals, fall or winter are the seasons of maximum weight gain and summer tends to be the season of minimum weight gain for children, and the time of maximum weight loss for adults. Several researchers also find that minimum weight gains, or even weight losses, occur simultaneously with maximum height gains (Bogin, 1979; Tobe *et al.*, 1994).

Seasonal variation in height growth

The cause of seasonal variation in height growth rate is not completely understood; however several investigators suggest that seasonal periodicity in sunlight may act on the human endocrine system. The effect of sunlight seems to be one that synchronizes the body's natural fluctuations in growth-regulating hormone activity so that all the necessary hormones are working simultaneously to speed-up or slow-down the rate of skeletal growth. The study of the sunlight effect published by Nylin (1929) was probably the first. He experimentally tested the sunlight effect on a sample of Swedish boys. He exposed one group of boys ($n = 45$) to 'sunlamp' treatments (using a lamp that produced both visible and ultraviolet light) during the winter months in Stockholm (from winter solstice to spring equinox) and compared their growth to a group of boys ($n = 292$) not receiving treatment. Situated at about 59° N latitude, Stockholm has only six hours of daylight at winter solstice. During the three-month period of treatment, the experimental group averaged 1.5 cm more growth in height than the control group. During the summer, the control group grew at a faster rate than the experimental group, so that over the entire year there was no difference between groups in total height gain. Only the time of year when the maximum gain occurred differed between groups.

Some years later, Marshall & Swan (1971) compared the monthly growth rates over a year's time of blind and normally sighted children living in southern England. Both groups showed rhythmic changes in growth rate of equal magnitude. The months of maximum growth of individual blind children were distributed evenly throughout the year. The maximum growth rates of the normally sighted children virtually all occurred between the months of January and June. The authors suggested that seasonal variation in day length might have affected the growth rates of normally sighted children. To further investigate this hypothesis, Marshall (1975) analyzed monthly measurements of height, taken over a two-year period, for 300 healthy children living on the Orkney Islands (59° N latitude). Marshall also recorded changes in day length, hours of insolation (hours during the day when the sun is not obscured by clouds or haze), rainfall, and temperature. No significant associations were found between monthly increments of height and any of the meteorological variables. Marshall concluded that there was no support for the 'day length' hypothesis.

However, two other studies lend support to an association between variation in sunlight and growth rate. Vincent & Dierickx (1960) found that healthy children living near the equator in Kinshasa (then Leopol-dville), Zaire, grew more rapidly in height in the dry season than in the rainy season. Diet, temperature, humidity, and sunlight variation were considered as possible influences. No evidence in favor of the first three was found. Although day length was slightly longer during the rainy season, there were far more hours of insolation (bright sunshine), and more opportunity for children to be exposed to the sun, during the dry season. The authors concluded that exposure to sunlight could regulate growth rate. My own early research (Bogin, 1978) focused on seasonal variation in growth. Starting in September, 1974 I measured monthly increments of height growth over a 14-month period for a sample of 246 healthy children, juveniles, and adolescents of high socioeconomic status living in Guatemala City (14° N latitude). I used a statistical technique called periodic regression and harmonic analysis to analyze the data. This technique can detect if a true seasonal pattern of growth is present in a series of measurements (Bogin, 1977). The sample of children that comprised 43 boys and 42 girls aged 5.0 to 7.9 years old showed the clearest evidence of a true seasonal pattern of growth in height. About 75 percent of the children grew at a significantly faster rate during the dry season than during the rainy season (Figure 6.3). Conversely, about 25 percent of these children grew at their fastest rate during the rainy season and their slowest rate in the dry season. Results such as these are found in most seasonal growth studies, but the reason why a minority of children follow a pattern that is just the opposite to that of the majority has never been explained.

It is possible to offer an explanation for the seasonal pattern followed by the majority of children. As in the African study, day length was longer during the rainy season, by about 1.5 hours, but there were significantly more hours of insolation during the dry season (1662 hours from October to April) than during the rainy season (962.8 hours from May to September). Both in equatorial Africa, and in Guatemala, therefore, the strongest association between sunlight and growth rate is not for day length, but for hours of insolation.

A relation between light and growth has been known since 1919, when it was shown that ultraviolet light could cure rickets, a disease of bone growth. A few years later the relationship between vitamin D and normal bone growth was demonstrated (see review by Cousins & Deluca, 1972). It is now known that cholecalciferol, vitamin D_3, is synthesized by the skin (and modified in the liver and kidneys to its active form) when people are exposed to ultraviolet light. The physiological action of vitamin D_3 is to increase the intestinal absorption of calcium and to control the rate of skeletal remodeling and the mineralization of new bone tissue (Rasmusen,

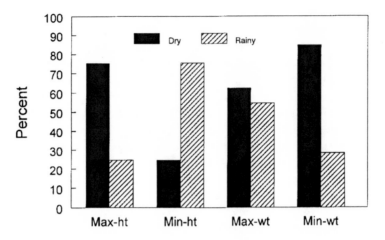

Figure 6.3. Seasonal variation in height and weight growth of healthy, high SES Guatemalan children. The sample includes 43 boys and 42 girls, aged 5.0 to 7.9 years old. The timing of maximum and minimum rates of growth in height follow a true seasonal trend. About 75 percent of the children have maximum rates of height growth in the dry season and minimum rates of height growth in the rainy season. Rate of weight growth does not follow a seasonal trend. About equal numbers of children have maximum weight gains in both the dry and rainy season. A significantly greater percentage of children have minimum weight increments, including weight loss, in the dry season.

1974; Vaughn, 1975). Vitamin D_3 is essential for normal bone growth, and thus growth in height.

Vitamin D_2, the form of the vitamin that was once commonly added to milk and other foods, has these same general properties, but is many times less potent in action (Wurtman, 1975). Haddad & Hahn (1973) working in St. Louis, Missouri and Stamp & Round (1974) working in London, England found no correlation between dietary intake of vitamin D_2 and levels of vitamin D_2 metabolites in the circulatory system of healthy white adult subjects. Both investigations did find significant month-to-month variations in the circulatory levels of vitamin D_3 metabolites. The amount of vitamin D_3 in the blood rose and fell in direct proportion with the availability of ultraviolet radiation from the sun. In a more recent study by Guillemant *et al.* (1995), measurements of height, weight, and vitamin D_3 of 28 male adolescents were taken in September, 1993 and March, 1994. The boys, living in or near Paris, France, ranged in age from 13.5 to 15.75 years old at the start of the study and all were in good health. This is not really a seasonal growth study since only one six month period of time was considered. Nevertheless, the boys gained an average of 2.9 ($\pm$1.3) kg and

3.3 ($\pm$ 1.2) cm over the fall and winter months. The vitamin D levels in their blood dropped significantly over the same time. The loss of vitamin D was due to both utilization of the body's stores of the vitamin for bone growth and other metabolic activities, and the reduced production of the vitamin during the winter months.

These studies show that the major source of vitamin D is the body's synthesis of the D_3 form, and not dietary supplements of the D_2 form. In both the St. Louis and London studies cited above, the amount of ultraviolet light penetrating the atmosphere and reaching the ground at noon was 15 times greater in June than in December (Wurtman, 1975). Variation in day length, outdoor exposure, and clothing requirements increase the difference between seasons in the exposure of people to ultraviolet light. While dietary vitamin D_2 may be of little importance, the amount of vitamin D_3 in cow's milk could be. Samuel Brody (1945, p. 211), whose work on growth theory was reviewed in Chapter 3, noted a strong seasonal variation in the concentration of vitamin D_3, as well as vitamin A, other vitamins, several minerals, and fatty acids, in cow's milk. Brody reports that in the northern hemisphere, the July levels of vitamin D_3 were at 200 percent of the yearly average, while the December levels were about 35 percent of the yearly average. Today, processed milk sold at food markets, at least those in the United States, has uniform levels of vitamin D_3 throughout the year. Such uniformity did not exist 25 or 30 years ago, when, in fact, much processed milk was fortified with vitamin D_2. It is possible, therefore, that the seasonal variation in the concentration of vitamin D_3 in cow's milk may have contributed to bioavailability of this vitamin in the diet of boys and girls studied prior to 1970, and their seasonal pattern of growth in height.

In the tropics, where variation in day length is minimal, the amount of ultraviolet light reaching the ground is a function of the amount of cloud cover blocking the sun's radiation. Physicists call the process whereby light energy is reduced when passing through any medium 'extinction'. The extinction of ultraviolet radiation occurs when sunlight entering a cloud is scattered by collisions with water droplets or suspended particles. It is estimated that a thick white cloud will extinguish 90 percent of the ultraviolet light entering it (Van de Hulst, 1957). Thus, at tropical latitudes, the measurement of hours of insolation provides a fairly accurate indicator of available ultraviolet light and rate of vitamin D_3 synthesis in the body. The studies of seasonal variation in growth conducted in Zaire and Guatemala found that the availability of ultraviolet light, and the opportunity to be exposed to it, are greater in the dry season than during the rainy season. Greater amounts and exposure to ultraviolet light may have increased the rate of vitamin D_3 synthesis, calcium absorption, skeletal remodeling, bone mineralization, and growth in height, although direct studies of these relationships have yet to be done.

Seasonal variation in weight

The causes of seasonal variation in the rate of growth in weight seem to fall into two categories, depending on the circumstances of the populations studied. The first category includes populations suffering from seasonal food shortages or disease, such as the people of The Gambia studied by Billewicz (1967) and Billewicz & McGregor (1982). In these cases, the cause of weight growth variation is clearly due to increases and decreases in energy balance during the year.

The second category includes healthy, well-nourished populations, and in this case the causes are not so easily discerned. For my own research, I reviewed the literature relating to seasonal variation in the rate of growth in height and weight. I found that 22 of the 29 studies showed that maximum increments in height and weight do not occur at the same time of the year (Bogin, 1977). The authors of some of these studies speculate that children have a natural, endogenous rhythm for growth in weight that is independent of both the seasonal rhythm of growth in height and of seasonal variation in climate.

In my own research (Bogin, 1979), I measured monthly increments of weight change of 85 healthy well-nourished children (43 boys and 42 girls) living in Guatemala. I found that minimum weight gains, and even weight loss, occurred for most individuals (90 percent of the boys and 80 percent of the girls) during the dry season, the time of maximum increments in height. The number of children losing, or not gaining, weight showed a tendency to increase from month to month during the dry season. In the last three months of the dry season (March, April, and May), the percentage of boys and girls losing or not gaining weight rose from 30.5 percent, to 37.0 percent, and finally to 57.5 percent. The pattern of weight change could not be explained by observable behavior related to diet, exercise, or disease. Perhaps the increased rate of metabolism required to grow in height at the fastest rate for the year required more energy than was supplied by the diet of the these, otherwise, healthy and well nourished children. The children may have lost body fat to supply the energy needed by the skeleton for growth. Unfortunately, I was young and inexperienced at the time of the study, and I did not take any measures of body composition. To my knowledge, monthly measurements of body composition have never been taken on a reasonably large sample of children.

Month of birth effect

Seasonal variation in the rate of height growth is now a well-established phenomenon, and its causes are probably known. A month-of-birth effect on the height of both young people and adults has been reported in a few studies. The validity of this phenomenon is less well established and its causes are not understood. The month-of-birth effect was noted by Henneberg & Louw (1990) in a cross-sectional study of 1165 boys and girls, six to 18 years old, attending schools in Cape Town, South Africa. The participants in the study were all judged to be healthy and well-nourished. Using appropriate statistical methods, the data for all age groups and both sexes were combined in order to assess the effect of month of birth on attained size. Those individuals born during the months of August to January were found to average about 0.7 cm taller and 800 g heavier than individuals born from February to July. Though small, these differences are statistically significant. In a follow-up study, Hennerberg & Louw (1993) measured 1522 boys and girls, six to 18 years old, living in the poorest region of the Cape Province of South Africa. Most of these participants showed evidence of chronic undernutrition. Even so, a significant month-of-birth effect was evident in the data, but the timing was shifted. Individuals born during the months of November to April were taller and heavier than those born during May to October. In both studies, the pattern of month-to-month deviations of height and weight is approximately sinusoidal, that is the deviations tend to rise and fall smoothly from the birth month of minimal size to the birth month of maximal size.

A large study of 507 125 human males, all 18 years old, living in Austria finds strong evidence for a month-of-birth effect (Weber *et al.*, 1998). The height of subjects was measured as part of the army induction process. The data were collected over a ten-year period, meaning that ten different years of birth are represented. A significant sinusoidal pattern in height at 18 years, with a peak in April and a trough October, was found. The difference between peak and trough amounts to 0.6 cm. That value is very similar to the difference found in the South African studies. The major contrast between these studies is that in the southern hemisphere the maximum values for height are clustered in the last half of the year, while in the northern hemisphere study the maximum effect is clustered in the first half of the year. The month of birth effect, then, is similar to the seasonal growth effect in that birth during the spring and summer months leads to greater average height later in life. Of course, spring and summer are six months out-of-phase between the northern and southern hemispheres. This points to some seasonally variable environmental factor, such as sunlight, as the cause of the month-of-birth effect.

Evidence for an environmental cause also comes from studies of non-human animals. Hennerberg & Louw (1993) also report on the weight of 804 German Shepherd dogs, of both sexes aged 72 days to eight years old. The dogs were of known pedigree and all were raised under highly similar conditions at the same South African kennel. A significant month-of-birth effect on weight was found for these dogs, and the effect was as strong as that for humans and followed the same sinusoidal pattern, and monthly timing, as for the well-nourished Cape Town human sample. A month-of-birth effect was not found in a study of rabbits born and raised in a research laboratory at the University of Pittsburgh (Dechant *et al.*, 1997). A major difference between the studies of the dogs and the rabbits is that the dogs were allowed access to the out-of-doors, and seasonal variations in climate, while the rabbits were confined indoors and reared under constant conditions for lighting, temperature, and other environmental variables.

Weber and colleagues speculate that the month-of-birth effect may be caused by sunlight influencing the pattern of human growth during the late fetal and early postnatal period. The exact mechanism by which sunlight could do this is unknown. It is known that growth in length is faster during the last trimester and first few months after birth than at any later time of life. The stimulus of greater amount and intensity of sunlight during the spring and summer months may act on both the mother and the fetus/neonate to program a pattern of growth that is different than that for individuals born during the fall and winter months. Hennerberg & Louw (1990) point out that there is no evidence of a seasonal effect on birth weight or birth length. So, the month of birth effect on growth must act on growth after birth, and possibly accumulate over time. Infants born during the spring and summer may grow a bit more in the first year or so following the birth year than infants born in the fall and winter. By six years of age the birth month effect is detectable, significant statistically, and seems to persist to age 18. The average size of the effect, at less than one centimeter, is small in biological terms and may have little practical impact on an otherwise healthy individual. Nevertheless, the discovery of this effect and the eventual understanding of its causes, may help researchers to better understand the control of human growth and may offer clinicians new treatments for infants and children with growth disorders.

Migration and urbanization

Throughout human history people have moved from place to place. When all humans lived as hunters and gathers, migration redistributed the population but the consequences for human biology were negligible since life in one place was probably much like life in another nearby place. The advent of agriculture, an event confined to the past 10 000 years of human history, changed the effect of migration on human biology. The discussion in Chapter 5 relating to the 8000 years of secular trend in Latin America, demonstrates that the advent of agriculture, along with the concentration of people into state-level societies, resulted in a decline in human stature and human health. Indeed, the formation of cities, both in the time of the agricultural revolution and later in the time of the industrial revolution, has been a major force shaping both human biology and culture.

Throughout recorded history, social, economic, political, and biological events encouraged the development of cities and the migration of rural people to urban areas. For instance, opportunities for wage-earning employment, education, and health care are more available, generally, in the city than in the countryside. Given this, it is not surprising that rural-to-urban migration is the most common type of migration that has occurred during recorded history (Smith, 1984), and this type of human migration is occurring more rapidly today than ever before, especially in the least economically developed nations of the world (United Nations, 1980). Demographic change during the past 200 years provides a glimpse at the extent of urban migration. In the year 1800, there were about 25 million people living in urban areas. In 1980, there were about 1.8 billion and by the year 2000 it is estimated that there will be 3.2 billion, which is a 128-fold increase in two centuries. By contrast, the natural increase of the world's population will be only 6.4-fold in the same two centuries (about one billion people in 1800 and 6.4 billion in 2000).

For the first time in human history, more than half of the population will live in urban areas by the end of the century, and 60 percent by 2020. More people will live in cities by the year 2025 than occupied the whole planet ten years ago (World Resources Institute, 1996). Migration redistributes the genetic, physiological, morphological, and sociocultural differences found in human populations. It is likely, therefore, that rural-to-urban migration would have some effect on the growth and development of migrants, and the recipient population. Livi (1896) and Ammon (1899 – both cited by Boas, 1912, 1922) were the first to publish studies of the growth of urban migrants. Livi found that the children of urban migrants in Italy were taller than rural sedentes (the non-migrating rural population). He believed the reason for this was **heterosis**, the marriage of urban migrants from different rural regions leading to 'genetic vitality' in their offspring. Ammon also found the children of migrants to be taller than rural sedentes, but he argued for the action of natural selection to explain this. Though it is not clear what type of selection pressure was supposed to be involved, perhaps he meant that, in the rigors of the urban environment, only the 'fittest' (the tallest?) would survive.

These views that a genetic mechanism, heterosis or natural selection, was at work stem directly from the erroneous belief that human types are genetically fixed, that types will not change when exposed to different environments. This belief was shattered by the publication of Boas' study of the 'Changes in the bodily form of descendants of immigrants' (1912), which was discussed in both the Introduction and Chapter 1. Boas found that neither natural selection nor heterosis could adequately account for changes in growth, rather these were due to biological plasticity in the face of the new urban environments. The now classic studies of Shapiro (1939) on the growth of Japanese children in Japan and Hawaii, and of Goldstein (1943) and Lasker (1952) on the growth of Mexicans in Mexico and the United States confirmed the nature of human developmental plasticity. Today, human developmental plasticity is taken for granted, but it was the migration research of Boas, and those following his lead, that established the validity of this phenomenon.

Shapiro's Japanese migrant study compared the growth of Hawaiian-born Japanese of Japanese immigrant parentage, Japan-born Japanese who migrated to Hawaii, and Japanese sedentes living in the same villages from which the migrants originated. The sedentes were mostly farmers and laborers in rural villages. The exact location of the migrants was not given, but 65% of the recent immigrants were employed in sedentary occupations (store clerks), and 76% of the Hawaiian-born were either students or sedentary workers. The immigrants appear to have been developing an urban lifestyle. The sedentes and the recent immigrants differed in a few anthropometric measurements, some increasing and some decreasing with migration. The largest differences were between the immigrants and the Hawaiian-born. The latter were taller and more linear in body build than their parents or the sedentes. Shapiro argued that, with migration, there were improvements in diet, health care, and socioeconomic status and that these conditions, associated with an urban lifestyle, were responsible for the growth changes.

The urban advantage in growth was not always present. Meredith (1979), Malina *et al.* (1981), and Wolanski (1990) reviewed studies from the period 1870–1920 that showed rural-living children in the United States and Europe were taller than their urban peers. For example, Steegmann (1985) found in a study of eighteenth century British military records, that the stature of rural-born recruits averaged 168.6 cm and the urban-born recruits averaged 167.5 cm, a statistically significant difference. By 1930 this pattern was reversed and urban children were consistently taller and heavier than rural children. Other sources of data (Eveleth & Tanner, 1976; Fogel, 1986; Bogin, 1988a) provide some ideas about changes in the rural and urban environment at the turn of the century. For example, prior to 1850, mortality rates for infants, children, and adults were higher in the cities than in the rural areas of Europe and the United States. Around the year 1900 the trend reversed, due to improvements in sanitation, water treatment, and food preservation in the city. Urban children benefited from these improvements, as reflected in their greater stature and weight, and more rapid maturation, compared with rural children.

A well-conceived study of rural-to-urban migration and growth was carried out in Poland. The results were published in Polish by Panek & Piasecki (1971) and were summarized in English by Eveleth & Tanner (1976). A new industrial town was created on the outskirts of Cracow in 1949. In 1965 the population reached 1599 persons per square kilometer, similar to that of many other European cities. Most of the population growth was due to migration from rural villages. The children and youth measured for the study had been born in the new city or had lived there at least 10 years. The children living in the 'new city' were, on average, five to six centimeters taller and one to two kilograms heavier than the rural sedentes. The urban children matured earlier, both for tooth eruption and age of menarche. In the post-World War II period, the conditions for life in the Polish countryside were very poor, with widespread malnutrition and high rates of infectious disease. In Polish cities condition of life were better, and children of urban migrants experienced significant improvements in growth.

In many ways, post-war Poland was a developing nation. As in Poland, the populations of the developing nations of the third world also show that healthy, well-nourished children living in the cities are usually taller and mature earlier, as measured by rates of skeletal development and age at menarche, than their rural age peers (Eveleth & Tanner, 1976, 1990). This was confirmed in my own research (Bogin & MacVean, 1981c). We found that urban schoolchildren, from low socioeconomic status families, living in Guatemala were taller than rural children of the same age. However, migrants to urban slums in less developed countries do not usually experience the benefits of the urban environment. In Asia, Africa and Latin America the slums are often on the outskirts of the cities. As squatter settlements they have no official access to city services and facilities. Not surprisingly, the growth of migrant children living in these slums is not significantly different from that of children living in the impoverished rural areas. Indeed, Malina *et al.* (1981) showed that children living in the urban slums of Oaxaca, Mexico, were significantly shorter and lighter than nearby rural children from a reasonably prosperous town. Johnston *et al.* (1985) showed that residents of an impoverished community on the outskirts of Guatemala City were no taller or heavier than rural populations in Guatemala.

The rural–urban continuum

Cities are not uniformly healthier places than the rural countryside. In today's world the migration of people, goods, and services flows along a continuous path from urban areas to rural areas and back again. The development of suburban regions that surround traditional cities, and ex-urban areas that extend some of the trappings of urban life to rural towns and villages, blurs the distinctions that once existed between the city and the countryside. Since about 1990, research about the effects of migration and urbanization has begun to focus on the rural–urban continuum. One example of this new research focus is the Cebu Longitudinal Health and Nutrition Survey, a study carried out in the Philippines between 1983 to 1986 (Adair *et al.*, 1993). Many types of biocultural data were collected from people living in the city of Cebu (about one million inhabitants then), areas immediately surrounding Cebu (called peri-urban areas), from people living in small towns out of the urban area, and from people living in isolated rural areas in the mountains and off-shore islands.

One aspect of the research focused on the health and growth of infants, aged two to 12 months, living in urban squatter settlements (slums) in Cebu or living in isolated rural areas. Adair and colleagues report that infants living in urban slums and in isolated rural areas have the lowest weight-for-age of all residence areas. The urban slum infants also have the lowest values for weight-for length. In contrast, the infants raised in isolated rural areas have the lowest length-for-age. Factors that contribute to the patterns of infant growth seem to be that infants raised in urban slums have the highest prevalence of infectious diseases and of diarrhea. Also, the mothers living in urban slums report lowest rates of breast feeding. These factors are associated most strongly with the inadequate growth in weight of the urban slum infants. In the isolated rural areas the factor associated with inadequate growth in length seems to be the quality of the diet. The rural diets may be deficient in one or more essential nutrients. Adair *et al.*

mention zinc deficiency as one possibility, but there may be others as well. The researchers conclude that urban and rural environments are not uniformly distinct, nor are they internally homogeneous. In cities, the harmful effects of urban environments on growth seem to be confined to the squatter slums.

Is there a biocultural selection for rural-to-urban migrants?

Though the environment for growth in the city, versus the countryside, may determine the size of children, it has also been proposed that a biological selection of individuals takes place for migration and for marriage. By biological selection it is meant that migrants may not be a random sample of the rural population. Migrants may be genetically taller, mature more rapidly, or differ in other physical aspects from rural sedentes. Positive assortative mating between such people may further differentiate migrants from sedentes.

Several studies provide tentative support for biological selection. Steegmann's (1985) historical study of eighteenth century British military recruits found that conscripts who migrated from their county of birth were significantly taller (169.1 cm) than recruits living in the county of their birth (167.6 cm). Recall though, that he also found that urban-born recruits were shorter than rural-born men. Steegmann pointed out that eighteenth century Britain was a developing country and that urban areas were characterized by food shortages and unhygienic living conditions. Thus migration to the city was probably not responsible for an increase in stature; rather taller men living in rural counties were more likely to migrate than shorter men. Whether these were genetically taller men, or individuals who had experienced better living conditions prior to migration, was not known.

In the United Kingdom, Martin (1949) found that among men inducted into the army, (1) migrants (men living in a county other than their county of birth) were taller and heavier than the national average; and (2) migrants were taller and heavier than the natives of the recipient counties. However, Martin also found that the sedentes of the recipient areas were taller and heavier than the sedentes of the exporting counties and that migrants to 'tall' counties were taller than migrants to 'short' counties. These two facts suggested that the receiving counties in general, and the 'tall' counties in particular, had better living conditions than the exporting or 'short' areas, but data for this and age at migration, were not known.

Based on a retrospective study of 14 different ethnic groups, Kaplan (1954) found that migration was selective for physical type. Growth differences between migrants and sedentes were found too soon after migration to be due to an environmental change. However, no account of age of migration or the pre-migration environment was given. Illsley *et al.* (1963) studied migrants into and out of Aberdeen, Scotland. They found that migrants were, on average, taller than rural or urban sedentes. Migrants had generally better health than sedentes. Migrant women had lower rates of low birth weight and perinatal death for their children. Finally, migrants were generally of higher socioeconomic status than sedentes.

The Illsley *et al.* study suggests what the true meaning of migrant selection may be that it is more likely to be selection for socioeconomic status than biological selection *per se*. More recent studies confirm this. Kobyliansky & Arensberg (1974) studied migrants from Russia and Poland to Israel. The migrants were of higher socioeconomic status and were taller than the Eastern European sedentes. Mascie-Taylor (1984) reviewed geographic and social mobility in England and found that the effects of selection are additive; migrants tended to be the taller individuals of any geographic area and the taller individuals within any social class. However, the higher social classes were also more mobile. Which is more important, stature or social class? Higher socioeconomic status can, by itself, lead to increased body size and rate of maturation. As MacBeth (1984) concluded in her review of this issue, differences in socioeconomic status confound any unique biological difference between migrants and sedentes. Further confounding the issue is the fact that there is greater tendency for tall individuals to achieve higher socioeconomic status (Tanner, 1969; Bielicki & Charzewski, 1983, and see the next section of this chapter). However, the predominant selection of migrants seems to be for socioeconomic status rather than for tallness.

A more recent study of rural-to-urban migrants in Guatemala strongly supports the SES effect. Kaplowitz *et al.* (1993) made use of longitudinal data from the INCAP four-village study. At 36 months of age there were no statistically significant difference in body length, or other body measurements, between individuals who would eventually migrate to Guatemala City and those who would remain in the rural villages. There were differences between eventual migrants and sedentes in measures of SES, education, and performance on several psychological tests. Eventual migrants had significantly higher values on most of these measures. These results are supported by migration research from other disciplines in the social sciences and economics (Rogers & Williamson, 1982), as well as the biological literature cited above.

Indeed, the best prospective studies, in which individuals are identified and measured before and after migration, find no evidence for biological selection. Lasker (1954) found that age of migration and length of time in the new environment were responsible for growth differences between migrants from Mexico to United States and Mexican sedentes. Similarly, no growth differences were found between eventual urban migrants and rural sedentes in South Africa (De Villiers, 1971), Switzerland (Hulse, 1969), Oaxaca, Mexico (Malina, *et al.*, 1982a), and Guatemala (Kaplowitz *et al.*,, 1993).

Migrants to urban environments are usually fatter, and have a different distribution of subcutaneous fat, compared with rural sedentes. Ramirez & Mueller (1980) found that Polynesians migrating from traditional villages to urban areas became fatter and significantly more of this fat was deposited at the subscapular skinfold site than at the triceps site. Acculturation to Western society and 'modernization' in behavior were associated with increased fatness and an increase in centralized obesity (greater fat deposition on the trunk of the body) in adult Dogrib Indians of Canada (Szathmary & Holt, 1983) and Mexican-Americans in Texas (Mueller *et al.*, 1984). Similarly, migration from the islands of Samoa to Hawaii and California was found to be associated with increased fatness and increased centralized obesity in adult Samoans (Pawson & Janes, 1981; McGarvey, 1984; Bindon & Baker, 1985).

All of these studies are of migrants from rural areas, and more traditional societies, to urban, more modernized societies. When researches examine rural–urban difference in modern industrial societies a different picture emerges. Between 1982 and 1984, Holle Greil measured 6000 men and women, ages 18–60 years, who lived in the former German Democratic Republic (East Germany). 'Ninety-six anthropometric measurements were made on each subject. Comparisons were made between urban and rural sub-samples and across the occupational groups stratified by physical stress or energy expenditure' (1991, p. 123). She found that both men and women living in large cities were taller than subjects living in small or medium size towns, who were, in turn, taller than subjects living in villages. Measures of skeletal robustness, such as biacromial and biiliac breadth, body weight, and fatness (as measured by skinfolds) followed the opposite pattern. Villagers were the most robust, heavy and fat, followed by towns folk and then large city dwellers. The level of physical activity associated with people's jobs was also associated with stature, skeletal robusticity, and fatness. Those men and women with more physically demanding jobs were found to be shorter, but more robust and fatter than people with less physically demanding employment. Greil did not test for interactive effects between urban–rural residence and job type.

In conclusion, it seems that the original work on the human growth response to migration of Boas (1912) was correct. The studies reviewed here confirm that the environment, as mediated by such factors as socioeconomic status, food availability, and health status, is the primary determinant of biological change in growth and development following rural-to-urban migration. As the human population of the planet continues to urbanize we may expect significant and large-scale changes to affect the biology, the health and the growth and development of people living in both cities and rural regions. One effect, already noted in much research, is that the growth of cities has come at the expense of deterioration of the quality of life in rural areas (Wolanski, 1990), as reflected in the different patterns of growth between rural and urban regions. However, I have also shown here that the lines between city and countryside are becoming less well-marked. The problems of the rural populations caused, in part, by unbridled urbanization are fast becoming problems for many city dwellers as well.

Socioeconomic status

Throughout this book, especially in the Introduction and Chapters 1, 2 and 5, the relationship between socioeconomic status (SES) and human growth has been discussed. The discussion, however, usually treated SES as a proxy for other factors know to influence growth, such as nutrition, disease, and workloads. In this section, I treat SES as a factor that, in and of itself, has a direct effect on human growth and development. To do so requires that we take a very critical look at human growth as a reflection of the biocultural environment into which the individual is conceived and developed.

As defined in Chapter 2, SES is a concept devised by the social sciences to measure some aspects of education, occupation, and social prestige of a person or a social group. In its most common application in the 'First World' industrialized nations, SES is measured by the years of formal education (schooling) and the occupation of an adult. In the 'Third World' developing nations, SES must often be measured by other criteria. In farming or herding communities, the size of land holdings or number of animals owned may be useful indicators of SES. Among the urban poor, the quality of the home, as indicated by the number of persons per room, the presence of running water and toilet facilities within the home, the ownership of various electrical appliances, and the type of cooking fuel used, are sometimes used as markers of SES (Johnston & Low, 1995). The SES of infants, children, juveniles, and most adolescents is ascribed to them based on the SES of their parents. Some societies have very rigid boundaries between people of different economic, educational, and occupational statuses. In these societies the boundaries establish well-defined social classes and a person's SES is, in many ways, constrained by that person's social class. Other societies allow varying degrees of mobility between social classes, and often that mobility is linked to the quantity and quality of formal education. An example of this was given in Chapter 1, the students of the Carlschule, an eighteenth century high school attended by sons of the nobility and of the bourgeoisie. The bourgeois graduates of this school were groomed to be employed by the nobility to take care of their business affairs. A similar relationship between education and social class exists in the United States today, in that graduates of certain prestigious university business schools are selectively employed by the largest corporations, or wealthiest individuals. Many of these business-school graduates themselves come from well-to-do families, and therefore retain the social class of their parents, but some were raised in middle-class families and eventually supersede the social class of their parents. Many other examples of upward, and downward, social mobility may be described in the relatively open societies.

Both in rigid and more open societies, social class and SES are powerful influences on human physical and psychological growth and development. Conversely, stature, body composition (fatness and muscularity), and rate of development influence the social, emotional, and economic status of children, youth, and adults. An appreciation of these biosocial interactions between growth and socioeconomic status (SES) has been realized only in the past 200 years. Following the studies of children attending the Carlschule in the eighteenth century and the work of Pagliani in Italy and Bowditch in the United States in the nineteenth century (see Chapter 1), it has been known that children of lower socioeconomic status (SES) are generally smaller and mature less rapidly than children of higher SES. A more recent example is the work of Bielicki *et al.* (1981), who studied the stature of 13 000 Polish military recruits, born in 1957 and measured in 1976, in relation to three environmental factors: occupation and education of the father (SES), population size of the conscript's home city, town or village (urbanization), and number of siblings in the conscript's family (family size). Mean stature decreased with lower SES, less urbanization, and larger family size. Though each factor was shown to have an independent and significant effect on stature, the strongest effect, as measured by a statistical analysis, was for SES, followed by family size, and then urbanization.

The human populations of southern Mexico and Guatemala have been studied to assess the influence of several variables on child growth. These populations include people of various ethnic groups, such as the Maya and Ladinos mentioned previously, people engaged in a continuum of occupations from subsistence agriculture (low SES) to professional and technical jobs (high SES), rural and urban residents, and citizens of two different nations. Using a multifactorial approach, the relative influence of each of these variables on the growth status of Mexican and Guatemalan children can be computed. The following analysis uses data from published sources for six groups of children. Three of the groups, rural Zapotec-speaking Indians, rural Ladinos, and urban Ladinos from the state of Oaxaca, Mexico, were studied by Malina *et al.* (1981). The Zapotecs are subsistence farmers, with some income from crafts and wage labor. The rural Ladinos are cash-crop agriculturists, and differed from their Zapotec neighbors in their more 'westernized' life style, clothing, agricultural techniques, and formal education. The urban Ladinos live in a colonia, a very low socioeconomic status neighborhood, and parents of the children work as unskilled or semi-skilled laborers. The three Guatemalan groups, urban Ladino children of very high SES, urban Ladinos of low SES, and rural Kaqchikel-speaking Maya children of low SES, were studied by me (Bogin & MacVean, 1978, 1981a, 1984). The high SES urban Ladinos are from some of the wealthiest families in Guatemala and are afforded all of the health, nutritional, and social benefits of their privileged economic status. The low SES urban Ladinos are children of low-paid semi-skilled laborers. The families of the rural Kaqchikel follow many traditional Maya cultural practices in terms of language, clothing, house construction, and many social behaviors. However, only four percent are agriculturists, the traditional occupation, the remainder are small business persons engaged in the production and sale of crafts and clothing, or are vendors of agricultural produce.

Means and standard deviations for the height, weight, arm circumference, triceps skinfold, and arm muscle circumference of each group are given in Table 6.2. The data represent average values for children aged 7.00 to 13.99 years old. Children of higher SES, from urban areas, and of Ladino ethnicity are generally larger, in most dimensions, than children of lower SES, from rural areas, or Indian ethnicity. The most significant difference between samples is for the triceps skinfold measure, an indicator of energy reserves (i.e., adipose tissue) and, hence, nutritional adequacy. Diet surveys conducted in Mexico and Guatemala find that intakes of protein and calories are below national standards for rural Zapotec and Maya populations (Bogin & MacVean, 1984).

A canonical correlation analysis was used to calculate the rank order of importance of each of the independent variables (parental occupation, rural–urban residence, ethnicity, and nationality) on growth. This analysis reduces groups of related variables to statistical 'factors'. In this case, the five anthropometric measurements (stature, weight, etc.) were reduced to

Table 6.2. *Means (X) and standard deviations (SD) for growth status variables of children from Oaxaca, Mexico and Guatemala*

Guatemalan samples

Growth variable		Urban Guatemalan (High SES)		Urban Guatemalan (Low SES)		Kaqchikel Indian (Very low SES)	
		Boys	Girls	Boys	Girls	Boys	Girls
Height (cm)	mean	138.40	138.37	130.53	131.00	126.76	125.50
	SD	(6.56)	(7.83)	(6.71)	(6.61)	(5.93)	(6.10)
Weight (kg)	mean	34.31	35.77	29.36	29.83	27.37	27.23
	SD	(6.56)	(7.75)	(5.39)	(5.33)	(4.07)	(4.33)
Arm circumference (cm)	mean	22.20	22.07	19.56	19.89	18.59	18.47
	SD	(2.59)	(2.37)	(1.80)	(1.81)	(1.53)	(1.43)
Triceps skinfold (mm)	mean	11.23	13.37	8.47	10.51	6.34	7.67
	SD	(5.03)	(4.61)	(2.87)	(3.44)	(2.34)	(2.66)
Arm muscle circumference (cm)	mean	18.67	17.86	16.86	16.54	16.60	16.04
	SD	(1.47)	(1.35)	(1.38)	(1.35)	(1.26)	(1.16)

Oaxaca samples

Growth variable		Rural Zapotec		Rural Ladino		Urban Colonia	
		Boys	Girls	Boys	Girls	Boys	Girls
Height (cm)	mean	125.59	127.13	129.31	129.14	127.63	127.47
	SD	(5.66)	(6.27)	(5.61)	(5.76)	(6.04)	(5.79)
Weight (kg)	mean	25.43	27.00	27.33	28.00	27.10	27.27
	SD	(3.26)	(3.69)	(3.59)	(3.86)	(4.23)	(4.13)
Arm circumference (cm)	mean	17.43	18.26	18.54	19.29	18.08	19.04
	SD	(1.36)	(1.43)	(1.43)	(1.41)	(1.66)	(1.60)
Triceps skinfold (mm)	mean	6.16	8.76	5.46	8.14	6.27	8.13
	SD	(1.41)	(2.01)	(1.64)	(2.33)	(2.24)	(2.63)
Arm muscle circumference (cm)	mean	15.49	15.50	16.84	16.73	16.81	16.50
	SD	(1.22)	(1.19)	(1.17)	(1.14)	(1.31)	(1.14)

Note: Ages 7–13 were combined to compute the means and standard deviations.

one factor, which may be called 'growth status'. The correlation between the 'growth status' factor and each of these independent variables was calculated, and from the values of the correlation a canonical score for each independent variable was derived. The absolute value of the canonical score is a measure of the relative influence of parental occupation, urban–rural residence, ethnicity, or nationality on 'growth status'. The canonical scores ranked in the following order: rural–urban residency, 1.68; parental occupation (SES), 1.43; Mexican–Guatemalan nationality, 0.98; and Indian–Ladino ethnicity, 0.90. Clearly, the importance of factors associated with the general physical, social and economic environment transcend the influence of ethnicity or nationality, and their possible genetic or cultural concomitants, as determinants of the growth of these samples of children. In this example, rural residence and low SES combine to limit the means that people have to obtain sufficient food for adequate growth.

In developing nations, such as Mexico and Guatemala, there are great differences in the quality of the environment for growth from rural to urban areas, and from low to high SES. In the developed nations of North America, Western Europe, Australia, and Japan, public and private systems of health care delivery, food distribution, and social services may significantly reduce environmental differences between rural and urban region and SES classes. Even so, the power of the socioeconomic effect on growth is still in evidence. For instance, variation in height due to region of origin and father's occupation (used as a measure of SES) was analyzed by Mascie-Taylor & Boldsen (1985) in a national sample of 33 000 11-year-old British children and their parents. Children and adults from Scotland, Wales, and northern England were found to be shorter than subjects living in central and southern regions of England. However, statistical analysis of the data showed that the regional effect could be explained, almost entirely, by variation in the occupation of the father. In central and southern England, fathers had significantly higher occupational status than in the other regions. These finding indicate that within Great Britain, geographic variation in climate, diet, and genetic variation is subordinate to socioeconomic status as the major determinant of growth in height.

A number of studies have established that in the United States and Western Europe, children, juveniles, adolescents, and adults of lower SES are, on average, shorter, have less muscle mass and skeletal mass, but more fat mass than individuals of higher SES (Garn & Clark, 1975; Fulwood *et al.*, 1981; Clegg, 1982; Malina *et al.*, 1983; Lasker & Mascie-Taylor, 1989; Gordon-Larsen *et al.*, 1997). Garn *et al.* (1977) found that poorer, leaner girls in the United States grow up to be poorer, fatter women who have babies of lower birth weight than women of higher SES. Because birth weight is a powerful determinant of subsequent growth, the authors wondered whether the SES differences found in later life might not be due to differences in prenatal growth and birth weight. The question was addressed by Garn *et al.* (1984) in a study of more than 3000 live-born singletons from a national sample of births in the United States. Using a composite score of parental education, occupation, and income to measure SES, the sample was composed of two SES groups: low, with scores from one (minimum) to three, and high, with scores from eight to ten (maximum). After matching, or correcting, for maternal pre-pregnancy weight and birthweight, it was found that low SES children were significantly shorter, weighed less, and had smaller head circumference than the high SES children. The differences were present at birth and increased each year after birth up to seven years of age. The postnatal SES effect was also found to influence mental development. Garn *et al.* selected a sample of low and high SES infants, all eight months old, with scores of 80 or higher on the Bayley Scales of infant development. Scores above 80 indicate satisfactory motor and mental development. General intelligence (IQ) tests were given to the children at four and seven years of age. At these ages, about 25 percent of the low SES children had low IQ scores, while only about seven percent of the high SES children had low scores. These findings showed that the SES differences in physical and cognitive growth were not due to prenatal and perinatal influences alone, rather the socioeconomic effects were cumulative to seven years of age. Of course, this finding is not unexpected since it is known, as shown in the previous chapter, that the infant and childhood periods of growth are highly sensitive to many environmental determinants of growth.

The National Child Development Study of the Great Britain is another longitudinal study of the influence of SES on growth in height and weight. Data from this study are based on the population of all infants born in England, Scotland, and Wales between March 3 and 9, 1958. Lasker & Mascie-Taylor (1989) published the mean height, weight, and **body mass index** (BMI) of these boys and girls at ages 7, 11, and 16 years stratified by the social class of the male head of household. Lasker & Mascie-Taylor (1996) also published mean heights at age 23 for these same samples. There are two to three thousand individuals in each age group. 'In Britain, social class is officially ascribed on the basis of the occupation of the male head of the household . . . the Registrar General's 5-fold class designations . . . are . . . social class I – professional; II – intermediate; III – skilled; IV – semi-skilled; and V – unskilled . . .' (1989, p. 1). The distribution of height, weight, and BMI of this sample, by age and social class for the boys, is shown in Figures 6.4, 6.5, and 6.6. BMI is calculated as

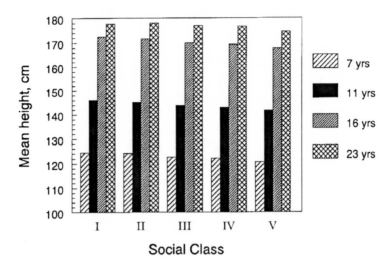

Figure 6.4. Mean height of boys born in England, Scotland, and Wales by social class of their father (or male head of the household). Data from the National Child Development Study (Lasker & Mascie-Taylor, 1989, 1996).

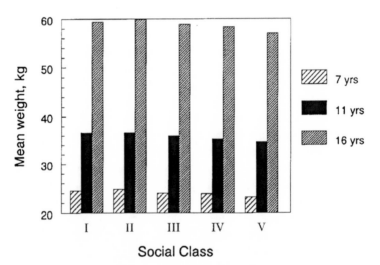

Figure 6.5. Mean weight of boys born in England, Scotland, and Wales by social class of their father (or male head of the household). Data from the National Child Development Study (Lasker & Mascie-Taylor, 1989).

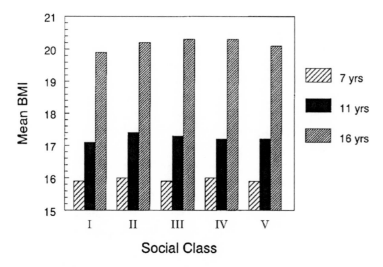

Figure 6.6. Mean BMI scores of boys born in England, Scotland, and Wales by social class of their father (or male head of the household). Data from the National Child Development Study (Lasker & Mascie-Taylor, 1989).

[weight in kilograms/(height in meters)^2]. Higher BMI scores indicate that an individual has relatively more weight-for-height than a person with a lower score. In the general population higher BMI score usually indicates more body fatness. In special segments of the population, such as highly trained athletes, or body-building enthusiasts, a higher BMI score may be due to increased muscle mass.

Lasker & Mascie-Taylor find that mean body size is significantly related to social class, and declines, generally, from social class I to V at each age. Sex (male or female) is also a significant influence on body size but there was no sex by SES interaction in any statistical analysis. Accordingly, the data for males shown in the figures are paralleled in the data for the females. The statistical impact of the social class effect is achieved by age seven years, and is then maintained to age 23 years. The notable exception to the stability of size by social class over time is the change in BMI between ages 11 and 16 years. The three lower social classes, III, IV, and V, have a significant increase in BMI compared with social classes I and II. The change is due to absolutely faster rate of growth in weight of the lower three classes.

Lasker & Mascie-Taylor (1989) also analyzed the association between body size of the boys and girls in the study and the social mobility of their father. They found that the study subjects with fathers who moved up one social class were large for their old social class but small for their new social class. The effect of the size of the child and social mobility of the father was found for the seven-year-olds, but was not statistically significant for 11- or 16-year-olds. The authors explain the association of child size at age seven and father's social mobility by stating, 'It cannot be that the growth of the child causes the male head of household to change type of occupation. It must be that socioeconomic factors that lead to the change in occupation (probably including standard of living) are established years before the job change when the young child is most susceptible to social class influence on growth in stature and weight' (p. 7).

Even within a seemingly homogeneous population, subtle differences in SES have a significant influence on growth. Johnston *et al.* (1980) analyzed the longitudinal growth records of 276 rural Mexican children. All the children came from families practicing traditional subsistence agriculture, with minimal opportunities for wage labor. The sample of children was divided into two groups, one with evidence of chronic undernutrition and growth failure, and another with satisfactory nutrition and growth status. Out of 38 variables measured for their impact on growth and nutrition, three were significant: socioeconomic status, father's linear size, and mother's linear size. Malnourished and poorly growing children came from poorer families, with less-well-educated parents, and had parents with smaller linear dimensions (height, biacromial breadth, etc.) than better nourished children. Since small body size is likely to be the result of poor living conditions (see previous sections of this chapter), the effect of parental linear dimensions may itself be due to the low SES environment of the parents when they were children. Garn *et al.* (1984) called this transgenerational influence of low SES the effect of 'recycling of poverty'.

SES factors associated with the growth of Maya living in the United States

In previous chapters my research with Maya refugee boys and girls living in the United States has been discussed. These boys and girls are significantly taller, heavier, more muscular and fatter than Maya of the same ages living in Guatemala. However, there is growth variation within the refugee sample as well. In this section the analysis focuses on differences in growth in height of the children within the Los Angeles sample of Maya. For this sample we (Bogin & Loucky, 1997) collected information for place of birth of the parents, for the child, and the length of time the child has lived in the United States. Figure 6.7 illustrates the distribution of height by age for all of the children in this sample. Imposed on the distribution are the linear regression and its 95% confidence limit lines (the confidence

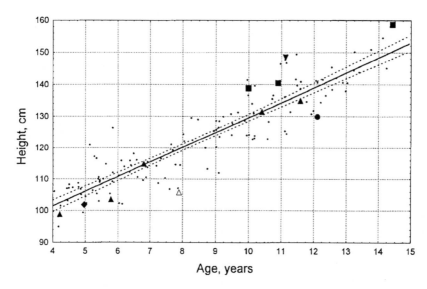

Figure 6.7. Distribution of height by age for the Los Angeles sample Maya children. Lines represent the linear regression and 95 percent confidence interval. Bold symbols indicate children from families described in the text.

limits define the range of heights which may be considered 'average' for a given age).

All of the Maya parents were born in Guatemala. Their children in the sample were born in three countries, Guatemala, Mexico, and the United States. The effect of child birthplace is estimated by forming two groups, those born in the United States and those born elsewhere. A multiple regression analysis of height on birth place and length of time in the US for the entire sample, after removing the effect of age, is not significant ($P = 0.21$ for birth place, and $P = 0.61$ for length of residence). A birth place effect may be noted if the analysis is restricted to the cases that lie outside the confidence interval. Of the 51 children above the confidence interval 30 were born in the US and 21 elsewhere. Of the 50 children below the confidence interval 21 were born in the US and 29 in Mexico or Guatemala. A chi-square test, assuming an equal number of children in all cells, shows that birthplace is a significant effect at $P = 0.07$. We interpret this result as a tendency for US, born Maya refugees to be taller than those born in Guatemala or Mexico.

The bold symbols in Figure 6.7 indicate individual children for whom we have some family ethnographic data. My coauthor, James Loucky, a social anthropologist, and I visited these families in their homes and gained some understanding about their life in Guatemala, decisions to migrate to the United States, and life in the United States. While we do not have enough ethnographic data for formal statistical analysis, we present some information about these children and their families as a first step toward a more detailed biological and ethnographic, i.e., biocultural, analysis of their SES.

The solid upward-pointing triangle symbol (▲) represents children from the 'M' family, which has resided in the United States for about three years, but has not made a successful adaptation to life in Los Angeles. They are very poor, even in comparison with other Maya in the community. The four older 'M' children were born in Guatemala, but the youngest child was born in a refugee camp in Mexico. The three oldest children are of 'average' height; the two youngest are at or below the lower confidence limit. The oldest 'M' daughter (age 10.5 years) does not attend school, rather she stays at home to care for the younger children while the parents are working. The two Maya men with whom we visited this home commented that it was a shame that the girl does not go to school. The parents could not speak English, or much Spanish, and were having difficulty finding work. A few weeks after our visit this family left Los Angeles to look for work elsewhere.

The 'M' children were measured in the family's apartment, which they share with the 'T' family. The 'T' family has been in Los Angeles longer than the 'M' family, and is, in fact, supporting the 'M' family. The open triangle (△) symbol points to the datum for height of the youngest daughter of the 'T' family. She was born in Los Angeles, but is the shortest girl for her age-group in the entire sample. The variation in stature of children from these two families shows that neither migration to the US nor birth in Los Angeles automatically result in taller stature.

The solid circle symbol (●) is for a boy who was born in Los Angeles, but returned to Guatemala when he was one year old. He lived there with his mother until he was 11 years old, at which time the family migrated back to Los Angeles. At age 12 years, he is about 10 cm shorter than the 'average' child of the Los Angeles sample and about 5 cm shorter than boys his age of the Guatemala village sample. We were not told in detail why this boy traveled back to Guatemala, but we were informed that the family is very poor and the boy and his mother needed support from their Guatemalan relatives.

The solid square symbols (■) denote the children of two sisters who have made successful social and economic adjustments to life in the Los Angeles area. One sister trained as a nurse and works as a health para-professional. Her husband is a skilled tailor working in the garment industry and earns more than double the official minimum hourly wage. The other sister is a beautician and works in an upscale shop. Both families rent their own apartments in a middle-class suburb of Los Angeles. The parents say they moved from South-Central Los Angeles to these more expensive neighborhoods in order to protect and educate their children in better schools. The parents are of average height for Maya raised in Guatemala. Two of the children are siblings who were born in Guatemala, the 10- and 14-year-olds. The 11-year-old is a cousin who was born in Los Angeles. The above average growth of these children may reflect the socioeconomic success of these families, combined with the investment the parents make in their offspring.

Higher SES alone is no guarantee of taller stature. The solid diamond symbol (◆) indicates a boy from a family that has lived in South-Central Los Angeles since 1981. The father of this family is economically successful. He owns a garment manufacturing shop and has contracts to make jeans for a major wholesaler. Before leaving Guatemala the father was a butcher and a community leader. In rural Guatemala this means that he devoted considerable time and money to sponsor *fiestas* and other community events. He continues this practice in Los Angeles and maintains his community leadership role. Our impression from visiting this family's home, and talking with the father, is that the family is very traditional in terms of many Maya values and behavior. Surprisingly, however, there was little mention of investment in his children in terms of education or their future. We also found during our visit that the parents of ◆ are of shorter stature than average for Maya adults. Given the data we have, it is impossible to decide if the shorter-than-average stature of this boy is only a genetic trait, or if the style of parental investment also contributes.

The downward-pointing solid triangle symbol (▼) is for a girl from another family that has a leadership role in the Maya community. The parents have only primary school educations and work as semi-skilled laborers in the garment industry. They are shorter in stature than the parents of the children discussed in the previous paragraph. However, they invest much of their resources into their children, who are expected to finish high school and go on to post-secondary education. The father of this family told us that,

> What is in the future depends on what my children do. Here the children will stay, eat well, live better than in Guatemala. But many adults are returning to Guatemala, because children here abandon their parents. So, it depends on your children here. If you have good work, and children study well, and you earn well, it is possible to rent or even buy a good house, and not have to think about returning to Guatemala.

In further conversation this man explained that the investments he makes in his children will pay-off for him when he is old and needs his children's support. At present, the above average height of this girl may well be a consequence of her parents' investments.

Parental investment and growth

The act of migration to the United States is based, in part, upon parental investment decisions. In biology, the study of parental investment is guided by life history theory (Stearns, 1992), centered on decisions regarding when to begin reproducing, how many offspring to have, how often to reproduce (i.e., birth spacing), and how much time and energy to give to each offspring. In both Guatemala and the United States, Maya parents say that children raised in the United States are more likely to survive, grow better, and be healthier than children raised in Guatemala. Maya women state that their infants and children are nearly twice the size as they would be if raised in Guatemala. The mothers of Indiantown often ascribe the effect to the infant formulas they use (Stebor, 1992). One mother explains,

> My daughter, Rosita, is four years old and is very small, I think she will be small all her life because she was so sick in Guatemala when she was a baby. She still doesn't eat well. Now look at my son who is almost a year old [born in the U.S.]. Already he is walking, which means his legs are very strong. He is twice the size of Rosita when she was a baby. I tell you the difference is milk [formula]   (ibid, p. 106).

Maya women, both those pregnant and those with infants, receive free or low-cost health care and nutritional supplements from the WIC (Women, Infants and Children) Program. Stebor reports that Maya women acknowledge the value of the WIC program, and justify the investment of time and money (lost wages) required to enroll in the program by pointing to their bigger, healthier babies and children. Maya infants in Florida, and probably Los Angeles, are fed more total food, including some breast-feeding along with formula feeding. The WIC program educates Maya mothers to follow hygienic practices when preparing and storing formula, and to use safe drinking water to mix the formula. One way that the SES effect works on growth is via parental education and here we may see that health education, combined with the use of milk formula, explains improved growth of the refugee children. These Maya refugee data also add more support to the 'milk hypothesis' discussed earlier in this chapter.

In rural Guatemala, prenatal and postnatal infant medical care, child care education, and safe drinking water are usually not available. Due to chronic undernutrition for the rural poor in Guatemala, Maya women may not produce a sufficient quantity of breast milk, and infant formulas or cow's milk are too expensive for most Maya to purchase. Poverty, low SES, and chronic undernutrion all contribute to poorer growth.

SES effects on rate of maturation

A girl's age at menarche is known to be a sensitive indicator of environmental conditions of growth, and girls from lower SES backgrounds usually have a later mean age at menarche than girls from higher SES backgrounds (Johnston, 1974). The subtleties of the SES–menarche relationship were reviewed by Bielicki & Welon (1982) for Polish populations. When parental education, family size, and nutritional status were controlled statistically, menarche was found to occur later in girls whose fathers lived in small towns than in big cities, and in girls whose fathers had lived in a city for less than 15 years, rather than for 15 to 20 years. Daughters of full-time farmers reached menarche later than daughters of part-time farmers living in the same villages. Daughters of men with an elementary school education had later menarche than daughters of men with elementary school and two years of vocational school education. These findings show, once again, how powerful the SES effect may be, even when living conditions vary only slightly between groups of people.

Two other studies of maturation and SES show the power of this relationship. Low *et al.* (1982) found, in the period 1961–3, significant differences in the sexual maturation (breast and pubic hair development, and age at menarche) of Hong Kong girls of high and low SES. A follow-up study during 1977–9 found no SES differences in sexual maturation. The change was entirely due to the more rapid sexual development of the low SES girls during the recent period. The authors pointed out that social and political changes, resulting in higher wages for unskilled laborers and the availability of low-cost, descent public housing, were the major environmental improvements between the earlier and later studies. Thus, socioeconomic improvements that are, at best, indirectly aimed at infants and children can have direct influence on human development.

The other study, by Hoshi & Kouchi (1981) is an assessment of changes in the age at menarche in Japan from 1884 to 1980. Mean menarchial age was about 15 years during the period 1884–1920. By 1940 the age had dropped to about 14.2 years, a decrease similar in magnitude to that found for European and North American girls. World War II reversed this trend, so that by 1952 mean age at menarche returned to 15 years. Although the war had ended by this time, these girls were conceived and born just prior to the war years, and passed through their environment-ally sensitive periods of infant and childhood growth during the war. There was a rapid decline in mean age at menarche after 1952, so that in 1959 the age was 13.3 years. Girls of that mean age were born just after the war ended, and grew up during the reconstruction period following the war. In recent years the decreasing trend has slowed, and in 1980 the average menarchial age was 12.4 years, which is very close to the average age for well-off girls from other populations. As previously described, the post-war period in Japan was one of tremendous social, economic, politi-cal, and dietary change. The rapid decline in the age at menarche immedi-ately after the war, when these changes were most drastic, and the slowing of this decline more recently, shows how closely linked are the socioeconomic development of a nation and the physical development of its children.

The meaning of SES

What are the direct causes of the relationship between socioeconomic status and human growth? It is often argued that SES is, in reality, only a proxy for better health care, which reduces childhood mortality and mor-bidity, and also results in increased growth (e.g., Malina, 1979). However, the SES effect is more subtle than this. Bielicki & Welon (1982) listed four primary factors relating SES and growth: (1) higher SES allows for better nutrition, (2) better health care, (3) reduced physical labor for children, and (4), greater growth-promoting psychological stimulation from parents, schools, and peers. The relation of the first three to growth has been adequately described in preceding sections of this discussion. Bielicki & Welon did not explain how psychological stimulation could increase growth, but they believed that growing up in an urban environment would expose people to more of these stimulants. At the time of their research there was little evidence in support for psychological growth stimulation. Today, there is still no convincing evidence for this, but there is consider-able evidence that negative emotional stress can delay, or even stop, growth in height. This evidence is reviewed in the next chapter in the context of the hormonal factors that regulate growth.

In a sense, Bielicki & Welon were right to think that the urban environ-ment would be the source of growth-promoting factors. The benefits of the city, in terms of the availability of food, health care and education, for those who can afford these, have already been enumerated. But something about the city goes beyond these obvious influences of human development. Matsumoto (1982) found that urbanization and income not needed for food purchases were associated with faster rates of growth. He analyzed changes in the rate of growth in height of Japanese children born in the years 1888 to 1962. The age at peak height velocity (PHV) during the adolescent growth spurt was used to assess the impact of several environmental factors on growth rate. As expected, World War II had a delaying effect on growth rate; on average, boys born in 1894 reached PHV at about 14 years, boys born in 1936, and therefore exposed to the war, reached PHV at about 14.9 years. After the war, the average age at PHV lowered rapidly and reached about 13 years for boys born in 1950–5 and 12.8 years for boys born after 1960. Similar changes occurred for girls, who generally reach PHV about two years earlier than boys.

Matsumoto evaluated the influence of six independent variables on rate of growth: (1) diet, (including separate categories for calories from starchy foods and calories from fat), (2) family size, (3) industrialization, (4) national income level, (5) urban population rate (number of people living in cities), and (6) Engel's coefficient (a statistic that measures the ratio of food expenses to total living expenses). Using the statistical technique of lagged correlations, the time difference between year of birth, the change in the independent variables, and the trend in the age at PHV was calculated. It was found that diet, family size, industrialization, and national income level had maximum effects on age at PHV after a 16 to 24 year lag from year of birth. As the average age at PHV occurred before age 16 in all birth cohorts, it was unlikely that any of these variables was a primary cause of the changes in rate of growth. In contrast, the lag between year of birth and urban population rate, Engel's coefficient, and age at PHV was eight to ten years. Thus, the increase in the number of people living in cities and the decrease in the amount of money spent on food relative to other living expenses, preceded the decrease in age at PHV.

Could these factors be direct causes of more rapid growth? Or, are these factors still only proxies for more direct causes? Population growth and increased population density are linked to a variety of physiological effects in human beings, including higher rates of leukemia (Muirhead, 1995) and lower rates of insulin-dependent diabetes mellitus (Staines *et al.*, 1997). In a broad sense, both cancer and diabetes are growth related in that cancer is unregulated cellular multiplication and insulin is a growth regulating hormone (see next chapter). There may be ways, as yet unknown, in which population density brings about cellular and hormonal changes that promote skeletal growth. The extra income families have, beyond that needed for food and other basic needs, could be spent on any number of material or social goods and services that might act as psychological stimulants. How these could promote faster or greater amounts of growth in height is a matter that requires careful experimentation and hypothesis testing, in other words, no one knows.

Risk focusing

It is known that within large cities, SES can act to stratify human populations into groups that are exposed to either positive or negative environments for growth. In the United States, for example, much research establishes a direct link between urban families living in poverty and a high prevalence within these families of low birth weight, chronic illness, lead poisoning, low quantity and quality of health care, short stature, obesity, lower performance on mental tests, increased school absenteeism, and school failure (Crooks, 1995). High SES families living in the same cities experience an opposite suite of consequences. In this, we see once again that economic, political, and cultural systems act to differentially allocate the benefits and risks for growth between SES groups. Epidemiologists refer to this process of differential allocation as **risk focusing**. Larry Schell (1997) defines risk focusing as '. . . the sociocultural process of allocating exposure to toxic or infectious materials to groups whose members enter the group partly because of previous exposure to those materials' (p. 70).

A model of risk focusing, with an emphasis on growth and development, is shown in Figure 6.8. The model shows that high SES allocates individuals to a low risk of exposure to negative stressors (the toxic or infectious materials), a low risk for poor physical and mental development, a high expectation for educational and occupational achievement, and, finally, a high expectation of continued high SES. In contrast, low SES is associated with high risks for both exposure to negative stressors and poor growth and development, a low expectation for achievement in education and occupation, and a recycling of low SES into the future.

A specific example of this type of risk focusing is the case of exposure to the toxic metal lead described by Schell (1997). Lead is one of many environmental pollutants that is known, or suspected, to have an effect on human development. Readers interested in some of these other pollutants, such as noise, polychlorinated compounds, and other toxic wastes may consult reviews by Schell (1991a, b) and Schell *et al.* (1992). Lead occurs naturally in the environment but at trace levels. Human activities concentrate lead to the point that some people may receive a single acute exposure, or chronic, lower-level exposure. Lead poisoning can cause growth failure, delays in cognitive development, and other developmental problems. Infants and children are at greater risk for lead poisoning because they may play where lead accumulates and because they absorb more lead

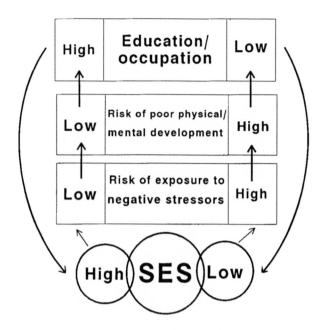

Figure 6.8. Model of risk focusing developed by Lawrence Schell (1997). In this model, SES factors allocate and focus risk of exposure to negative stressors, leading to disability and poor qualifications for employment, lower occupational status, and low SES.

from both their gastrointestinal tract and their respiratory tracts than adults. Low SES families in the United States, and many other urban locations in the world, often live in neighborhoods with high lead concentrations in the soil and air. The lead comes from old, dilapidated buildings where leaded paint was used and from leaded automobile gasoline which left lead dust in the soil. The lead dust is still present even though leaded gasoline was banned for sale in the United States in the late 1980s. The low SES families are constrained to live in these high risk neighborhoods due to the cost of rents in other, lower risk, areas.

The Centers for Disease Control of the United States has set a value of 10 micrograms/deciliter (μ/dl) of lead in the blood as the action level, meaning that lead levels of this amount or greater have known detrimental effects. Schell reports that national surveys show that of all infants and children (one to five years old) living in the central cities of the USA, an average of 36.7 percent of African-Americans exceed the action level. In contrast, only 6.1 percent of white infants and children living in the same central cities, but in different neighborhoods, exceed the 10 μg/dl action level. In the

United States, people with darker skin color, including African-Americans and Latinos, are disproportionately represented in the high risk neighborhoods due, in part, to 'racial' discrimination in employment and housing. Schell points out that the high lead burdens of African-Americans is not only due to recent exposures, but is multigenerational. Lead is transferred from mother to fetus through the placenta, which may result in ever greater lead burdens in succeeding generations.

The case of risk for lead exposure in the cities of the United States shows that the risk-focusing model is an example of a classic positive feedback relationship. High risks for exposure mean that infants and children may be physically or mentally impaired. This may elevate the risk for poor educational achievement and, later in life, poor opportunities for employment. This, in turn, may force people to continue to live in the same high-risk neighborhoods, and recycle negative risks and low SES into the next generation. Of course, poverty increases the risk for many other adverse consequences, some of which have been discussed previously in this book. Deborah Crooks (1995) published a very useful illustration of these risks with reference to school achievement, an important component of SES in many societies. Her illustration is reproduced here as Figure 6.9.

Schell concludes '. . . that culture can introduce stressors rather than buffer [against] them . . . [and] . . . political and economic factors, may constrain the choices that people can make in responding to their environment' (p. 67). It is in this political and economic context that we may see the real meaning of the SES effect. In societies that are stratified by social class, SES differentially allocates people to receive either the benefits or risk of urban environments, and for that matter, any environment.

In recent years, only three nations, The Netherlands, Norway, and Sweden, seem to have eliminated differences in growth between children from higher and lower SES levels. This was accomplished in all three nations by providing high-quality national programs for health care, nutrition, and other social services, to all segments of the society (Roede, 1985). Most of these programs were established during the recovery period after World War II. Prior to the War, SES differences in growth did exist. In Norway, Brundtland & Walløe (1973) and Brundtland *et al.* (1980) found that for children born after 1955 there were no differences in average age at menarche, or the average height of boys and girls from different SES backgrounds. Children from lower SES families weighed more than children from higher SES families, a common finding for developed industrial nations. Gunilla Lindgren (1976, 1995) found that at ages seven, 10 and 13 years, there were differences in height between SES groups for Stockholm boys and girls born in 1933 and 1943. For boys and girls born in 1955,

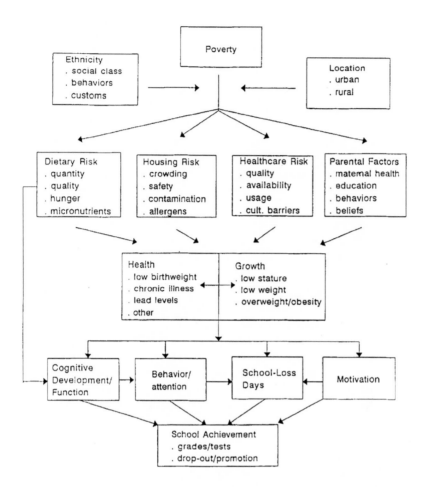

Figure 6.9. Biosocial risks for children living in poverty related to health, growth and school achievement. Model developed by Deborah Crooks.

however, there were no SES-related differences in average height and weight, or the average age at peak height velocity, peak weight velocity, and menarche. Socioeconomic differences, as measured by education, occupation, and income, do exist in these Scandinavian nations, but federal systems of guaranteed health care and social support services provide people of all SES levels with an equal share in the environmental opportunities for growth and development. This was the case at least until 1980, when Lindgren (1995) reports that significant differences in height and head circumference were found between 5.5-year-old children of the highest and lowest SES groups in Sweden (the lowest SES group being small, on average). The reason for the reappearance of an SES effect is not completely known. A longitudinal study of living conditions during the years 1968 to 1991 and its relations to adult stature in Sweden offers some possibilities. The researchers (Nystroem & Lundberg, 1995) interviewed 4574 Swedish men and women over 24 years of age about their family's SES during the years of growth and development, episodes of economic hardship, family size, dissension within the family and break-up of the family. Reduced adult stature is associated with lower childhood SES, economic hardship, and large families, but SES has the largest effect. So, even in societies that have consciously tried to eliminate SES effects on human health and development, and have done so for a time, the SES effect may return. Lindgren observes that 'In many European countries – including Sweden and the UK – the gap between rich and poor is widening, so also the inequality of health as mirrored in life expectancies and prevalence of different diseases in different socioeconomic strata . . .' (p. 420).

Size and social mobility

There is another side to this issue of socioeconomic effects and growth, which is that within all social classes taller individuals tend to move up in SES and shorter individuals tend to move down or remain stable in SES. Some aspects of this phenomenon were discussed above in the section 'Migration and urbanization', especially the observation that, on a population basis, migrants tend to be taller and of higher SES than sedentes. On an individual basis, Scott et al. (1956) found that taller women from Aberdeen, Scotland worked at more skilled jobs than shorter women. Furthermore, when taller women married, their husbands worked at more skilled jobs than the husbands of shorter women. These findings were confirmed by Schreider (1964b) in a study of British women showing that, regardless of the occupation of the father, taller women tended to marry men with more highly skilled jobs and shorter women tended to marry men performing semi-skilled or unskilled labor.

Taller people achieve higher educational status than shorter people, as was shown by Parnell (1954) for university students in England and other European countries and by Kimura (1984) for Japanese university students. Stature-related social mobility even occurs within families. Susanne (1980), working in Belgium, and Bielicki & Charzewski (1983), working in Poland, found that within families the taller siblings were better educated than shorter siblings. In Belgium, there was a higher correlation between a young man's stature and his level of education, than between the same individual's level of education and his father's occupation. In the Polish study, the average difference in stature between the better educated and the less well educated brothers ($n = 116$ pairs) was 1.26 cm, a statistically significant difference. The results from both studies indicates that tallness leads to a higher level of education. Better education, of course, is likely to lead to more skilled and higher paying employment, and, consequently, higher SES.

These studies show that the tall tend to rise in SES and the short tend to sink. Stature, by itself, does not determine occupation, education, or socioeconomic attainment, but there is a strong social bias in favor of the tall which may help facilitate their SES climb. Throughout the world there is a general bias in favor of taller men by women in courtship and mating. This bias operates among the Mehinaku Indians of Brazil, a traditional culture based on subsistence farmers and fishing. Gregor (1979) found that taller Mehinaku men had more wives and lovers, more wealth, and more social prestige than shorter men, and the latter were often the objects of social and sexual ridicule. It is interesting to note that the range of stature for Mehinaku men observed by Gregor was 151.8 cm to 175.9 cm, meaning that the tallest men would be considered about average in height if United States values of tallness and shortness were applied. Gregor reviewed ethnographic data from several other cultures and found that only the Crow of North America and the Mbuti pygmies of Central Africa specifically mention tallness as a disadvantage.

Numerous studies, from a diversity of disciplines, find that the taller, non-obese man or woman is given preference in many arenas linked with SES, such as the perception of intelligence, academic performance, and social skills, as well as initial job hiring and perception of both current and future job performance (Hensley & Cooper, 1987; Loh, 1993; Lindeman & Sundvik, 1994; Anonymous, 1995). The empirical data show that height and weight have a small, but statistically significant, effect on measured academic performance, and both have an initial effect on wages when a person is hired for a job, at least in the higher income countries. The cognitive and school effect is almost entirely due to rate of maturation – faster maturation during the growing years results in both greater stature and cognitive performance (Lindgren, 1995). However, stature has an independent effect on preference for employment and salary. The statistical effect of height on income of adult males in the United States was analyzed in a study by Ekwo and colleagues (1991). The researchers used data from the second National Health and Nutrition Examination Survey of 1976–80. Anthropometric and socioeconomic data were available for a nationally representative sample of 4563 men aged 20 to 65 years old. The correlation between height and family income was 0.13, which is small but statistically significant ($P = 0.001$). The researchers analyzed the data for the confounding effects of age and 'race' (whites versus blacks) and found that, 'White males in the highest income category were significantly taller, heavier, more adipose, had higher levels of education, and were more often married than their lowest income counterparts. Among non-whites, martial status and education alone differentiated highest from lowest income nonwhite males' (p. 181).

The reason for the height bias in education, earnings, marriage, and many other socioeconomic realms seems to be that height is a strong predictor of social dominance (Hensley, 1993; Ellis, 1994). The way that social dominance is played out in society is usually in terms of both overt and subtle preferences for people who have characteristics of the desired type, that is the stereotype that a society defines as attractive and successful. One way in which the height bias operates in the United States was shown by Keyes (1979). Based on result from a questionnaire survey, Keyes found that men under 175 cm in stature, about the average height for men in the United States '. . . invariably wished they were taller'. Keyes reviewed other research and found that job status and economic rewards could account for this desire. In two studies, conducted three years apart, of graduates from the University of Pittsburgh, there was a 12.4 percent difference in initial starting salary favoring men 188 cm tall versus men 180 cm tall. The salary difference favoring *cum laude* graduates was only four percent. Additional research on 5000 Army recruits measured in 1943 found that in 1968, those over 183 cm earned eight percent more than those below 168 cm, even after the influence of IQ, educational level, marital status, and occupation were statistically controlled. In another study, 140 personnel officers of companies involved in retail sales were asked which of two equally qualified job applicants would they choose – one who is 185 cm tall or one who is 165 cm. The taller applicant was preferred by 72 percent of the personnel officers, 27 percent had no preference, and one respondent chose the shorter applicant.

Another example of how social bias, social dominance, and height seem to be related comes from a study titled 'Height as a measure of success in academe' (Hensley, 1993). The author of this study analyzed a random sample of 83 male faculty and 52 male heads of department in the United States. Hensley reports that, overall, academics are taller than the average American of the same age and sex, but the degree of tallness varies by academic rank. The mean difference in stature from lower to higher rank are: assistant professors, 3.2 cm taller than the average male, associate professors, 3.8 cm taller, and full professors, 5.0 cm taller. The sample of department chairmen averaged 5.4 cm taller than American men of the same age. A satirical perspective of the height bias in academia was offered by the British comedy troupe 'Monty Python'. The text of their television sketch called 'Archeology Today' may be found at the web site http://www.dcscomp.com.au/sdp/sketches/archeolo.htm.

The height bias operates in the non-academic world at least as strongly as it does within the Ivory Tower. A report in *The Economist* (Anonymous, 1995) that cites a 1980 survey of American Fortune 500 companies found that more than 50 percent of chief executives were over 183 cm tall, but only three percent were less than 170 cm tall. Another arena of bias is politics, as all but three winning American presidential candidates have been taller than the loser – so much for political philosophy! Winners of political contests are perceived as being taller than they actually are, and conversely, political losers also 'lose' height. A Canadian study interviewed 177 voters about their perceptions of the heights of several candidates for the 1988 Canadian Federal elections. The subjects were asked to estimates the heights of the candidates both before and after the election. The losers were judged to be shorter after the election, while the winner was considered to be taller than he was before the election. The authors of this study (Higham & Carmet, 1992), state that their results confirm previous research showing that people in higher social status positions are perceived as being taller than they really are. In addition, their new research shows that the perception of social status and height is dynamic, in that election outcomes can rapidly alter judgments about stature.

It is possible to speculate about the historical and psychological reasons why the 'bigger is better' bias exists in many societies. There is considerable evidence from research on non-human social animals, especially primates, that larger individuals enjoy higher status in their social hierarchies (Ellis, 1994). The rewards of higher status are very tangible, as higher ranking males and females have preference to desirable foods, more security from predation and from harassment from other members of their social group, and are often have greater reproductive success. So, the social advantages of larger physical size may predate the evolution of *Homo sapiens*.

The human species adds some further complications to the advantages of larger size. A social bias for desirable types, in terms of stature and body composition, exists in virtually all human cultures. While most societies favor tallness, not all favor fatness, especially levels of adiposity that are defined as obese in the industrialized nations. Moreover, people judged to have excessive height are often at a social disadvantage. In the United States, for example, the ideal height for men is about 188 cm (Anonymous,

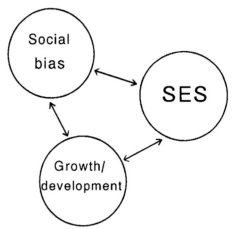

Figure 6.10. Growth and development as both a cause and a consequence of social and economic effects. Double-headed arrows indicate two-way interactions within a cyclical process.

1995), and there is little evidence for greater economic, social, and educational rewards for men above that ideal. Male heights above 193 cm, and female height above 175 cm are incompatible with standard sizes of clothing, the design of most furniture, and the space provided in many automobiles, which may require higher expenditures of money in order to compensate. Men, and especially women, who are deemed 'too tall' face both overt and subtle types of discrimination in employment and social life. School-age boys and girls who are taller than average are often expected by their teachers to act more mature, be more responsible, and have higher academic performance than their shorter class-mates. Teasing and outright physical abuse may be heaped on the 'too tall' child. The effects of these types of stresses on taller than average boys and girls has not been well studied.

The concrete result of these biases, just as for the discrimination stemming from racial prejudice, is that individuals or groups of the accepted 'type' are more likely to receive better care, in the widest sense, than individuals or groups of the undesired type. A positive feedback relationship between growth and socioeconomic status results from the social bias, better environmental conditions lead to larger size, taller individuals tend to rise in SES, and higher SES leads to better environmental conditions. An opposing cycle exists for those from lower SES, or shorter individuals from any social class. The result is that differences in physical size between individuals are both a consequence and a cause of socioeconomic effects on growth (Figure 6.10).

7 Genetic and endocrine regulation of human growth

Children tend to resemble their parents in stature, body proportions, body composition, and rate of development. It may be assumed that, barring the action of obvious environmental determinants of growth, these resemblances reflect the influence of genes that parents contribute to their biological offspring. Genes do not directly cause growth and development. Rather, the expression of a genetically inherited pattern of growth is regulated by many proteins that genes produce, and the entire process is mediated by several biological systems, especially the endocrine and neurological systems. A marvelous example of the interaction of genes, proteins, and the endocrine system may be seen in the action of **homeobox genes and *Hox* genes**. The description and elucidation of homeobox and *Hox* genes is one of the most important advances of molecular biology of the past two decades.

The homeobox is a sequence of 180 DNA base pairs that codes for a 60 amino acid segment of a protein. First discovered in the genome of the fruit fly, *Drosophila*, homeobox sequences are found in all eukaryote organisms so far examined. These highly conservative DNA sequences – the same homeobox sequences are found in organisms as diverse as hydra, nematodes, all arthropods (the group that includes insects) and all chordates (the group that includes human beings) – produce proteins that regulate the expression of other genes '. . . and control various aspects of morphogenesis and cell differentiation' (Mark *et al.*, 1997, p. 421). *Hox* genes are a category of homeobox genes that encode transcription factors (Holland & Garcia-Fernàndez, 1996), which are proteins that initiate and regulate the conversion of the DNA code to the RNA sequence that is used to make amino acid polypeptide chains.

In multicellular animals, homeobox genes act to delimit the relative positions of body regions, for example the head, thorax and abdomen of insects, or the general body plan and limb morphology of vertebrates. Homeobox genes seem to have their greatest impact during the earliest stages of development. The proteins that homeobox genes produce are needed to regulate the expression of other DNA to '. . . sculpt the morphology of animal body plans and body parts' (Carroll, 1995, p. 479). The DNA affected by homeobox proteins will, in turn, produce other proteins that mediate cellular differentiation, growth, and development. These 'down stream' proteins do not act alone. Some of them must combine via a process called molecular zipping before they have any effect on a given segment of DNA (McKnight, 1991). These and other proteins need an appropriate environment to have any effect. The biochemical environment of the egg cell and, a bit later in time, of the mother's womb and the placenta, provide a host of factors needed for growth, including nutrients and hormones.

Throughout life, the endocrine system often provides the necessary biochemical environment for gene action. The human adolescent growth spurt, for example, requires adequate amounts of two hormones, growth hormone and testosterone (boys) or estradiol (girls), to be secreted into the bloodstream. Without these two hormones the genes that regulate growth of skeletal, muscle, and adipose tissue will not increase in activity enough to produce the growth spurt. The endocrine system also responds to the influence of many environmental factors that affect human development. Situated between the action of genes and the external environment, the endocrine system serves as a mechanism that unifies the genes we inherit and the environments in which we live to shape the pattern of growth of every human being.

This chapter describes some aspects of the nature of the genetics of growth, the endocrine system, and the interaction of genes, the environment, and hormones on human development. These are each very active areas of research with very large literatures. Moreover, knowledge in these areas of research is changing rapidly. Only a few aspects of the research being conducted in genetics and endocrinology can be covered in this book. Accordingly, the emphasis will be on research relating to topics discussed in previous chapters, especially those that relate to the biocultural nature of human development.

Genetics of human development

Before discussing further the newest and most exciting developments in the molecular biology of growth, such as homeobox genes, an overview of evidence for genetic influences on growth is in order. Some historical background was given in Chapter 1. Here the focus is on the general concept of heredity and growth, a theme that was addressed by Daryl Bock (1986). He analyzed longitudinal data from boys and girls, raised under favorable environmental conditions, participating in the Fels Research

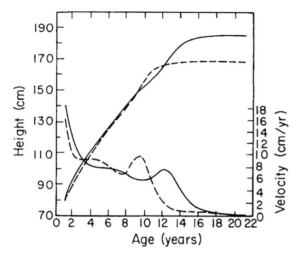

Figure 7.1. Distance and velocity curves for the growth in height of the tallest girl (solid lines) and earliest girl (dashed lines) participating in the Fels Research Institute Study of growth (after Bock, 1986).

Institute Study (which was described in Chapter 1). As of 1983, 214 boys and 234 girls had been measured to maturity. The families of the Fels sample live in the state of Ohio, are largely of middle socioeconomic class and of European cultural background. Thus, the sample is not representative of all children of the state of Ohio or of the United States, however the Fels study provides a wealth of data about the growth and development of normal children.

Bock analyzed the data for several individuals from this study, cases representing the extreme variants in the normal range of growth and development. His purpose was to describe variation in the inheritance of patterns of growth. In Figure 7.1, the height and velocity of growth in height of the tallest girl and the earliest maturing girl (defined by the rate of skeletal maturation) are compared. Both girls were tall for age throughout infancy and childhood, being above the 95th percentile for height for all girls in the Fels sample. The early maturing girl entered the adolescent growth period in her seventh year, reached peak height velocity at 9.4 years, and stopped growing by age 13 years, when she reached the 75th percentile of height for girls living in the United States. The skeletal maturation of this girl, as estimated from hand–wrist radiographs, was advanced over her chronological age by about three years. Thus, her tallness during childhood was due, in large part, to her advanced maturation.

The tall girl entered adolescence at about 10.0 years, reached peak height

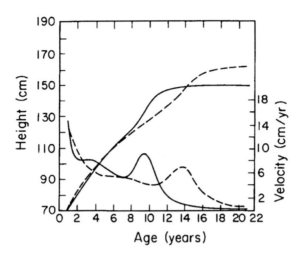

Figure 7.2. Distance and velocity curves for the growth in height of the shortest girl (solid lines) and the latest girl (dashed lines) participating in the Fels Research Institute study of growth (after Bock, 1986).

velocity at 12.2 years, and ceased growing at about 18 years at a height of 185.1 cm, which is at the 99th percentile. The parents of this girl were also tall, compared with other adults of like sex and age, which lead Bock to state that the growth of this girl is a '. . . typical example of intrinsic tall stature of familial origin'. By intrinsic Bock seems to mean that the girl's tallness is a genetically determined characteristic, inherited from her parents.

Presented in Figure 7.2 are growth curves for the shortest girl and the latest maturing girl (again, defined by rate of skeletal maturation), that Bock analyzed from the Fels study. During childhood both girls were of similar stature, being about at the fifth percentile for the Fels sample. The short girl entered adolescence just a little later than the earliest maturing girl (shown in Figure 7.1), stopped growing in her fourteenth year, and reached 149.0 cm, which is below the fifth percentile. This girl was advanced in skeletal maturation from age eight to age 16 years. Bock explained that her short stature is a family characteristic, inherited from her parents. Her stature is influenced by both a reduced amount of growth in size, combined with a developmental pattern for early maturation.

The late maturing girl entered adolescence about two years after the short girl and, with this extra time for prepubertal growth, reached 160.1 cm, which was just under the 50th percentile. Her skeletal maturation was delayed in comparison with her chronological age. The delay fluctuated between one-half year and two years throughout her childhood and adolescence. This slow rate of development allowed for a prolonged growth period and attainment of average stature.

Bock described these four cases of growth as 'unusual', meaning that they represent the limits of the range of normal variation in amounts and rates of growth for the sample of the Fels Research Institute. The four girls were raised under favorable environmental conditions, and showed no evidence for acute or chronic diseases that influence growth and development. Given this, and measurements of height of the parents of each girl, Bock's analysis indicates a major genetic component in the determination of size and rate of growth in these girls.

A similar study was published by Prokopec & Lhotská (1989) based on a sample of 81 boys and 78 girls. The subjects, all from Prague, were measured annually from birth to age 20 years. The Preece–Baines growth curve was fit to the longitudinal data of each subject. From these fitted curves for all the boys and girls, the tallest, the three shortest, the three slowest maturing, and the three fastest maturing of each sex were selected. None of these extreme cases was known to have any major chronic or acute diseases. Neither the subject's history of common childhood diseases, nor the occupation of the fathers had an effect on positive or negative growth and development. In contrast, the **mid-parent height** did predict the stature of offspring. Mid-parent height is the average of the stature of the mother and the father. The positive impact of mid-parent stature on offspring growth attests to the role of heredity, although parents and children living in the same household also share a very similar environment. Inspection of the Preece–Baines curves showed that tall or short stature at age 20 could be predicted from stature at age four years. The predictability of adult height from stature at age four attests to the early establishment of individual patterns of growth and their stability over time.

Studies such as these lend support to the concept of a **genetic potential** for body size. Some researchers extend this concept to genetic potentials for body composition and body proportions. The term 'genetic potential' usually means that every human being has a genetically determined upper limit to adult stature, the ratio of leg length-to-sitting height, and other anthropometric dimensions. This concept is so well entrenched in the field of human growth research that it is almost always used without definition or justification from research (that is, without reference to the literature) in scientific and popular articles. Interested readers may peruse recent issues of journals dealing with human biology, physical anthropology, pediatrics medicine, and related fields, where they will find the phrase 'genetic potential' used in this way. The problem with this casual usage is that the assumption on which it is based is not true. In Chapters 5 and 6 I reviewed research showing that adult stature, body proportions, and body composition are highly plastic. The short stature and the relatively short legs of past generations of the Japanese or the Maya, compared with Europeans, were once assumed to be genetically determined traits, but now we know that they are indicators of the quality of the social, economic, and political environment.

In a more general sense, the whole concept of 'genetic determination' is seriously flawed. That notion implies that the flow of information about how any human trait is developed, be it height, body fatness, personality, or intelligence, originated in the DNA and then unfolds into the phenotype. Within this scenario, one may allow for a greater or lesser amount of environmental influence on the phenotype, but the flow of information basically begins with the DNA and moves one way. In the following sections of this chapter I will show that the role of DNA in human development is much more complex, and often much less direct, than this. I am not going to enter into a debate at the low level of 'genes versus environment'. Rather, my primary goal is to explain that the interactions between genes, hormones, and the environment may flow in all directions. A secondary goal is to indicate where we lack knowledge concerning the factors that produce the wide range of human phenotypes. Taken together, these goals should lead to a much more critical and tempered use of the concept 'genetic potential'.

Homeobox genes, evolution, and growth

What is the 'major genetic component' for growth? How does it guide growth and development so as to produce a human being, rather than a chimpanzee, or a whale, or an oak tree? In the broader perspective of evolutionary time, what sort of genetic changes are required to produce new patterns of growth? Specifically, were new genes needed to evolve the new human life cycle stages of childhood and adolescence? In Chapter 4 I indicated that no new genes were needed, rather changes in the regulation of the expression of existing genes within the primate genome were all that was needed. However, new genes also evolve. Some additional discussion of homeobox genes may provide a theoretical understanding of how the evolution of gene regulation and the evolution of new structural genes comes about.

In a review of homeobox genes and evolution, Carroll (1995) writes that one of the most primitive living chordates, called amphioxus, has a single homeobox sequence that appears to be ancestral to all vertebrates. The jawless fish, such as lampreys and hagfish, seem to have two homeobox clusters. Jawed vertebrates (modern fish, amphibians, reptiles, birds, and mammals) have four homeobox clusters. Some of the DNA in these homeobox clusters arose by gene duplication, that is the DNA of ancestral organisms was copied to make two sets of identical DNA. But, much of the DNA in the modern vertebrate clusters is unique to vertebrates and represents new genetic material. The proteins manufactured by the vertebrate homeobox genes regulate the formation of the distinct regions of the vertebral column (cervical, thoracic, lumbar, sacral, and caudal), the formation of the limbs, and the formation of teeth. Carroll points out that the vertebrate hindlimb evolved from the pelvic fin of fish and the forelimb from the pectoral fin. In most amphibians, reptiles, and mammals the fore- and hindlimbs are nearly identical in gross morphology. Carroll states, 'It is difficult to explain how the fore- and hindlimbs of tetrapods are so similar to each other when they evolved from different fish fins. It has been suggested that a major change in *Hox* gene expression brought about a serial homology of the forelimb and hindlimb' (1995, p. 484–5). By this he means that the manner in which homeobox genes regulate limb formation has evolved to sculpt the isomorphic appearance of the limbs.

The great apes and human beings have fore- and hindlimbs that are distinct in shape and function. Apes have relatively long forelimbs and are brachiators, while humans have relatively long hindlimbs and are bipedal. There is clinical and physiological evidence that homeobox gene regulation plays an important role in the evolution of these differences. A homeobox gene that seems to regulate the timing and extent of local growth rates in human hands and feet was discovered by Muragaki and colleagues (1996). A mutation in the amino-terminal end of the HOXD13 region of a human homeobox gene sequence causes the condition called synpolydactyly. People with this condition have their carpal bones of the hands and the tarsal bones of the foot transformed into short carpal-like and tarsal-like bones. Some people also have deformed or duplicated digits (several X-ray photographs are given in the original article). The mutation that causes these phenotypes is an expansion of a polyalanine region of the homeobox. Normally the HOXD13 DNA codes for 15 alanines (one of the amino acids) in sequence. When this polyalanine region is expanded, the extra alanines produced result in limb malformation. In a subsequent paper (Goodman *et al.*, 1997), workers from Muragaki's laboratory were able to show that the number of extra alanines, varying from seven to 14 in the subjects studied, is directly related to the severity of the malformation. This is very strong evidence for the direct role of homeobox genes in normal, as well as abnormal, growth and development.

This mutation, and the malformed hands and feet it produces, do not explain the evolution of human bipedalism from ape brachiation. But, it is through the study of mutations that we often come to understand how genetic systems work normally. It is quite possible that other regions of HOXD13 regulate the growth of ape versus human limbs. It is already known that the HOXD13 and HOXD11 regulate the rate of cell division in the proliferative zone of growing cartilage of the chick tibia and fibula (Goff & Tabin, 1997). Goff & Tabin propose that all *Hox* genes are growth promoters, but that some are more effective than others. Since several *Hox* genes may compete with each other for the target site of their regulatory action, '. . . the overall rate of growth in a given region is the result of the combined action of all of the *Hox* genes expressed in that region competing for the same target . . .' (ibid, p. 627). Goff & Tabin's model may account for the growth differences of limbs, and other body segments, between closely related species, such as chimpanzees and human beings. It is also possible that this model might explain how the timing, duration, and number of life cycle stages are regulated.

Twin studies as an approach to the genetics of growth

Twin studies offer a traditional methodology, used long before homeobox genes were discovered, for delineating the influence of heredity on growth. A typical strategy is to compare the growth of monozygotic (MZ) and dizygotic (DZ) twin pairs. MZ twins are the product of a single fertilization, and have a virtually identical genetic make-up. DZ twins result from separate fertilizations and are about as similar genetically as any two full siblings. Both MZ and DZ twins share similar uterine environments, and often share similar postnatal environments, so the differences between them in growth and development during fetal and early post-natal life are, in principle, due to genetics. Many studies of MZ and DZ twin pairs indicate that there is, indeed, a statistically strong genetic component in human growth. The power of the genetic effect seems to be greatest for skeletal dimensions (e.g. height, limb lengths, and craniofacial dimensions). The genetic effect is less powerful for girths (e.g., arm or thigh circumference), and it is weakest for skinfolds (Sharma & Sharma, 1984).

Tables 7.1 and 7.2 contain data that compare the correlations in height and weight between MZ and DZ twin pairs at ages from birth to eight years. The data are from the Louisville, Kentucky Twin Study, which began in 1962 and by 1979 had recruited and studied more than 900 twins (Wilson, 1979). If amounts and rates of growth are totally controlled by the genotype, then correlation coefficients for MZ twins, who are genetically

Table 7.1. *Correlation coefficients for height between monozygotic (MZ)*
and dizygotic (DZ) twin pairs from birth to age eight

Age	Total *n*	MZ	DZ	
			Same sex	Different sex
Birth	629	0.62	0.79	0.67
3 months	764	0.78	0.72	0.65
6 months	819	0.80	0.67	0.62
12 months	827	0.86	0.66	0.58
24 months	687	0.89	0.54	0.61
3 years	699	0.93	0.56	0.60
5 years	606	0.94	0.51	0.68
8 years	444	0.94	0.49	0.65

From Wilson (1979).

Table 7.2. *Correlation coefficients for weight between monozygotic (MZ)*
and dizygotic (DZ) twin pairs between birth and eight years

Age	Total *n*	MZ	DZ	
			Same sex	Different sex
Birth	992	0.63	0.68	0.64
3 months	766	0.74	0.66	0.40
6 months	819	0.81	0.63	0.39
12 months	828	0.88	0.55	0.37
24 months	779	0.88	0.53	0.50
3 years	713	0.88	0.52	0.54
5 years	606	0.85	0.48	0.62
8 years	444	0.88	0.49	0.46

From Wilson (1979).

identical, should be equal to 1.00, a perfect positive correlation, at all ages.
The correlation coefficient for DZ twins, who share an average of 50
percent of those genes that are free to vary, should be equal to 0.50 at all
ages. Of course, this assumes that parents are randomly selected from the
population of potential mates. This is usually not the case, for instance,
positive assortative mating for height occurs in many human populations.
Even so, the data in Tables 7.1 and 7.2 show that by one year of age, and
thereafter, the correlation coefficients predicted from the genetic model are
very nearly found for this relatively large sample of twins.

The coefficients from birth to six months of age, however, do not
correspond with genetic expectations. At birth MZ twins are less

concordant in size compared with DZ twins. One reason for this may be that MZ twins often share a monochorionic placenta during the prenatal period. Falkner (1978) reviewed data from seven studies of twin placentation, and found that about 70 percent of MZ twins have monochorionic placentae. Vascular anastomoses (arterial and venous connections) between the parts of the placenta supplying blood to each twin occur in monochorionic placentae. This results in a transfusion of blood, and the oxygen and nutrients carried by the blood, between the twins. The transfusion of blood is usually not equal, which means that the twins do not receive an equal maternal blood supply, possibly resulting in undernutrition and hypoxia (low oxygen availability) for the disadvantaged twin. In the condition of dichorionic placentation, each twin receives a separate supply of maternal blood and nutrients.

Falkner (1966) found that in a sample of 92 MZ twins, the within-pair difference in birthweight averaged 326.0 grams in monochorionic twins and 227.8 grams in dichorionic twins. Many subsequent studies confirm that monochorionic twins are more discordant in birthweight than dichorionic twins. These studies also show that monochorionic twins are generally of lower birthweight than dichorionic twins (Reed *et al.*, 1997). These effects on birthweight in monochorionic MZ twins suggests that one or both of the fetuses is exposed to a deficient maternal blood supply and may have to adjust its intrauterine growth rate to adapt to placental insufficiency. One consequence of this type of uterine environment would be that the role of fetal genetics would be reduced leading to the lower correlation coefficients in length and weight at birth seen in Tables 7.1 and 7.2.

The values of the correlation coefficients at birth for DZ twins, displayed in Tables 7.1 and 7.2, are higher than expected from a simple genetic model of growth. DZ twins are more likely to have separate placental connections with the mother's vascular system. In this placental condition, it is more likely that each twin receives a relatively equal share of oxygen, nutrients, and other substances from the mother. This may explain, in part, the relatively high concordance in size of the DZ twins at birth.

Robson (1978) reviewed birthweight statistics for several populations of people living in developed nations, and found that for single fetus or DZ twin pregnancies, up to 66 percent of the variance in birthweight is due to maternal environment. That environment includes the quality of the mother's diet, smoking habits, alcohol usage, and other variables of this type that influence the quantity and quality of the maternal blood supply to the fetus. Robson estimated that genetic factors account for about 34 percent of the variance in birthweight, and of this only ten percent of the variance is due to fetal genotype. Thus, the higher than expected concordance in the birthweight of DZ twins is likely to be due to the shared maternal environment, which sets some common limits to the growth of both twins. After parturition, growth is determined by the interaction of the environment with the unique genotype of each DZ twin. This probably explains why the correlation coefficients for DZ twins in the Louisville, Kentucky Twin Study become less concordant as the twins get older.

In some instances, prenatal environment influences on growth have effects that last for many years after birth, and even act transgenerationally (discussed in detail in Chapter 2). These environmental effects may obscure the contribution of genes to the determination of size. Wilson (1979) examined a sub-sample of MZ twins from the Louisville study. This sub-sample consisted of ten MZ twin pairs with the largest differences in birth weight. The lighter twin averaged 57 percent of the birth weight of the heavier twin and the average absolute difference between twins in birth weight equaled 1064 grams. The smaller twin was usually of low birth weight and required special postnatal care. Wilson found that by six years of age the relatively large initial disparities were progressively reduced but not eliminated. The mean difference for weight at six years was 2.19 kg for height 1.85 cm, indicating that the lighter twin at birth was still ten percent lighter and two percent shorter, on average. For these 10 pairs of MZ twins the within-pair correlation for height was only 0.72 compared with the correlation of 0.93, or higher, for all MZ twins in the Louisville study (Table 7.1).

Similar long-term differences in growth between MZ twins of markedly different birthweight have been reported by Falkner (1966, 1978). In one case, full-term MZ twin boys had birthweights of 1460 grams and 2806 grams and birth lengths of 43 cm and 50 cm respectively. The smaller twin was considered small for gestational age, and required special neonatal care. Inspection of the placenta showed that it was monochorionic, with only about 40 percent of the placenta supplying maternal blood to the smaller twin. During the first year of life, the growth rate of the smaller twin exceeded that of the larger twin, but the catch-up in growth was incomplete. Differences in height, weight, and other physical measurements persisted at all ages to 16 years of age, e.g., heights and weights at age 16 were 161.9 cm and 50.6 kg for the smaller twin and 167.3 cm and 58.5 kg for the larger twin.

The implication of these observations is that twin studies may provide clear indications of the genetic control of growth, but only when the environment for growth is, essentially, equal and favorable, meaning that the environment does not inhibit the growth of one or both twins. More than two-thirds of MZ twins share a monochorionic placenta, and many of these twins do not experience equally favorable prenatal environments. Few twin studies systematically account for the effect of placental type or the subtle consequences it may have on growth and development. An additional caveat is that few studies evaluate the effect of assortative mating (e.g., similarity in height and other dimensions) of the parents. Unless researchers evaluate these potential biases twin studies should be interpreted with some caution.

Correlations in growth between biological relatives (non-twins)

Studies of familial correlations in growth may help clarify the role of genes and the environment. Familial correlations for serial measurements of stature were analyzed by Byard *et al.* (1983), using data from the Fels Research Institute that included measurements of pairs of relatives, e.g., siblings, parent and offspring, cousins, uncle–nephew, aunt–niece, etc. Each of the pair had been measured once a year from one to 18 years of age. Correlations were calculated based on age-matched measurements (e.g., father's height at age 15 and his son's height at age 15). Multivariate analysis of the correlations found that degree of relatedness explained most of the variation in stature. That is, first degree relatives had higher correlations than second or third degree relatives. However, the effect of the common environment between first degree relatives, usually living in the same household, could not be separated from their genetic similarity. For example, between the ages of one and 15 years, correlations between siblings were always higher than those between parents and offspring. Theoretically, both parents and their offspring and siblings share about 50 percent more of their genes than the amount shared at random between any two unrelated members of a breeding population. Both full siblings and parents and offspring, then, should have approximately equal correlations in stature. However, the siblings lived together in the same households, which may have resulted in a 'commonality of environment' effect, increasing the value of sibling correlations. A similar pattern of familial correlations of growth was found by Russell (1976) and Mueller (1977), who also interpreted it as due to the effect of a more similar environment for growth shared by siblings than by parent–offspring pairs.

Other studies show how familial correlations, although theoretically a measure of genetic similarity, are equally a measure of the environment. For instance, the power of the environment to influence the value of sibling correlations in size was demonstrated by Mueller & Pollitt (1983). They used data gathered by Dr. Bacon Chow (Chow, 1974) from a study of the effects of nutritional supplementation of pregnant women on the subsequent growth of their offspring (Dr. Chow died before analysis of the data could be completed). The study included measures of the prenatal and postnatal growth of siblings, who were living in a rural Taiwan village characterized by high rates of chronic undernutrition. Each woman in the study contributed two infant participants. During pregnancy with the first child the mother was untreated, while during pregnancy with the second child she was given either a high calorie supplement or a placebo. No supplement was given to the children directly, so any nutritional intervention relating to the growth of the children was mediated by the mother prenatally or during lactation.

There were 108 pairs of siblings whose mothers received the high calorie supplement and 105 pairs of sibs whose mothers received the placebo. Correlations at birth in weight, length, head circumference, subscapular skinfold, and the index of weight/length^3 between sibs in the placebo group were significant and all were near the value of 0.50. Sibs in the supplemented group had birth size correlations that were '. . . unusually low and often insignificant' (Mueller & Pollitt, 1983, p. 11). The low correlations were due, presumably, to nutritional supplementation of the mother, which produced more favorable prenatal growth in the sib exposed to the supplement. The differences in sibling correlations between the two groups virtually disappeared by age 2.5 years. Apparently, the maternal mediated effect of the high calorie supplement was limited to prenatal life and infancy. After weaning, the generally adverse nutritional environment of the village was a stronger influence on the growth of all children.

Longitudinal studies, such as the Fels Research Institute and Bacon Chow study, are rare due to the time and expense required to collect data of this type. A more common approach to familial correlation research is to calculate parent–child correlations in stature. One such study, by Martorell *et al.* (1977), used data from the INCAP 'Four Village Study' of malnourished rural Guatemalan families. Correlations between mid-parent height and child stature (or length) were obtained for children aged six months, one year and then yearly up to age seven. The authors hypothesized that in this chronically malnourished sample, the stature of both the parents and the children would have been stunted. Furthermore, different degrees of malnutrition would have been experienced by different individuals. As a result, it was expected that the correlations between mid-parent stature and child stature for this sample would be lower than the values predicted from a simple genetic model, and lower than values from better nourished populations in developing countries. It was found, however, that correlations for the Guatemalan sample did not differ from samples from the United States or Northern Europe. The authors considered the notion that variability in stature in the Guatemalan sample is as much a product of genetic influences as it is in the developed nations. However, the authors rejected this proposal, and showed instead that socioeconomic and nutritional status were correlated across generations. That is, parents who had relatively better living conditions (e.g., housing, nutrition, etc.) when they were children were more likely to provide a better environment for their own children. Thus, environmental and genetic factors contributed to the parent–child correlation in stature, and it was not possible to quantify the unique contributions of either.

Familial correlation and heritability estimates for stature in a West African population were calculated by Roberts *et al.* (1978). The sample studied included the people of two villages in The Gambia, where traditional subsistence agriculture and rural lifestyles were practiced. The authors found that correlations for stature between husbands and wives were low and not statistically significant, indicating that there was no assortative mating for height. Correlations between parents and offspring, and between full siblings, for this sample were lower than those found for European or North American samples of middle to upper socioeconomic status. Moreover, the correlations between full siblings were lower than correlations between parents and children. The heritability for stature was estimated to be about 0.56 (1.00 being a perfect heritability and 0.00 indicating no heritability).

To help put these African results in perspective, Byard *et al.* (1983) found, for a United States population, that sibling correlations were higher, generally, than parent–offspring correlations and that the heritability of stature was about 0.68. Roberts *et al.* suggested that their findings reflect a relatively larger environmental influence on stature than in the American or European studies. High rates of infant mortality (up to 50 percent of newborns died by age five years), malaria, droughts and food shortages, and other 'rigours of the traditional way of life in West Africa' (p. 23) all influenced the growth of the villagers. The authors emphasized that family members of different generation, and also older and younger siblings, growing up under equally harsh or equally good environmental conditions, would tend to have higher correlations, while those growing up under dissimilar environments would tend to lower familial correlations and heritability estimates.

Similar findings are reported by Dasgupta *et al.* (1997) for a sample from rural West Bengal, India. The sample comprises 504 individuals, 110 parent pairs (mother and father), 187 of their sons, and 133 of their daughters. The families are of middle status Hindu caste, all are farmers, and the mean per capita monthly household expenditure is Rs. 150, or about US$5.00. All the coefficients of correlation calculated are less than theoretical expectation of a simple genetic model. Correlations for weight were lower, generally, than those for height. This is expected since weight is known to have a stronger environmental determination than stature. The lowest stature correlations are for brother–brother and father–son correlations for stature, $r = 0.14$ and 0.17 respectively. The highest values are for the sister–sister and mid-parent–daughter correlation for stature, $r = 0.48$ and 0.45 respectively. In general, mother–child correlations are higher than father–child correlations. This has been found in several other studies and is usually ascribed to the '. . . persistent effect of the intrauterine environment [and to] greater maternal care of children' (p. 8). Overall, the lower than expected correlation coefficients indicate that factors other than genetic inheritance account for the majority of height variation in this rural, low SES population.

Lasker & Mascie Taylor (1996), whose analysis of data from the National Child Development Study of Great Britain, which was discussed in the previous chapter, investigated parent–child correlations in stature of that sample. They calculated correlations when the offspring were 16 and 23 years old. Their findings are given in Table 7.3. Allowance for social class of the father lowers these correlations, but not by more than five per cent in any case. Social class achieved by the offspring at age 23 had no statistical effect on the correlations. I discussed in Chapter 6 how mean stature is related to social class in this sample, but it seems social class has little influence on these correlations. The values of the correlations are consistently larger in this study than in the studies from Africa and India just discussed. One would expect this, as the standard of living in Britain is much higher than in these other regions. Yet, the correlations are still less than the theoretical value of 0.5 expected from a simple genetic model. Moreover, mother–offspring correlations are higher than father–offspring correlations. Further evidence, it seems, of the persistent maternal effect due to the prenatal environment and the tendency of mothers to care for children more so than fathers. The small reduction in the correlation values when adjusted for father's social class attests to the influence of other non-genetic factors, which are not well controlled in the National Child Development Study.

Lasker & Mascie-Taylor try to place there results in the context of parent–offspring correlation studies in general. They point out that the correlations reported in the literature, which range from about 0.01 to 0.52,

Table 7.3. *Parent–child correlations at age 16 and 23 years from the National Child Development Study of Great Britain (Lasker & Mascie-Taylor, 1996)*

	16 years	23 years
Father–son	0.36	0.41
Father–daughter	0.43	0.41
Mother–son	0.41	0.47
Mother–daughter	0.47	0.46

are sample specific, reflecting different degrees of genetic variability within the sample, the effects of different environments, different sampling techniques, and different age ranges among the subjects. It is difficult to come to any concrete conclusions about the meaning of these correlations when the sample may be so ill defined. Even when the sample is very well defined, such as the Fels Research Institute Study, it may be unrepresentative of the larger population both for its ethnic make-up and the high degree of self-selection of the participants (Garn & Rohmann, 1966). Family correlation studies, then, are like twin studies in terms of the serious methodological limitations. As with twin studies, great care must be exercised when interpreting the findings of these studies.

Correlation in growth between adopted children and their adopted families

Adopted children should, in principle, have no more genetic resemblance to their adopted parents than any two people selected at random from the general population. Correlations in physical measures between adopted children and their adoptive parents should be close to zero. If the correlation coefficients are significantly different from zero, then researchers may get some idea of the influence of the common family environment on physical growth. Garn *et al.* (1976, 1979) analyzed correlations in growth between adopted pairs of siblings, adopted and natural siblings, and parents and their adopted children. Correlations in fatness, measured as skinfolds, between biologically related parent–child pairs were higher ($r = 0.20$) than those for parents and their biologically unrelated adopted children ($r = 0.10$). Between the ages of five and 18 years, biological siblings had higher correlations in fatness ($r = 0.27$) than unrelated adopted siblings ($r = 0.19$). But, the point to emphasize is that for all pairs of relatives the correlations coefficients are significantly greater than zero. This means that some factor or factors other than genetic relatedness is producing a significant degree of resemblance between people growing up in the same household.

The effect of living in a common family environment, called the cohabitational effect by Garn *et al.* (1976), was probably responsible for the higher than expected correlations in fatness between non-biologically related kin. Garn and colleagues also found a significant correlation in stature between genetically unrelated family members in these same studies. One reason for the significant correlations in growth status between genetically unrelated family members is diet and activity (exercise). Cohabitation has been shown to result in significant correlations in energy intake, energy expenditure, serum vitamin levels, and blood lipids in people with no biological relationship (Garn *et al.*, 1979). These correlations indicate, once again, the difficulty of identifying unique genetic effects on growth from this type of statistical methodology.

The effects of genetic abnormalities on growth

Although it is often difficult to separate the hereditary and environmental contributions to the phenotype, it is known that mutations to the DNA a person inherits may produce various kinds of abnormal growth and development. Understanding what happens when the genetic material is disrupted by mutation allows us to better understand the role of normal DNA in growth and development. Achondroplasia is a condition that results in short stature due to impaired growth of the legs and arms (there are other growth consequences as well). About $1/15\,000$ live births in European populations are affected, making achondroplasia a frequent cause of skeletal dwarfism. In 1994 a French research team announced the discovery of the gene defect responsible for achondroplasia (Rousseau *et al.*, 1994). The gene has been mapped to the short arm of chromosome four, and the DNA change is a point mutation that alters the amino acid make-up of a protein called fibroblast growth factor receptor-3 (Bonaventure *et al.*, 1996). The change in that receptor protein causes achondroplasia and several other clinically related disorders (e.g., thanatophoric dysplasia, type I and II, and hypochondroplasia). To date, 100 percent of patients with achondroplasia have the identical DNA mutation.

Studies of people with unusual karyotypes (the number and type of chromosomes inherited by an individual) provide evidence for such genetic effects. Normal human karyotypes are 46,XY for males and 46,XX for females, 46 being the total number of chromosomes and X or Y being the types of sex chromosomes. A classic study by Tanner *et al.* (1959) examined people with sex chromosome anomalies, including individuals with

47,XXY (Klinefelter's syndrome) and 45,X (Turner's syndrome) karyotypes. People with the 47,XXY condition are phenotypically males, and taller on average than normal 46,XY males. People with the 45,X condition are phenotypically female, and much shorter, on average, than normal 45,XX females. Tanner *et al.* found that the body proportions (e.g., the ratio of leg length to stature) and rate of skeletal development of 47,XXY boys was like that of normal 46,XY boys. They also found that the rate of skeletal development of 45,X girls, up to puberty, was like that of normal 46,XX girls. Consequently, the authors concluded that genetic factors on the Y chromosome produce the male pattern of growth in body proportions and skeletal development.

An X chromosome effect on growth was proposed by Garn & Rohmann (1962). They used a longitudinal sample of hand–wrist radiographs and dental radiographs to study ossification rate (number of bony centers present), ossification timing (age at the appearance of a center), and tooth calcification in siblings. The sample numbered 318 brother–brother, sister–sister, and brother–sisters pairs. Garn & Rohmann hypothesized that rates of skeletal and tooth development are genetically controlled and some of these genes are linked to the sex chromosomes. They also proposed that pairs of sisters, who share the same paternal X chromosome, should have greater concordance in rates of development than pairs of brothers, who have only a 50 percent chance of sharing the same maternal X chromosome, or brother–sister pairs, who share no paternal sex chromosomes. It was found that the correlation between pairs of sisters in skeletal and dental development (averaging about 0.52) was significantly greater than the correlation between pairs of brothers or bother–sister pairs (averaging about 0.35). Garn & Rohmann interpreted these correlations as evidence for X chromosome genetic control for rates of development.

The research strategy of searching for genetic determinants of growth by describing the size and development of individuals with sex chromosome anomalies has been used by many other investigators, including a series of studies by Shirley Ratcliffe and her colleagues and by J. Varrela and his colleagues. Ratcliffe's recent work focuses on a longitudinal study of boys and girls with sex chromosome abnormalities who were identified at birth (1995). In particular, her work concentrates on individuals with supernumerary sex chromosomes (XXX, XYY, XXY). She finds that an extra Y chromosome has no detectable effect on prenatal growth, as measured by weight, length, or head circumference at birth. In contrast, an extra X results in smaller size, and the XXX karyotype significantly reduces all three birth dimensions. During infancy and childhood, XYY and XXY boys grow faster than XY boys, and during adolescence the XYY boys have a greater peak height velocity than XY boys. Boys with either an extra X or Y chromosome end up taller than XYs, and virtually all of the difference in stature is due to leg growth. XXX girls grow less than XX girls at all ages and end up significantly shorter. Slower growth in sitting height (length of trunk and head) of the XXX girls is largely responsible for the stature difference. Both XXY boys and XXX girls have significantly reduced head circumference at birth and at all later ages. Head circumference reflects growth of the brain, and Ratcliffe finds a significant positive correlation between head circumference and scores on the Weschler Intelligence Scale (an 'IQ' test) when the subjects of her study were seven to 14 years old – those subjects with a smaller head circumference also had a lower 'IQ' compared with subjects with larger head circumferences. In summary, Ratcliffe finds that a supernumerary Y chromosome increases growth in leg length and stature. An extra X chromosome reduces growth of the brain during fetal life, with no catch-up after birth. In girls an extra X also reduces sitting height and stature.

The work of Varrela and colleagues focuses on adults. Varrela *et al.* (1984a) recorded 25 anthropometric measurements from 48 adult women with the 45,X karyotype (Turner's syndrome), 24 of their mothers and sisters and 95 control women to quantify the anthropometric differences associated with the chromosomal abnormality. The 45,X women were relatively smaller than their 46,XX mothers and sisters, and the control sample, for all the length measurements (height, sitting height, leg length, etc.) and pelvic breadths (especially bitrochanteric), but relatively larger than their mothers, sisters and the control sample in other measurements of breadth, circumference and fatness. The authors believed that the cause of abnormal growth in the 45,X condition appears to be primarily genetic, though it is unclear if this is due to missing genes that contribute directly to growth in size, or to the impaired metabolic activity of 45,X cells. In support of the second possibility, Varrela *et al.* cited other research showing that *in vitro* cultures of 45,X cells grow more slowly than culture of normal 46,XX cells.

Varrela *et al.* (1984a) also found that the pattern of growth of 45,X girls may have been influenced by environmental factors. The subjects and controls of their sample were taller, and larger in several dimensions, compared with samples of 45,X women and normal women measured for earlier studies. The authors believe that this difference was due to improvements in living conditions since the time of the earlier studies, including higher socioeconomic status, better health care and nutritional status, and smaller family size, all of which are environmental changes that are known to be associated with increased growth. Another indication of the sensitivity of the growth of 45,X women to the environment is that Varrela *et al.*, and Brook *et al.* (1977) in another study, found that correlations in size between adult 45,X women and their mothers or fathers (about 0.49 and 0.56 in the studies cited) were significantly higher than correlations between 45,X women and their sisters (averaging 0.10 in the Varrela *et al.*, 1984a study). If the genetic cause of growth deficits in the 45,X condition is correct, then the father-45,X daughter correlations should be the lowest, since most affected daughters apparently lack the paternal, rather than the maternal, X chromosome in their karyotype. One estimate is that 80% of the X chromosomes in 45,X females are of maternal origin (Sanger *et al.*, 1971). Given that the father's sex chromosome contribution is small, the significant correlation between father-45,X daughter growth indicates an autosomal (non-sex chromosome) location of growth controlling genes or non-genetic influences on the size of the daughters. Furthermore, the correlation in growth between parents and their 45,X daughters were higher than those between 45,X girls and their sisters. This contrast may reflect differences in parental treatment of the normal daughter and the Turner's syndrome child. If so, the possibility exists that there are major environmental determinants of the growth of 45,X children.

Varrela (1984) measured the size of 29 adult men with Klinefelter's syndrome (47,XXY) finding that they were larger than normal controls in stature, arm length, leg length, triceps and subscapular skinfold, but smaller than the normal controls in biacromial diameter, bideltoid breadth, wrist breadth and most head dimensions. One interpretation of these findings is that there are genes on the X chromosome that control linear growth, though the direction of the growth control is not the same for all body parts. However, Varrela cites evidence that 47,XXY males are deficient in testosterone, but have normal plasma levels of growth hormone. It is known that during adolescence, growth of the vertebrae and of the shoulders is dependent on androgenic hormones, such as testosterone, while leg growth is controlled more by growth hormone (Tanner *et al.*, 1976a; Prader, 1984). The effect of the hormonal milieu within the body associated with the extra X chromosome in Klinefelter's syndrome results in a relative increase in leg length and a relative decrease in biacromial (shoulder) breadth. This produces a more 'feminine' appearance compared with chromosomally normal males. The growth pattern of normal and affected males is not traceable to specific alleles for growth in size per se, rather it appears to be a hormone-mediated effect.

In another study, Varrela *et al.* (1984b) measured body size of eight adult 46,XY women with complete testicular feminization. Although the 46,XY karyotype is usually associated with the male phenotype some 46,XY embryos and fetuses lack a sensitivity of their tissues to androgenic hormones (Kelch *et al.*, 1972; Prader, 1984) and they develop into women, with normal female appearance. Varrela *et al.* found that in height, leg and arm length, and most other bony dimensions, the 46,XY women were larger than their mothers and sisters and a sample of normal controls. In body proportions the 46,XY females were not significantly different from normal females. The authors concluded that there are genes on the Y chromosome with a general size increasing effect, and this is one reason that normal men and 46,XY women are larger, on average, than normal women. They also suggested that the development of male body proportions requires a tissue sensitivity to androgenic hormones and not just the presence of Y-linked genes. For, in the case of testicular feminization, the body proportions of the 46,XY women are under the control of estrogens (or 'female' hormones), and these women achieve normal female body proportions.

To test the growth promoting effect of the Y chromosome further, Varrela & Alvesalo (1985) recorded 25 anthropometric measurements from seven men with the 47,XYY karyotype, and compared these with measurements from four normal male relatives and 42 normal control males. The 47,XYY males were significantly larger than the normal men in stature, sitting height, leg length, and bistyloid wrist breadth. The authors noted that the serum levels of growth-promoting hormones in the 47,XYY males were about equal to the levels in normal 46,XY males, except for a 'slight' elevation in testosterone. Varrela & Alvesalo point out that no amount of testosterone alone could produce all of the growth differences between 47,XYY and normal males. As mentioned above, testosterone acts mainly to increase sitting height and shoulder width and does not strongly influence leg length. Moreover, other studies of the growth of 47,XYY males show that they are taller than average during childhood, before the pubertal increase in testosterone secretion occurs (Ratcliffe, 1976). Men with XXYY karyotype are also taller than average (Lee, 1996). All of these observations led many in research to conclude that the Y chromosome contains alleles that directly increase growth in linear dimensions, perhaps by a mechanism that increases cell proliferation (mitosis).

This line of reasoning is supported by two more recent studies of the effect of Y chromosomes on the growth of bone and teeth. The Y chromosome carries a gene called *SRY* (an acronym for 'sex-determining region Y'), which is known to be involved in formation of the male testis from the undifferentiated embryo gonad. *SRY* does not do this directly, as a gene located on chromosome 17 called *SOX9* is also needed. Some pathologies of both sexual and skeletal development are linked to mutations in the *SOX9* gene. The skeletal disorders are of the type related to inadequate cell proliferation (Foster *et al.*, 1994). Taking the dental approach, Alvesalo (1997) notes that it is known that the Y chromosome promotes growth of both tooth enamel and dentin. The dental influence of the X chromosome seems to be restricted to enamel formation. Since dentin, but not enamel, growth is influenced by cell proliferation, the gene or genes responsible for this must be associated with the Y chromosome. It is likely that these are the *SRY* and *SOX9* genes, as well as others. Alvesalo suggests that the Y chromosome related genes have pleiotropic effects throughout the body, and regulate the phenotypic expression of sexual dimorphism in the size and shape of many dental and skeletal features, including stature.

With studies such as these, and the work with homeobox genes, genetic auxologists are coming closer to understanding the role of DNA in the regulation of growth in the genetically normal individual. There are additional reports of progress. Genes implicated in the growth of the skeleton, organs, and fat tissue have been mapped to specific chromosome regions. The best of these studies are done with non-human mammals. One team of researchers used the offspring from a cross between the European wild boar and the domesticated Large White pig to map such genes to a region of chromosome 4 of these animals (Andersson *et al.*, 1994). Large White pigs were probably derived from the wild boar, but the domesticated pig is significantly larger, has a longer intestine, and less abdominal fat than the wild boar. The pigs also grow more rapidly than the boars. These phenotypic differences were produced by intensive artificial selection by pig breeders for a more commercially valuable animal. The offspring of the pig–boar crosses are intermediate in size and rate of growth compared to their parents. Localizing these growth traits to a region of chromosome 4 shows a clear genetic basis to the artificial selection. Andersson *et al.* point out that most of the gene regions linked to pig fatness have also been mapped to the human chromosome 1. Potentially, some of these genes may be involved in the regulation of human obesity. To date, no single human 'obesity gene' has been found, and Andersson *et al.* caution that a single gene, or even a major gene, does not exist since human obesity is a '. . . polygenic syndrome with a strong environmental influence . . .' (p. 1774).

The same may be said for most aspects of human growth and development. All of the genetic research, including the correlation studies of the growth of parents, children, and siblings, and the sex chromosome studies, indicate that the genetic control of growth is **polygenic** (many genes) and those genes are likely to be scattered among the autosomes rather than concentrated on the sex chromosomes. It is also clear that differences in growth attributable to genetic factors exist between individuals, and by extension, may exist between biological populations. At the population level, such genetic differences probably account for only a small amount of the variation in size and rate of growth. Strong environmental influences, in combination with the action of the endocrine system as a mediator between the genotype and phenotype of a person, account for a much greater fraction of the variation.

Endocrinology of growth

Hormones are organic substances synthesized in specific body tissues, often called endocrine glands. The glands secrete their hormones into the bloodstream, where they circulate to specific, and distant, sites of action. An example, discussed in the previous chapter, was the hormone cholecalciferol, or vitamin D_3, which is synthesized in the deep layers of the skin. It travels through the bloodstream to the liver and kidneys, where its biological activity is enhanced, and then to its two primary sites of action: (1) the intestine, where it promotes calcium absorption, and (2) bone, where it regulates skeletal metabolism and bone growth.

There are several major hormones that have an effect on growth and development, and these are discussed below. In addition, there is a group of substances known as **growth factors** that have effects on growth, both independently and interactively with each other and with hormones. Growth factors are synthesized by specific cells within a wide variety of body tissues. For instance, human liver and fibroblast cells produce substances known as the insulin-like growth factors (IGFs), which promote cell division in bone, muscle, and other tissue. IGF synthesis can be stimulated by growth hormone, from the pituitary gland, and both growth hormone and IGFs may need to be present to have an optimal influence on growth (Preece & Holder, 1982; Prader, 1984). D'Ercole & Underwood (1986) described how growth factors, like hormones, may be carried by the circulation to their sites of action (the classic **endocrine action**), may act directly on the cells that synthesize them, an **autocrine action**, or may affect nearby cells in the same tissue, a **paracrine action**. There is also an **intracrine action**, whereby a hormone may act on the nucleus of the cell which manufactures the hormone. These four types of action are illustrated in Figure 7.3.

Hormones are a type of protein, and like all proteins they are manufactured by the transcription of DNA to RNA and then the construction of one or more polypeptide chains of amino acids. DNA, or genes, do not act directly on the organism that carries them, rather the DNA provides the template to manufacture substances such as hormones that have a

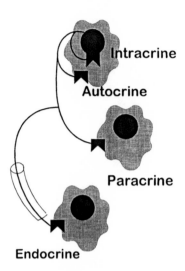

Figure 7.3. Four modes of hormone action. Growth hormone (GH), which is secreted from the pituitary, is used as an example of intracrine action on the nucleus of the secretory cell. The other three are autocrine action, in which GH is secreted and then binds to the cell membrane leading to new cell activity; paracrine action, in which GH binds to nearby cells; and endocrine action, in which GH is carried by the blood vessels to more distant sites of action (from Harvey *et al.*, 1993).

biological action. It is important to understand that the production and secretion of a hormone are only part of the process by which endocrine substances have an effect on the body. The target tissues for a hormone must be sensitive to its presence. Tissue sensitivity may be influenced by several factors, including aspects of the environment in which the organism lives, the sex and age of an individual, binding proteins that may carry the hormone to its site of action, the presence of biochemical receptors at the tissue level that recognize the hormone, and the production of 'secondary messengers', so called because some hormones do not cross cell membranes and require intermediary substances to carry their 'message' into the target cells. Here we can see how the 'external and internal environment' influences hormone action.

The following account is limited to the more elementary aspects of the endocrinology of human growth. The reader should be aware that there are additional complexities of the endocrine system. Moreover, endocrinology is a very active area of research, and our current state of knowledge may be subject to substantial revision.

The major hormones of human growth and maturation

The actions and interactions of hormones and growth factors provide a system of fine control for the regulation of growth and development. A central feature of this system is the hypothalamic regulation of the pituitary gland. Figure 7.4 illustrates the location of the hypothalamus and pituitary at the base of the brain. Blood vessels directly connect the hypothalamus to the anterior pituitary gland, and allow for neurochemical communication from the hypothalamus to the pituitary. The hormones of the hypothalamus stimulate or inhibit the release of the pituitary hormones, and these are released into the general circulation where they act on specific target tissues throughout the body. Figure 7.4 also outlines the major hormones produced and secreted by each endocrine organ and indicates the target tissues of the pituitary hormones. The following discussion is confined to those hormones involved in growth and maturation.

Thyroid hormones

Thyrotropin releasing hormone (TRH), secreted by the hypothalamus, stimulates the release from the pituitary of thyroid stimulating hormone (TSH). The pituitary TSH acts on the thyroid gland to promote the release of two metabolically active hormones, thyroxin and triiodothyronine. A negative feedback relationship controls the release of TSH and the two thyroid hormones into the bloodstream. Negative feedback is a control mechanism that involves self-regulation of a mechanical or biological system inhibition. A mechanical example is temperature regulation by a thermostat. A thermostat is a device, as in a home-heating system, a refrigerator, or an air conditioner, that automatically responds to temperature changes and activates switches controlling the equipment. In a heating system, a drop in temperature results in the closing of a switch that turns the heater on. When the temperature reaches the setting on the thermostat the switch opens and the heater is turned off. Rising temperature, then, inhibits the production of more heat. Negative feedback works in an analogous manner in biological systems, such as when rising levels of one of the thyroid hormones sends a signal back to the hypothalamus or the pituitary to terminate the release of TRH or TSH.

In Figure 7.5 three negative feedback loops are illustrated showing how thyroid activity is controlled. This model may be applied, generally, to the other hypothalamic–pituitary–target tissue hormones. The ultrashort feedback loop involves autocrine action of TRH on the hypothalamus. In this case, rising levels of TRH within the hypothalamus may suppress the

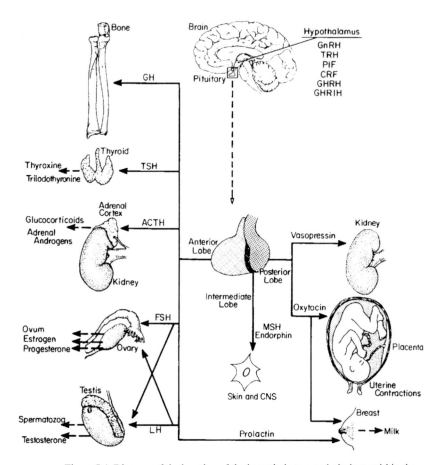

Figure 7.4. Diagram of the location of the hypothalamus and pituitary within the brain, and a schematic illustration of the target organs and tissues of the pituitary hormones. Abbreviations are: GnRH, gonadotropin-releasing hormone; TRH, thyrotropin-releasing hormone; PIF, prolactin-release-inhibiting factor; CRF, adrenocorticotropin-releasing factor; GHRH, growth hormone-releasing hormone; GHRIH, growth hormone release-inhibiting hormone; GH, growth hormone; TSH, thyroid stimulating hormone; ACTH, adrenocorticotropic hormone; FSH, follicle stimulating hormone; LH, luteinizing hormone; MSH, melanocyte stimulating hormone (from Schally *et al.*, 1977).

activity of TRH secretory cells directly. A second form of feedback control involves rising systemic levels of pituitary TSH, which may decrease the release of TRH through a short feedback loop, another example of para-crine action. Finally, an increase in the blood levels of thyroxin and triiodothyronine may suppress TRH secretion through a long feedback

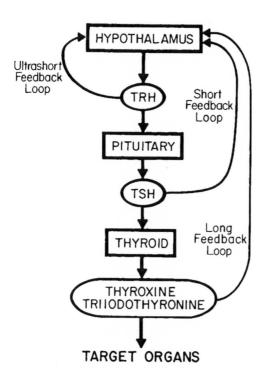

TARGET ORGANS

Figure 7.5. Feedback circuits for the control of the
hypothalamic–pituitary–thyroid system.

loop, based on endocrine action. All three avenues of feedback control may
work simultaneously to 'fine-tune' the level of thyroid hormones in the
bloodstream.

A fine level of control is needed, since thyroxin and triiodothyronine
have powerful metabolic actions. Thyroid hormones are needed for normal
growth in stature, the development of normal body proportions, formation
of bone from cartilage, and formation of the teeth. A deficiency of these
hormones (hypothyroidism) during infancy and childhood results in
growth retardation and mental impairment, and in the extreme case, the
child suffers from a form of dwarfism called cretinism. Thyroid hormones
seem to have an important role in the maturation of brain enzyme systems
and myelination, the covering of nerve fibers with a fatty insulation which
speeds up the transmission of nerve impulses. This accounts, in part, for
why hypothyrodisim in infants results in mental impairment.

Sizonenko & Aubert (1986) reported that thyroxin has been detected in
the human fetus at 78 days of age, and serum levels rise, generally, through-

out the prenatal period. By two weeks postpartum, the infant reaches adult levels of serum thyroid hormone activity. The exact nature of the role thyroid hormones play in cell division, and growth in size, remains to be determined, though it seems that normal levels of thyroxin favor protein synthesis and thus serve to provide the 'building blocks' needed for the growth of all body tissues. Thyroid hormones have a direct effect on protein synthesis and cell division, and also work along with GH and the IGFs to promote growth (Weiss & Refetoff, 1996).

Adult-onset hypothyroidism is associated with slower metabolic rate, resulting in a decreased activity of involuntary muscles, including heart and gut muscles, in voluntary muscle weakness, and in weight gain. Hyperthyroidism (an excess of the thyroid hormones) speeds up metabolic activity and results in fast heart rate, nervousness, increased appetite, and weight loss.

Gonadal hormones

The pituitary secretes two hormones that regulate gonadal activity, luteinizing hormone (LH) and follicle stimulating hormone (FSH). They were first discovered to have an influence on the maturation of the female reproductive system, hence their names are derived from ovarian functions. Nonetheless, these same hormones also have a significant effect on the male reproductive system and the secretion of hormones from the testes that have an influence on growth and maturation. The pituitary gonadotropins have but a single hypothalamic releasing hormone (Schally *et al.*, 1977), called gonadotropin-releasing hormone (GnRH).

In women, FSH and LH stimulate the growth of the ovaries and the release of ovarian hormones. The ovarian hormones are called, collectively, the estrogens. In men, FSH promotes the development of the seminiferous tubules and initiates the production of spermatozoa. LH stimulates secretion of hormones from the testes, collectively called the androgens. Androgens are required to complete the formation of mature spermatozoa and these hormones also have an influence on the growth of bone, muscle, fat, and other tissues. In men the serum levels of androgens or estrogens regulate the secretion of GnRH, LH and FSH through a, generally, negative feedback relationship (similar to that for the thyroid hormones). In women, negative feedback regulates the levels of androgen production while serum estrogen levels exert a positive feedback influence on the hypothalamus and pituitary. Positive feedback works to increase or amplify the levels of activity of a system. In women, positive feedback between the ovaries, the hypothalamus and the pituitary results in the menstrual cycle. During the pre-ovulatory phase of the menstrual cycle GnRH, LH, and FSH stimulate the ovary to release estrogens, which in turn further stimulate the hypothalamus and pituitary to secrete more GnRH, LH, and FSH. The progressive amplification of these hormones in the bloodstream leads to an LH surge at mid-cycle that results in ovulation. Negative feedback is restored during the post-ovulatory phase, which ends with menstruation. Then the cycle repeats.

Gonadotropins, estrogens, and androgens have major influences on human development. LH and FSH may be detected in fetal pituitary tissue as early as ten weeks of age (Sizonenko & Aubert, 1986). In the male fetus, LH and FSH probably provoke protein transcription by the SRY gene on the Y chromosome. As mentioned above, proteins made by the SRY gene eventually direct the undifferentiated gonad to develop into the testis, which will produce and secrete androgens, especially testosterone. However, the source of the prenatal testosterone may not be from fetal tissues entirely. It is known that a chromosomal male embryo lacking the hypothalamus or pituitary develops into a phenotypically male fetus, though with somewhat underdeveloped penis and testes. A placental hormone, human chorionic gonadotropin, is likely to be capable of stimulating differentiation of the testis and production of testosterone. This placental hormone may play a role in the development of the normal fetus as well.

The role of LH and FSH in the development of the female embryo is not well understood. In chromosomally normal female embryos, estrogen is not required for the normal development of the female reproductive system during prenatal life. Estrogen is necessary for the final maturation of reproduction following puberty, including the ability to ovulate and lactate (Smith & Korach, 1996).

At birth, and up to about two years of age, serum levels of LH, FSH, and gonadal hormones are higher than at any time prior to the onset of puberty. Noting this fact, Grumbach *et al.* (1974) suggested that the sensitivity of negative feedback in the system of hypothalamic–pituitary–gonadal regulation is not well developed in the infant. Research with non-human primates by Plant and colleagues, discussed in Chapter 2, and recent human research (Hayes & Crowley, 1998), confirms Grumbach's initial hypothesis. The delay in maturation of the feedback system may be related to the growth and development of the child during infancy and later in life. The relatively high levels of estrogen and androgens in the bloodstream of the infant are correlated with the rapid velocity of physical growth, neurological development, motor control and cognitive advancement that take place during infancy.

The rate of production of gonadal hormones and growth velocity fall to relatively low levels by about two years of age, when the negative feedback control becomes highly sensitive. Low levels of gonadal hormones, and a decelerating rate of skeletal growth, are maintained until gonadarche (puberty), when a new positive feedback control develops. The switch from negative to positive feedback means that rising levels of gonadal hormone secretion result in an increased secretion of GnRH, LH, and FSH. During adolescence the rate of growth increases rapidly, that is, the growth spurt occurs. Prader (1984) and Preece (1986) find that the rising levels of testosterone and estrogens are significantly correlated with growth rate during adolescence, and they suggested that increased secretion of these hormones is likely to be one cause of the adolescent growth spurt. They were correct, as much more is now known about the role of the gonadal hormones at the time of adolescence. Figure 7.6 presents a model of the action of androgens and estrogens on skeletal growth. The model was developed by Smith & Korach (1996).

The effect of gonadal hormones on personality and behavior has also been an area of much research. It is commonly believed that rising levels of testosterone can increase feelings of aggression and libido in both men and women. On the other hand, rising levels of estrogens are commonly thought to 'feminize' (i.e., reduce aggression) the behavior of both men and women. A series of studies by J. Richard Udry and colleagues have investigated these popular notion to see if they have scientific validity. The levels of serum testosterone is a strong, positive predictor of sexual motivation and behavior in both adolescent boys and girls (Udry *et al.*, 1985; Halpern *et al.*, 1993; 1997). However, there are also strong effects of social factors that moderate sexual activity. For example, a higher frequency of attendance at religious services significantly delayed age at first coitus. The researchers report that testosterone is not the primary cause of personality differentiation at puberty – testosterone does not cause male patterns of aggression in human beings (Udry *et al.*, 1988). Rather, as with sexual activity, aggression is moderated by several social factors. The same is true of so-called feminine behaviors in both boys and girls. Udry (1994) uses these data to develop a model for the development of human genders (as opposed to human sexes). It is a biosocial model, and requires inputs from both the internal endocrine environment and the external social environment of the individual in order to account for the development of personality and behavior.

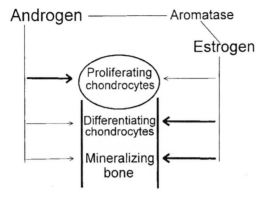

Figure 7.6. A model for the relative actions of androgens and estrogens on the growth of a long bone. Androgen hormones can be converted to estrogens via the action of the aromatase protein. The epiphyseal growth plate of the long bone is indicated as a sketch, with the regions of chondrocyte proliferation, chondrocyte differentiation, and bone mineralization labeled. The width of the arrows indicates the relative importance of androgens and estrogens on these regions and processes. Redrawn from Smith & Korach (1996).

Adrenal hormones

The role that adrenal hormones play in human growth and development, in relation to adrenarche and the evolution of the childhood period of human development, was discussed in Chapter 4. The adrenal hormones have other effects on growth as well. The cortex of the adrenal produces two classes of hormones, glucocorticoids and androgens. Glucocorticoids, such as cortisol, are involved in the body's ability to maintain homeostasis when faced with physical or emotional stress. Adrenal androgens are produced by the zona reticularis of the adrenal cortex, which is relatively large and active during fetal life, but undergoes involution after birth. Adrenal androgen levels are low throughout childhood until adrenarche, at about age six to nine years, when secretions begin to increase steadily until a plateau is reached late in the fourth decade of life (Wierman & Crowley, 1986).

The control of adrenal androgen production may involve the activity of glucocorticoids, and their pituitary stimulating hormone called adrenocorticotropic hormone (ACTH). Anderson (1980) proposed that rising levels of the glucocorticoids during the late fetal period act to suppress adrenal androgen production. Years later, stimulation by ACTH overcomes the inhibiting action of the glucocorticoids and produces adrenarche.

Compounding the problem of understanding the evolutionary importance of adrenarche is the fact that its physiological function is not known completely. Katz *et al.* (1985) found a positive correlation between levels of adrenal androgens, skeletal maturation, and fatness in adolescent boys that was independent of their serum levels of gonadal testosterone. In a related study, Zemel & Katz (1986) calculated the statistical contribution of testosterone and the major adrenal androgen (dehydroepiandosterone sulfate) to growth velocity in height for a sample of 181 male adolescents. Based on a regression analysis, it was found that both hormones had a significant effect on height velocity. However, the relative influence of testosterone was about five times greater than that of the adrenal androgen. Hediger & Katz (1986) found that increased adrenal androgen levels in skeletally mature young women were correlated positively with fatness and associated with a concentration of subcutaneous fat on the trunk of the body. Taken together this series of studies by Katz and his colleagues suggest that adrenarche may be one of the endocrine events that promotes maturation and determines, in part, adult body proportions and body composition. As discussed previously, these sorts of changes mark the transition from the childhood to the juvenile stage of the life cycle.

Adrenal hormones have a strong influence on the distribution of body fat. The placement of fat on the body, in turn, is associated with several metabolic diseases, including diabetes and heart disease. Centralized body fat, such as that in the abdominal region, is a 'risk factor' for these diseases. Centralized fat seems to exert an influence on the hypothalamic–pituitary–adrenal axis. In adults with high amounts of abdominal fat there is a lower secretion of pituitary growth hormone and of gonadal androgens and estrogens, due to suppression of hypothalamic GnRH and of pituitary LH and FSH. Women with centralized obesity seem to have increased secretion of adrenal androgens, resulting in a more masculinized body in terms of fat patterning and muscularity (Bjorntorp, 1997). The cause and effect relationship in all of this is not clear, because it seems to take both elevated cortisol levels and lower gonadal hormone and growth hormone levels to direct the fat to be stored at abdominal sites. Clinical depression and anxiety, as well as alcohol abuse and smoking, also play a role in these hormonal and fat deposit outcomes. Indeed, Bjorntorp reports that emotional stress may precede the physiological effects, and that the emotional disorder may cause the disruption in the hypothalamic–pituitary–adrenal axis. Experimental studies with monkeys supports this interpretation. The emotional stress leads to over-secretion of adrenal cortisol and other hormones, which seems to cause the more generalized hormonal imbalance, the centralized fat deposits, and the metabolic diseases.

Growth hormone and growth factors

Throughout prenatal and postnatal life the availability of growth hormone and other growth factors are necessary to maintain and promote physical growth and development. Growth hormone (GH) is synthesized and secreted by specialized cells in the anterior pituitary gland. A hypothalamic hormone, growth hormone releasing hormone (GHRH), stimulates the synthesis of GH and, along with other agents, causes the release of GH into the bloodstream (Barinaga *et al.*, 1985). Another hypothalamic hormone, growth hormone-release-inhibiting hormone (GHRIH), sometimes also called somatostatin, has an antisecretory effect on GH. At specific locations in the body, GHRIH also inhibits the secretion of TSH, glucagon, insulin, and several digestive acids and enzymes (Schally *et al.*, 1977). Experimental and clinical research show that secretion of GHRIH and GHRF are inversely correlated, and indicate that the direct feedback control of GH secretion is maintained within the hypothalamus (Argente *et al.*, 1992). The overall control of GH secretion is complex, with many inputs from the internal and external environment of the body. Figure 7.7 illustrates some of the more important aspects of GH regulation.

Unlike the other pituitary hormones so far discussed, GH does not seem to affect a single target tissue, but appears to have a general growth-promoting effect throughout the body. Early research by Cheek (1968) found that GH is needed for the body to retain nitrogen, sodium, chlorine, potassium, phosphorous, calcium, and other elements that make up body tissue. Cheek also reported that GH is needed for muscle cell division. Isaksson *et al.* (1982) demonstrated experimentally that GH stimulates long bone growth directly. The authors injected GH into the cartilage growth plate of the proximal tibia of rats, in which the pituitary gland had been surgically removed. Rat tibia receiving the GH injection grew significantly more than rat tibia receiving a saline injection. Green and his colleagues (Morikawa *et al.*, 1982; Nixon & Green, 1984; Green *et al.*, 1985) found that GH promotes the differentiation of preadipose stem cells into adipose cells. Given the widespread action of GH in the body, the name 'growth hormone' is an accurate description of the function of this endocrine product.

Another class of growth-promoting substances is called the insulin-like growth factors (IGFs); in older literature they are called somatomedins. The IGFs are similar to insulin in molecular structure and in biological action (Preece, 1986). Insulin, a hormone produced and secreted by cells of the pancreas, stimulates protein synthesis and the growth of cartilage cells. It is well known that GH acts on the pancreas to increase the synthesis of

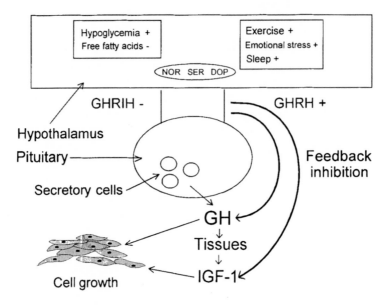

Figure 7.7. Some of the major features of the growth hormone (GH) regulation and the hypothalamic–pituitary–GH axis. GH is secreted by specialized cells in the anterior pituitary. GH enters the general circulation and is carried to its target tissues throughout the body. Growth hormone binding proteins (not shown) are needed for GH to have its effect on the target tissues. At the tissues, GH promotes the secretion of insulin-like growth factor 1 (IGF-1) by cells in the tissues. Both GH and IGF-1 may produce cell growth. The regulation of GH secretion is controlled by neurochemical signals from the hypothalamus. GHRIH inhibits GH, as indicated by the minus sign, while GHRH stimulates GH secretion, as indicated by the plus sign. The secretion of both GH and IGF-1 work to inhibit further GH release via feedback inhibition. Several environmental variables may influence GH release. Low food intake leading to hypoglycemia increases GH release. An increase of free fatty acids in the bloodstream lowers GH release. Exercise, short-term emotional stress, and sleep increase GH release. The impact of these environmental variables on GH is mediated by several neurochemical transmitters in the hypothalamus, including noradrenaline (NOR), serotonin (SER), and dopamine (DOP).

insulin, and also increases the serum levels of the IGFs. It was once thought that the unique origin of IGFs was from cells located in the liver, but it is now known that a wide variety of cells produce IGFs (Clemmons & Van Wyk, 1984). In humans there are two major types of IGFs: IGF-1, which may be the type that, in concert with GH, regulates postnatal growth, and IGF-2, which appears to be the type that controls some aspects of prenatal growth. Production of IGF-2 is stimulated by placental lactogen, a hormone similar to GH and prolactin (Sizonenko & Aubert, 1986).

Preece & Holder (1982) reviewed experimental and clinical studies of the role of GH and the IGFs on growth. They argued that the evidence suggested a cascade effect: hypothalamic GHRF stimulating the release of pituitary GH, which circulates in the bloodstream stimulating the production of the IGFs at the tissue level, where they have autocrine and paracrine actions to promote cell division. Preece & Holder cite several endocrine pathologies as evidence for the role of IGFs as the direct growth promoting substance, in particular the conditions Laron syndrome and acromegaly. One of the features of Laron syndrome is dwarfism. Some types of dwarfism are due to low levels of, or an absence of, GH, but Laron dwarfism results from low levels of IGFs despite normal or high levels of GH. Adult height is at or below 142 cm for men and 136 cm for women. The cause of Laron dwarfism is a genetic mutation that results in a deficiency of the cell receptors for GH (Rosenbloom *et al.*, 1992). Without the GH receptor, cells cannot be stimulated to produce IGF-1, and the result is impaired cell growth. Acromegaly is a condition of abnormal exessive growth characterized by a slow, but continuous increase in size of the bones of the face, hands, and feet throughout life, that often leads to gross disfigurement in adulthood. The professional wrestler and actor Andre the Giant (star of the 1987 film *The Princess Bride*) suffered from acromegaly. At the time of his death in 1993, at age 46, he was 224 cm tall and weighed 236 kg. Levels of GH and IGFs are both above normal in acromegalic patients; however, there is a higher correlation between IGFs and the clinical progress of the disease.

Clearly, the IGFs, along with IGF receptors, IGF binding proteins (proteins that carry IGFs to their target sites), and IGF proteases (proteases are involved in DNA transcription) have an important role in normal and abnormal growth. However, many studies confirm that GH alone can stimulate cellular growth. As mentioned above, studies of living animals and cultured cells find that GH administration promotes the differentiation of stem cells (i.e., undifferentiated cells) into specific types of tissue cells. For example, GH promotes prechondrocytes in the growth plate into cartilage cells (Nilsson *et al.*, 1986), and pre-adipose cells into adipose cells (Isaksson *et al.*, 1982; Morikawa *et al.*, 1982). Following differentiation, these cells are sensitive to the influence of IGFs, which promote growth in cell number (hyperplasia) by mitosis.

A **dual effector model** was proposed by Green *et al.* (1985) to model the coordinated actions of GH and the IGFs. In the 'dual effector' system, the differentiation of cells is controlled by GH and the selective multiplication of young differentiated cells controlled by specific IGFs. The effect of IGFs on cell growth must be selective for only the newly differentiated cells, since a general response of growth by all cells in a tissue would result in uncontrolled hyperplasia, leading, perhaps, to cancerous growth. A host of growth factors, including six to nine IGF binding proteins, tissue specific inhibitory factors, vitamin D_3, IGF-2, and many cell cycle regulators, such as the molecules p21 and p53, are involved in the inhibition of mitosis in all cells except the newly differentiated cells. The interplay of these growth factors and growth inhibitors seems to account for many human diseases, especially cancer. Accordingly there is a great deal of research activity in this area. Readers interested in greater detail may consult a recent molecular biology textbook to learn more about the control of cell differentiation and cell division.

One prediction of a dual effector model is that growth will occur in cycles, with alternating phases of stem cell differentiation followed by the expansion of newly differentiated cells by hyperplasia and hypertrophy. Growth cycles, or pulses, have been found when growth in length is measured on a weekly or daily basis. Periods of little or no growth, called stasis, are followed by bursts of rapid growth, called 'mini-growth spurts' or saltations (Hermanussen *et al.*, 1988; Lampl, *et al.*, 1992). Figure 7.8 illustrates the nature of these mini-growth spurts in terms of growth of the lower leg. The discovery of these periods of stasis and saltation replaced the traditional belief, based on annual or quarterly measurements, that growth is a smooth and linear process. The discovery also showed that short-term rates of growth cannot be extrapolated to annual rates, an important finding for medical intervention. Also important was the discovery that different segments of the body may be in stasis or growth at different times. A Japanese study (Ashizawa & Kawabata, 1990) of two children (ages 7.5 and 6.6 years) measured daily for one year found that all growth in stature was confined to the lower limbs, except for a seasonal pulse of trunk growth in the Spring (April–May).

How cycles of growth stasis and pulses of saltation are controlled is not known, but these growth patterns do lend support to the 'dual-effector' hypothesis. Growth stasis would occur during the differentiation stage, and saltations would occur during the expansion stage.

GH, IGFs, and rate of growth in normal children

Sizonenko & Aubert (1986) reviewed research showing that GH levels rise continuously throughout fetal life, peaking at the maximum lifetime level at about 35 to 40 weeks of gestation. The high value of GH in serum at this time may be due to two causes. First, the use by the fetus of placental lactogen as the 'growth hormone', rather than GH secreted by the fetal pituitary. Second, the negative feedback system for the regulation of GH

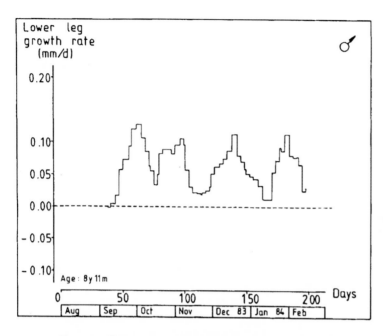

Figure 7.8. Growth velocity data for the lower leg ('knee height') of a boy aged 8 years 11 months, measured once weekly. 'Mini growth spurts' occur from week to week, and a longer term cycle of increases and decreases in the size of the 'spurts' is also apparent (after Hermanussen *et al.*, 1988).

secretion is not fully functional during fetal life. After birth, GH levels in serum decrease, so that at about one month of age they reach adult values, and at about three months of age the GH regulatory system seems to mature. Throughout the rest of life, GH is secreted in a pulsatile fashion, primarily at night. However, physical and emotional stress are capable of producing episodes of GH secretion at any time of day.

There is a significant positive correlation between serum levels of IGF-1, and IGF binding protein-3, and size (both height-for-age and weight-for-age) in prepubertal boys and girls (Jones & Clemmons, 1995). Preece *et al.* (1984) found for boys, that levels of IGFs, and possibly GH, were higher during adolescence than during childhood or adulthood. In fact, IGF levels were highest at the time of peak growth velocity in height; however, the variation of IGFs to growth rate during the total span of the adolescent spurt was not statistically significant. Rosenfeld *et al.* (1983) measured IGF levels during adolescence in boys and girls, finding that peak IGF values usually occurred about a year after a child achieved peak height velocity. From these studies, it seems that GH and IGFs play, at best, a limited role in the adolescent growth spurt. Prader (1984) argued that normal levels of GH and IGFs are needed to maintain normal adolescent growth, but that the pubertal growth spurt is more directly related to the marked increase in the secretion of testosterone or estrodiol, and not GH. The coordinated action of GH, IGF-1, and the gonadal hormones during adolescence is discussed below.

Other growth factors

In addition to the IGFs, which appear to be regulated, in part, by GH release, there is ever-increasing recognition of other growth factors which are specific for different tissues. Studies of the action of these growth factors was reviewed by D'Ercole & Underwood (1986). Epidermal growth factor and fibroblast growth factor appear to act like IGFs, in that they stimulate cell division of newly differentiated epidermal or fibroblast cells. Platelet-derived growth factor is released from platelets during blood coagulation, and may have an important function in response to injury and during cell differentiation. Bone-derived growth factor appears to be capable of stimulating cell division in cartilage and bone cells. Nerve growth factor may have little or no mitogenic (cell division) influence, but seems to be needed to differentiate nerve cells and to guide the elongation of axons toward their target sites. There are other known or postulated growth factors. How they all act and interact during the growth of the individual person, or any experimental animal, is not well understood, nor is it known if they are unique substances or members of a family of closely related autocrine and paracrine factors.

Other pituitary hormones, such as prolactin, melanocyte-stimulating hormone, vasopressin, oxytocin, and their hypothalamic releasing or inhibiting factors are necessary for normal metabolic activity, the maintenance of the placenta and fetus, birth, and other life-sustaining processes. Nonpituitary hormones also play a similarly vital role. In a broad sense, therefore, these hormones are essential for growth, but a detailed discussion of them is outside the scope of this book.

Hypothalamic, pituitary, and gonadal regulation of gonadarche

Hormones play key roles in the regulation of animal and plant life histories, particularly in the timing of transitions between prematurational stages and in the scheduling of reproduction. Furthermore, hormonal mechanisms are subject to information about the external and internal environment of the individual   (Finch & Rose, 1995, p. 1).

The transition from the juvenile stage to the adolescent stage of the human life cycle heralds the onset of reproductive maturation. A complex interplay between hormones, the central nervous system, genes, and the environment underlies this transition. One of the first endocrine signs of gonadarche (puberty) in boys are increases in the night-time secretion of gonadotropin-releasing-hormone (GnRH) by the hypothalamus, LH by the pituitary, and testosterone by the testis. More than 20 years ago, Boyer *et al.* (1974) and Judd *et al.* (1977) measured the serum levels of these hormones in normal boys by drawing blood every 20 to 30 minutes over a 24-hour period. In boys with an average bone age of eight 'years' or older, hormone levels began to show an increase during the night, while the boys were asleep. Serum LH levels increased about four-fold and testosterone levels increased about five-fold over prepubertal baseline values. During the daytime, all endocrine levels remained at baseline values. These night-time episodes of GnRH, LH and testosterone release preceded development of primary or secondary sexual characteristics. Weitzman *et al.* (1975) found that by middle to late adolescence, when the secondary sexual characteristics were well on their way towards maturity, the secretion of LH occurred in 'pulses' during the day as well as during the night. Testosterone levels remained relatively high at all hours in these post-pubertal boys. This endocrine pattern is also found for adult men.

Boyer *et al.* (1976) did not find a night-time increase in LH or FSH in adolescent girls, but did find a peak in serum levels at about 3.00 p.m. Jakacki *et al.* (1982) found that one endocrinologically normal 11.8-year-old girl, of two girls studied, had significant night-time LH secretory pulses. Whether secreted by day or by night, both studies found that, as for boys, a pulsatile secretion of GnRH and LH is necessary to initiate and maintain normal gonadal secretions of estrogens (or androgens in boys). Experimental studies show that a constant infusion, or single daily doses, of LH administered to LH-deficient children, monkeys, or rats will not promote gonadal maturation, but doses administered in a pulsitile fashion will do so (Marshall & Kelch, 1986).

These findings suggested to some researchers (e.g., Grumbach *et al.*, 1974) that the onset of night-time and/or pulsatile secretions of LH and gonadal hormones marked the change from the childhood pattern of hypothalamic–pituitary regulation of the gonads to the pubertal pattern of reproductive system control. That is, during childhood relatively low levels of circulating gonadal hormones acted to inhibit the secretion of releasing factors from the hypothalamus and gonadal-stimulating hormones from the pituitary. At the onset of puberty, one of two changes was believed to occur. Either the feedback system was reset, so that much higher levels of gonadal hormones were required to inhibit the hypothalamus, or gonadal hormones acted to stimulate the hypothalamus and pituitary to release even more gonadotrophic hormones.

A marvelous series of experimental studies with Old World monkeys, especially the rhesus, and clinical studies with humans strongly supports and refines the first hypothesis. Tony Plant and Ayesh Perera (Perera & Plant, 1992; Plant, 1994) find that the hypothalamic–pituitary–gonadal axis fully matures during fetal life. This maturity includes the pulsatile secretion of GnRH, which shows all the features of adult levels of activity. By late infancy, however, the GnRH 'pulse generator' (the term used to refer to the incompletely understood mechanism that controls GnRH release) is suppressed. The cause of the suppression is definitely localized to the hypothalamus, and not the pituitary or the gonads. A brief discussion of this process and an illustration of it was presented in Figure 2.11. At gonadarche, the suppression mechanism is removed, and the already mature pulse generator resumes the pattern of activity that it already demonstrated during the fetal and early infant stages of development.

Human evidence for this developmental sequence of hypothalamic activation–suppression–reactivation was published by Jakacki *et al.* (1982). They used a sensitive assay method to measure the gonadotropin secretions of 15 prepubertal children with bone ages less than ten 'years'. Blood samples were taken every 20 minutes for periods of three to 11 hours. Nocturnal levels of LH and FSH were higher in all children and a relatively low pulsatile secretion of LH was detected in eight of the children, some with bone ages as young as five 'years'. The authors concluded that '. . . puberty is heralded by the amplification, rather the initiation, of a circadian pattern of gonadotropin secretion' (p. 457). In retrospect, these findings showed that the hypothalamic–pituitary–gonadal system does not change its feedback relationships, rather it seems that low levels of GnRH, LH, FSH, and gonadal hormone secretion during childhood represent the inhibition of the reproductive endocrine system by some, then unknown, mechanism. At about eight to ten years of age, the inhibiting mechanism relinquishes control and the hypothalamic–pituitary–gonadal systems start to approach mature levels of operation. A mathematical model of this developmental sequence is shown in Figure 7.9.

The mechanism that inhibits the mature functioning of the GnRH pulse generator is still not completely understood. Perera & Plant (1997) have examined the morphological structure of the hypothalamus and the neuronal architecture of its GnRH secretory cells. When sections of this tissue are properly stained and viewed by electron microscope, they find that that there is a greater density of nerve cells and synaptic connection in the

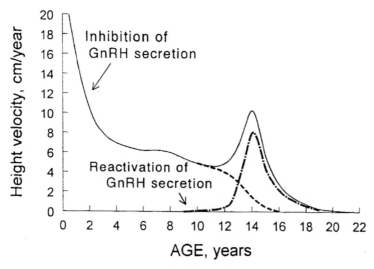

Figure 7.9. A model of human growth based on the activity of hypothalamic gonadotropin releasing hormone (GnRH) secretion by the hypothalamus. The thin solid line is an idealized velocity curve of human growth for boys. This curve can be described mathematically by two linked functions: a prepubertal and a pubertal function. The prepubertal function is associated with the phase of quiescent gonadal hormone activity due to inhibition of GnRH secretion by late infancy. The prepubertal rate of growth reaches zero and is 'switched-off' by 16 years of age, as indicated by the heavy broken line. The pubertal function (heavy chain–dash line) is 'switched-on' at about nine years of age, when the GnRH pulse generator is reactivated. As the frequency and amplitude of GnRH pulses increase, there is a rise in the pulsatile secretion of pituitary luteinizing hormone and gonadal testosterone. Spermatogenesis and an increase in rate of growth take place. The size and shape of the adolescent growth spurt is the product of interaction between the prepubertal and pubertal functions. The sequence of events for girls is similar (modified from the concept of Stutzle *et al.*, 1980).

medial basal hypothalamus of juvenile rhesus monkeys than in adults. Other regions of the hypothalamus did not differ between juvenile and adult monkeys. The authors suggest that the pulse generator suppression mechanism may be explained by plasticity in the neuronal inputs of the GnRH regions of the hypothalamus. How the timing of these plastic changes is controlled is still unknown.

Although the causes of the onset of puberty are not known the consequences of hypothalamic maturation for the gonads and for physical development are well documented. Preece (1986) pooled the results of several studies and found that, for girls, the amplification of pulsatile secretion of GnRH results in an increase of serum levels of LH and estradiol during puberty; for example, estradiol levels rise from about 50 picomoles/litre at age seven years to over 400 pmol/l by age 15 years. Preece also analyzed studies of children from whom both growth and hormone measurements were taken, to verify relationships between levels of gonadal hormones and physical maturation that had been long suspected but never proven. It was found that the increase in estradiol levels is highly correlated with secondary sexual development, for example maturation of the breasts. In an earlier study, Preece *et al.* (1984) discovered a similar high correlation between testosterone levels and genitalia development in adolescent boys. Using Tanner's (1962) system of rating genital maturation stages, Preece *et al.* found that at genitalia stage 2, when the first signs of enlargement of the penis and testes are clinically detectable, the serum concentration of testosterone averaged about 7.0 nanomoles/liter. At genitalia stage 5, when the penis and scrotum have an adult size and appearance, serum testosterone concentration averaged about 19 nmol/l.

Prader (1984) and Preece *et al.* (1984) review studies that show that a pubertal increase in testosterone secretion is needed for a normal growth spurt to take place in boys. Of course, normal levels of other growth-promoting hormones, such as thyroxin and growth hormone, must be present for a normal growth spurt to occur. For about two years before the peak height velocity (PHV) of the adolescent growth spurt, rising testosterone levels are positively and significantly correlated with increasing rates of growth in height. For at least three years after PHV, rising testosterone levels are negatively and significantly correlated with rate of height growth. Preece (1986) states that these findings suggest the principal effect of testosterone in early adolescence is to stimulate bone growth, but in later adolescence its main action is to promote epiphysial fusion, which slows growth. It is also likely that in the later stages of puberty testosterone exerts an inhibiting effect directly on the growth of cartilage and bone cells, since boys with higher levels of testosterone terminate their growth spurt more quickly than boys with lower levels of testosterone (Preece *et al.*, 1984).

Prader (1984) reviewed endocrine studies of the control of the adolescent height growth spurt in girls. Girls produce androgens in the adrenals and by conversion of ovarian estrogens to androgens, which have been shown to stimulate bone growth. However, girls lacking these androgens or lacking tissue sensitivity to androgens, such as girls with testicular feminization (see discussion of such cases in the genetics section of this chapter) experience a normal growth spurt. From this, Prader concluded that, in otherwise normal girls, the high levels of estradiol secreted during female puberty are directly responsible for the growth spurt. Presumably, estradiol acts like testosterone to stimulate bone growth in early puberty and to suppress cell division and facilitate epiphysial closure in later puberty.

More recent work shows that the rise in serum levels of the gonadal hormones during adolescence is directly responsible for the increased levels of GH, IGF-1, and IGF-Binding Protein-3 (IGFBP-3 is one of several binding proteins for IGF-1). High levels of androgens and estrogens cause irreversible changes in the hypothalamic–pituitary regulation of both GH and IGF-1 (Belgorosky & Rivarola, 1998). Androgens and estrogens also have direct effects on bone growth, by acting on cell division and maturation at the growth plate. Both types of sex hormones are needed for normal growth in boys and girls. Boys and girls without any estrogen or without receptors for estrogen never have an adolescent growth spurt. They have delayed bone maturation, unfused epiphyses, continued growth into adulthood, tall stature, low bone mineralization, and little or no secondary sexual development (MacGillivary *et al.*, 1998). All of these growth abnormalities are post-gonadarche events, as fetal survival, placental growth, and sexual differentiation are normal.

Melatonin and leptin: no role in the control of human puberty

Though the mechanism by which GnRH, LH, and gonadal hormone secretion is inhibited during childhood and stimulated with the onset of adolescence is not known, some studies have indirectly indicated that melatonin, a hormone produced and secreted by the pineal gland, may be involved. The pineal gland is located roughly at the center of the brain, near the inferior dorsal margin of the third ventricle, and contains or produces several compounds. Melatonin is the only one of these compounds for which a biological function is known (Tamarkin *et al.*, 1985). Wurtman & Axelrod (1965) described how photic information received by the eyes is carried by the nervous system to the pineal. The electrical information received by the pineal is transduced by cells in the pineal into the endocrine product melatonin. The melatonin is secreted into the general circulation to be carried to its sites of action. In seasonally breeding animals, changes in day length during the year correlate with levels of melatonin in circulation and the onset or suppression of gonadal maturation.

In relation to human puberty, Wurtman & Axelrod reported that for many years there had been speculation that high levels of melatonin in children might serve to suppress reproductive maturation. For instance, pineal tumors have long been known to be associated with precocious puberty in boys. Such children have an unusually early puberty, usually occurring before the age of eight years, and also have an abnormal pattern of growth. They are relatively tall during childhood, but this is due to an early pubertal growth spurt. The early spurt means that growth in height ends at a young age, often before the teenage years begin, which means that as adults these individuals are short.

Today it is known that melatonin plays an important role in reproduction, sleep, and, possibly, aging in many mammals. Some researchers still believe that melatonin plays a role in the onset of puberty. The action of melatonin on sleep and aging captured much public attention and led to a small industry of popular publications about the 'miraculous' nature of melatonin and the manufacture and sale of over-the-counter drugs containing melatonin. It is reported that a melatonin-based contraceptive for humans in undergoing clinical trials (Silman, 1993). Melatonin is clearly involved in the regulation of reproduction in seasonal breeding animals, such as horse, sheep and hamsters. Silman, who has been a champion of melatonin's role in human puberty for many years, explains that 'Humans are not seasonal breeders . . . due to an *impairment* of the retino-pineal pathway . . .' (p. 3, emphasis added). This somewhat humorous, but biologically misguided statement is just one example of the exaggerated nature of much melatonin research. Humans have a completely functional pathway from the optic nerve to the pineal gland, and variation in light levels do cause changes in the production of melatonin by the human pineal. The reason why humans are capable of reproduction throughout the year, while some mammals are seasonal breeders, lies in the evolutionary adaptations that each species makes to its environment.

The scientific understanding of the action of melatonin on human physiology is not well understood. Numerous studies note a correlation between changes in melatonin secretion, sexual maturation, and reproduction in humans. In one review of the research, Reiter (1998), who has long been involved in pineal research, states that '. . . The transition from Tanner stage 1 to Tanner stage 5 of sexual maturation is associated with a significant reduction in nocturnal melatonin levels, but a cause–effect relationship has not been established . . . Menopause is associated with a reduction in melatonin . . . [but] . . . In males of the same age melatonin levels also drop with no significant alteration in reproductive physiology' (p. 103).

Succinctly stated, there is no evidence that melatonin causes the onset of puberty, and there is weak evidence that melatonin directly regulates human reproductive physiology.

Leptin, discovered in 1994, is a protein that is transcribed by the so-called obesity (*ob*) gene. It is called the obesity gene because laboratory mice that have a mutation on the *ob* gene which prevents the manufacture of functional leptin become morbidly obese. Another strain of mouse has a mutation in the gene that codes for the leptin receptor, which also results in obesity, even in the presence of sufficient leptin in the bloodstream. The role of leptin in these cases of rodent obesity seems to be that leptin signals to the brain when there has been sufficient food intake to maintain body weight and body fatness. Leptin is secreted from adipose cells and seems to travel to the hypothalamus, where it regulates the hunger and satiety control centers (http://www.rockefeller.edu/pubinfo/ob.nr.html).The role that leptin and human analogues of the mouse obesity gene play in human obesity is under investigation.

It is also hypothesized that leptin is somehow involved in the regulation of human puberty. A correlation exists between levels of body fat and the onset of menarche in girls (Frisch, 1974). Could it be that as body fat is added during the juvenile years, more leptin is released into the blood-stream, until a critical level is reached? If this is true, then once past that threshold level of leptin, the hypothalamus is triggered to release its inhibi-tion on the GnRH pulse generator, and gonadarche occurs. This is an attractive and clearly delineated hypothesis. Unfortunately, the research to date finds no support for it. Leptin levels definitely reflect the level of fat mass of the body, rising with more fatness. However, a study by Arslanian *et al.* (1998) finds no significant difference in leptin levels between prepuber-tal and post-pubertal boys and girls. Nor were there sex differences in leptin levels after controlling for body fatness. Similar negative findings for a leptin–puberty effect in humans are reported by Palmert *et al.* (1998), who find no change in leptin levels at the time of gonadarche of boys and girls, and by Plant & Durrant (1997), who find no relationship of leptin levels to the timing of puberty in male rhesus monkeys.

Synthesis: genetic, endocrine, and environmental mediation of growth

Figure 7.10 is a summary of the relationship of GH and IGF-1 levels in the bloodstream to growth in stature. The straight line is the linear re-gression, that is, the mean value, of IGF-1 level for a given amount of GH secretion based on more than 100 clinical measurements of children, juveniles, and adolescents. The labels 'Short', 'Normal', and 'Tall' indicate boys and girls whose stature falls within the clinically accepted range. Note that as stature increases within this range, the levels of GH and IGF-1 increase. Exceptionally tall stature of people with acromegaly is due to an overproduction of GH and IGF-1. Growth hormone deficiency ('GHD' in the figure) results in exceptionally short stature. One type of clinical obesity is associated with an excessive production of IGF-1 in the

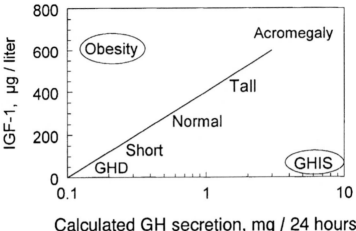

Calculated GH secretion, mg / 24 hours

Figure 7.10. Relationship of the calculated level of growth hormone (GH) secretion and the measured level of insulin-like growth factor-1 secretion to growth in stature. 'Short', 'Normal', and 'Tall' indicate stature within the clinically accepted range. 'GHD' indicates growth hormone deficency and 'GHIS' indicates growth hormone insensitivity syndrome. See the text for discussion of these growth variants and others indicated in the figure. Redrawn from the GH Insensitivity Syndromes Newsletter-KHP 336, 1993, Kabi Pharmacia Peptide Hormones, Stockholm, Sweden.

presence of a low–normal secretion of GH. Finally, a clinical condition called growth hormone insensitivity syndrome ('GHIS' in the figure) results from an excessive secretion of GH with very low levels of IGF-1 (below the 0.1 centile). Despite the production of GH, patients with GHIS are unable to utilize it or sense its presence in the blood. Due to this insensitivity, almost no IGF-1 is secreted by tissues and the GH feedback regulation system is deranged. Two clinical aspects of GHIS are very short stature ($< -3\,$SD units below mean height for age), and delayed growth of the skull and face. The unusual growth of the head and face is due to the fact that the bones of the base of the skull are exceptionally dependent on the GH/IGF-1 axis, and without these hormones facial development is extremely delayed.

The high correlation of GH and IGF-1 to variation in human growth, both normal and clinically pathological, attests to their central role in human growth and development. However, more importantly it helps us to understand how variation in body size, body composition, and body proportions between individuals and populations are brought about. At the center of the mechanism behind growth variation is the endocrine system. Of course, many more hormones, binding proteins, and growth factors are part of this mechanism. As hormones are protein molecules, they are ultimately coded for by DNA. Missing, additional, or otherwise mutated DNA may result in any of the growth disorders, a few of which are shown in Figure 7.10 or were mentioned in previous sections of this book. Many other disorders are described in pediatric medical texts.

Indeed, enough is known about the role that hormones play in pathological growth that hormone therapy is now a standard part of clinical medicine. There are medical specialists who treat patients with stunted growth, or excessive growth, by using hormones as medicine. The basic science knowledge that I have outlined here is part of the basis for these medical treatments. Further research, based on the results of clinical experience, helps to confirm the basic science and also reveals gaps in our knowledge which require additional research.

The interaction between DNA and hormones is two-way, meaning that the action of many hormones is to induce the DNA to transcribe itself into proteins. Indeed the hormone–DNA–protein path is probably more typical than the alternative DNA–protein–hormone path. For example, without the proper hormonal environment provided by the mother's body, a newly fertilized egg cell will not develop into a viable embryo. More to the point, the hormonal environment within a human being is exquisitely sensitive to the external environment in which that person lives. An adverse physical, social, or emotional environment can significantly alter the secretion of hormones, which, in turn, alters the regulation and the activity of DNA–protein synthesis. Some important examples of the interplay between hormones, the environment, and genetics are described in the following sections and in the last chapter of this book.

Why are pygmies short?

African pygmies, including such cultural groups as the Mbuti, Aka, Twa, and Efe, are some of the shortest populations in the world (Table 5.1). Several endocrine mechanisms have been identified as probable causes for their short stature. Rimoin *et al.* (1968) and Merimee *et al.* (1968) found that serum levels of GH and IGF-2 were normal in the sample of Central African pygmies they studied. However, Merimee *et al.* (1981) found that IGF-1 levels were significantly lower in serum samples drawn from pygmies than from taller peoples living near the pygmies. A nutritional cause for the low IGF-1 levels (see below) and for short stature could not be demonstrated. Rather, it seems that a genetic defect in the cellular mechanisms for the production, release, or cellular reception of IGF-1 is the cause of the short stature of the pygmies. A non-human animal model for pygmy growth was proposed by Eigenmann *et al.* (1984). They found that the small size of toy and miniature poodles is due to a defect of IGF-1 production. Both types of poodles have normal serum concentrations of GH and IGF-2, but significantly lower concentrations of IGF-1 than found in standard poodles. The smaller toy poodles have levels of IGF-1 lower than the miniature poodles. The endocrine similarities between the poodle models and the human pygmy situation with respect to IGF-1, adds support to a genetic hypothesis for the determination of pygmy stature.

It is also known that pygmies have low levels of the GH-binding protein (GHBP) and, from laboratory studies, that pygmy tissues are unresponsive to IGF-1. Taken together, the hormone–gene regulation of pygmy growth is complicated and the cause of pygmy short stature is not clearly understood. Merimee (1993) believes that all of the endocrine and growth characteristics of pygmies are the result of a low number of GH receptors (GH-R) in the cells of most tissues. In this sense, says Merimee, pygmy growth is similar to the short stature of people with Laron syndrome. Merimee reports that unlike Laron syndrome, pygmies have normal levels of GHBP and GH-R until age three years, and their growth in height is essentially normal as well. After age three years the level of GHBP, and by inference GH-R, decreases. According to Merimee, pygmy boys grow much like a control group of non-pygmy Africans until puberty, but pygmy boys have no adolescent growth spurt. Pygmy girls grow more slowly than the control group and have a much reduced adolescent spurt. Merimee emphasizes that the absent or reduced adolescent growth spurt is the main cause of the short stature of pygmies as adults.

Merimee's interpretation of the growth data are based on 23 cases of longitudinal data collected by Van de Koppel & Hewitt (1986) and a cross-sectional sample of more than 600 pygmies (Merimee *et al.*, 1987). One difficulty with all these data is that the individuals measured are not of known age, rather age is estimated. Robert Bailey (1991) published growth data for 51 Efe pygmy boys and girls of known age from birth to age 60 months. Bailey compares the length or height of the pygmy sample with the NCHS reference curves for height. At birth pygmies average 2.71 *z*-scores below the reference, and the height deficit increases to a negative 4.16 *z*-scores at 60 months. *z*-scores are a statistical value used to express the degree of standardized deviation that exists between a measured value and a reference value. The pygmy *z*-scores indicate that they are very short at birth and the degree of stunting increases with age. These data, of course, contradict those of Merimee. Bailey states that the 'age problem' accounts for the discrepancy, while Merimee views Bailey's data as 'unusual'.

Bailey is still collecting longitudinal data in order to see if his sample of Efe will experience an adolescent growth spurt. If there is no growth spurt, then pygmies will be an exception to the otherwise universal presence of the spurt in all human populations. However, Bailey interprets the data published by Van de Koppel & Hewitt as showing, '... that some girls and boys in their early teen years gained over 10 cm in 1 year' (p. 119). Experimental research with cell cultures derived from Efe pygmy subjects show that these cells show normal growth responses to estradiol, testosterone, vitamin D3, and triiodothyronine (Geffner *et al.*, 1996). These observations raise the possibility that the increased secretion of gonadal hormones at puberty could produce a significant acceleration in growth. Whatever the case turns out to be, further study of patterns of growth in pygmy populations will be of tremendous aid in helping us to understand how human growth is regulated, and possibly the types of endocrine and genetic changes required to evolve the human adolescent growth spurt.

A final question about pygmy size relates to the question, is pygmy short stature an adaptation? Jean Hiernaux (1974) argued that pygmy small size is an adaptation for optimal heat loss in the moist, hot forested environment of central Africa (some details of this explanation were presented in the previous chapter). Part of Hiernaux's argument rested on his belief that in addition to small stature, pygmies have a low weight to height ratio and long arms and legs in proportion to stature. All of these proportions would either help to dissipate excessive heat produced by the body or minimize the production of body heat. These assertions about pygmy body size and proportions were reassessed by Shea & Bailey (1996). They find that almost all the differences in body shape between pygmy and non-pygmy African populations are accounted for by a simple scaling relationship with stature. Shea & Bailey use several types of ontogenetic scaling analysis to come to this conclusion, but the relatively unsophisticated method of weight/height used in Table 4.1 also shows that pygmies and the African Turkana have the same body shape. In other words, pygmies show no unusual proportions that can be attributed to adaptation to the tropical forest. Shea & Bailey suggest that the small size of the Efe, and other pygmies, is an adaptation to limited food resources in the central African forests. Some support for a nutritional explanation of pygmy short-stature comes from research with the Mountain Ok of Papua New Guinea. The Mountain Ok are another short stature population; mean height of men is 152.7 cm and of women it is 146.7 ($n = 150$ subjects). Jessica Schwartz *et al.* (1987) studied their endocrine status and found that serum levels of GH, IGF-1, and IGF-2 were all within the normal range. The authors point out that these findings underscore the uniqueness of the low IGF-1 levels associated with short stature of the African pygmies. A restudy by Baumann *et al.* (1991) found that the levels of GHBP are significantly lower in the Mountain Ok than in a taller control sample. So, the short stature of the Ok may have an endocrine–genetic cause. However, it is likely that the small size of the Mountain Ok also has a nutritional cause. J. A. Lourie *et al.* (1986) surveyed 79 percent of the Mountain Ok population for a variety of anthropometric indicators of health and nutritional status. The authors find much evidence for chronic calorie undernutrition, for instance relatively small skinfolds and arm circumferences in people of all ages. Another indication of a nutritional cause for their short stature is that Mountain Ok living in villages within a day's walk of a mining town (gold and copper are mined in the region) are significantly taller, heavier, and fatter than Mountain Ok living in more remote villages. Opportunities for wage labor and the greater availability of food in mining towns are likely reasons for the differences in growth that were found.

Links between nutrition, hormone balance, and growth

It is often tempting to explain an unusual case of growth on a mutated gene that produces a deficit or excess of a hormone, such a GH, IGF-1, or GHBP. However, it is well known today that malnutrition, both undernutrition and overnutrition, can upset the endocrine balance in the body and this, in turn, can lead to a deranged regulation of many genes and gene systems. Obesity-induced adult-onset diabetes in humans is one example of overnutrition leading to an imbalance of the hormone insulin in the body. Restriction of food intake and loss of the excessive weight (mostly fat) usually restores insulin homeostasis. Nutritionists working with farm animals, such as hogs and cattle, make use of a balance between an adequate supply of nutrients and hormones to stimulate economically optimum rates of growth in these animals. Many of the principles applied in animal husbandry are equally pertinent to humans (Davis & Reeds, 1998). Undernutrition during childhood results in slow growth, delayed maturation, and, if prolonged into adolescence, short stature in adulthood. A link between undernutrition and hormone imbalance was noted by Pimstone *et al.* (1968). They found that GH levels were higher than normal in children suffering from kwashiorkor, a severe form of protein undernutrition, and from marasmus, a severe form of protein–energy malnutrition. Both groups of undernourished children were significantly shorter and lighter than expected for their ages. Upon recovery in the hospital, GH levels remained high when calories only were added to the diet, but fell towards normal values when protein and calories were added.

It may seem paradoxical that severely malnourished children should have high levels of GH but delayed growth. The elevation of GH is due to impaired release of IGFs and the lack of negative feedback of rising levels of IGFs on further GH release. Grant *et al.* (1973) showed that children with marasmus have relatively high GH levels and relatively low IGF levels. Three boys and two girls, aged 11 to 32 months old, were studied following admission to a South African hospital. After recovery for nine days in hospital, IGF levels had risen, while GH levels fell in the patients. The authors speculated that starvation reduces IGF secretion and activity as a means of saving body stores of protein for the maintenance and repair of tissues, at the expense of cellular growth.

These findings were repeated in many studies since 1973. In a controlled experimental study, Clemmons *et al.* (1981) found that fasting for 48 hours significantly decreased IGF-1 levels. The subjects were seven adult men suffering from obesity, who had volunteered to fast for ten days as part of a weight reduction experiment. In another study, Isley *et al.* (1983) placed five normal weight people (three women and two men) on a total fast, an energy-deficient diet, and a protein-deficient diet. The subjects were on each type of diet for five days, followed by five more days of recovery and then a further five days on a diet until all subjects had experienced each type of diet. The authors found that IGF-1 levels in serum dropped on each type of diet, and fell steadily each day so that they reached their lowest values after five days. IGF-1 levels rapidly returned to normal values with refeeding.

Many studies of human populations throughout the world also show that there is a clear association between nutritional status, hormone levels, and growth. A study of three groups of undernourished Malaysians by Wan Nazaimoon and colleagues (1996) is an example. A sample of 246 boys and girls, ages four to 15 years old, living in three jungle settlements was measured for stature and levels of IGF-1, IGFBP-3, thyroid stimulating hormone (TSH), and thyroxin (T4). Compared with the NCHS reference data for height, all of the Malaysians measured were very short, and clinical examinations confirmed they were malnourished in terms of total food intake. Iodine deficiencies are known from two of the villages and measurements confirmed low levels of T4 in these villages. The researchers found that IGF-1 and IGFBP-3 in the bloodstream were correlated strongly and positively with stature – shorter boys and girls had lower hormone levels. Amounts of T4 also correlated significantly with IGF-1 levels in subjects under age 11 years old. The authors explain that the path of causality in these results demonstrate that an inadequate supply of calories will lower levels of IGF-1 and IGFBP-3 and retard growth. When iodine intake is also inadequate, there is a further reduction in IGF-1 production and a further reduction in height.

All of the studies reviewed here show that the synthesis and secretion of several hormones essential for growth is dependent on both an intact genotype and a satisfactory nutritional environment. Changes in nutritional status can enhance or suppress the expression of DNA that codes for hormones. The interaction between nutrition, genes, and hormones, is complex, but well enough understood to see that simplistic notions of either genetic determinism or environmental determinism of growth are totally inadequate.

Psychological influences on endocrinology and growth

'Can thought influence body build?' William H. Mueller asks this question in the title of his 1995 paper. Much of Mueller's research focuses on the factors that influence body composition, especially fatness, and the effects of body composition on health. In an on-going study, Mueller and his colleagues find that thought (i.e., emotional states) may well influence body build (i.e., the physical size and shape of the body). Mueller quickly reviews the obvious; cases of emotional disturbance leading to anorexia or overeating and their consequences on body composition. The less obvious connections are explored as well, such as the evidence for direct pathways between stress, alterations in hypothalamic activity, and the accumulation of centralized body fat (fat on the trunk of the body). In his own research, Mueller finds that 15- and 16-year-old girls with high levels of body fat are more likely to hide feelings of anger and resentment than girls with less body fat. Boys of the same age with a high index of central fat distribution are more likely to express anger in aggressive, socially inappropriate, ways. High levels of body fat, especially with a centralized fat pattern, are risks for diabetes and heart disease. Aggressive responses to feelings of anger also entail many risks for health. Mueller concludes that the evidence available, while incomplete and tentative, indicates that thought does influence body build (and maybe health), and body build influences thought.

It has long been known that the quality of the emotional and psychological environment in which a child lives can influence his or her endocrine balance and growth. Emotional influences on adrenal function were described earlier in this chapter. **Psychosocial short stature** is a well-studied condition that shows how the endocrine system mediates the relationship between psychological factors and physical growth. Psychosocial short stature is a clinical condition of retarded growth in stature that cannot be ascribed to an organic problem with the child, but rather to behavioral disturbance and emotional stress in the environment in which the child lives. A diagnosis of psychosocial short statue is confirmed when the child is removed from that environment and growth is spontaneously restored. Many behavioral and emotional factors can lead to psychosocial short stature, but in many cases there is a clear hormonal connection. In a review of his clinical experience with psychosocial dwarfism Rappaport (1984) stated that '. . . the most consistent biological finding was the decrease of circulating somatomedin [IGF-1] activity . . .' (p. 44).

One of the first observations of a relationship between the psychosocial environment and human development was made by Harry Chapin in 1915. Chapin, a physician, visited ten orphanages and hospitals in the New York City region that cared for abandoned infants. These institutions provided an acceptable level of care in terms of hygiene and feeding, yet in nine of the ten all of the infants under two years old died. In 1942 Harry Bakwin proposed that the cause of this extraordinary mortality was emotional deprivation. In the first half of the twentieth century, the medical community in the United States believed that 'excessive' physical contact of an infant and its care-givers was deleterious. In the orphanages and hospitals, where the staff were likely to be overburdened with many infants, physical contact was reduced to a minimum. Bakwin believed that the deprivation of physical contact led to negative emotional state and death.

René Spitz (1945) carefully investigated the causes of poor growth experienced by emotionally disturbed infants and children. His studies focused on orphans confined to foundling homes and other institutions. Spitz compared the development of infants in a foundling home with infants raised in the nursery of a penal institution for delinquent girls. Inmates of the latter facility were the natural mothers of the infants. Both institutions provided an acceptable standard of housing, sanitation, medical care, and diet for the infants. The children in the penal nursery had more physical and social stimulation, due to their full-time care by their own mothers, or full-time substitutes. In the foundling home, care was provided by one nurse, and infants were confined to their cribs, without human contact, for most of the day. Over the two years of study Spitz found that the foundling home children became progressively delayed in their physical and mental development compared with the nursery infants and a control group of home-reared infants. At about age 3.0 years the foundling home children had average heights and weights expected for children aged 1.5 years. The developmental status of the nursery infants did not differ significantly from a control group of infants raised at home. Moreover the mortality rate for infants in the foundling home was 37 percent, while in the nursery group no child had died. Spitz used the term 'hospitalism' to describe the syndrome of poor physical and mental development and high mortality experienced by institutionalized children. It took some time for Spitz's research to make a significant impact, but eventually a new medical paradigm for infant and child care was proposed by pediatricians. By the 1950s, Benjamin Spock, author of an immensely popular 'baby book', advised parents to hold and cuddle their infants in ways that had been seen as indulgent just a few years earlier.

The critical importance of physical contact in early development was the focus of the famous experimental studies conducted by Harry Harlow and his colleagues (Harlow & Zimmerman, 1959; Kerr *et al.*, 1969). Harlow established several types of rearing environments for rhesus monkeys living in a laboratory. Some infants were raised by their mothers, and other infants were separated from their mothers, but given access to inanimate surrogates. One type of surrogate was a wire frame with a bottle and nipple positioned so that the infant monkey could cling to the wire and feed. Another type of surrogate was a wire frame covered with a soft textured cloth. Infant monkeys preferred to cling to the soft cloth surrogate and would even give up feeding for the opportunity to touch and caress the cloth. Behavioral and emotional development was impaired in the monkeys exposed to both types of surrogates, but the impairment was more severe in those with wire rather than cloth-covered surrogates.

The implications of the research by Spitz and Harlow were applied by Tiffany Field to the needs of preterm infants. Many preterm infants must be given extraordinary medical care in order to survive. Poor growth and development of those infants that did survive was a common outcome. Often, the medical care required that the infant be isolated from physical contact with the mother or other care-givers. Field conducted a series of experiments which showed that tactile stimulation could ameliorate much of the poor growth and development. The stimulation could be provided by placing an infant confined to an incubator on a sheepskin fur pad or by allowing the infant to be fondled through 'glove hole' access into the incubator. As little as 15 minutes of gloved touch resulted in 50 percent faster rates of growth in the isolated infant (Field, 1988).

War and growth

Warfare creates many orphaned infants and children and much emotional stress. In a fascinating example of the serendipitous nature of scientific discovery, Widdowson (1951) found that the psychological well being of German children, orphaned during World War II, directly affected their growth. Children in two orphanages, 'Bienenhaus' and 'Vogelnest', took part in a year-long nutrition supplementation experiment. Children of both orphanages were fed their usual ration for six months. In the second half of the study, the children of Vogelnest were given the usual ration plus additional, unlimited servings of bread, jam, and orange juice, but no supplement was given at Bienenhaus. Unexpectedly, during the first six months, when both groups received identical rations, the children of Vogelnest put on more weight, an average of 1.4 kg, than the children of Bienenhaus, an average less than 0.5 kg. Six months later the situation was reversed. Despite receiving extra food, the children of Vogelnest gained less weight (average gain of about 1.2 kg) than the children of Bienenhaus (average gain greater than 2.5 kg). However, eight of the children had gained an average of about 4.2 kg in the supplemented orphanage. Quite coincidentally, they had been transferred from Bienenhaus to Vogelnest just at the time when the supplement was given to Vogelnest. The records also revealed that the headmistress transferred between orphanages at the same time. Further inquiry by Widdowson found that the headmistress, a women referred to as Fraulein Schwarz, was a stern disciplinarian, who severely punished children for even minor infractions of behavior. She delivered these punishments at meal time, possibly to provide an example to the assembly of children gathered for the meal. The few children with favorable growth under the care of Fraulein Schwarz, were her 'favorites' who received no punishments, and when she transferred between orphanages she took her favorites along. Only these children, provided with positive psychological stimulation, showed the benefits of the extra food. Widdowson concluded that the emotional environment of the orphanages had a greater influence on physical growth than did the nutritional environment.

A serious limitation of the German orphanage study is that carefully controlled observations of behavior, food consumption, and energy expenditure did not take place. The results of the study cannot, therefore, be accepted at face value. Indeed, later studies challenged the idea that the emotional environment has a direct influence on growth. Sills (1978) reviewed the case histories of 185 patients hospitalized with the diagnosis 'failure to thrive', a syndrome characterized by stature and weight below the third percentile for age, but without obvious organic cause. Detailed interviews with the parents, and home visitations, revealed that in more than half the cases parental neglect and deprivation of the child were the causes of the retarded growth. Withholding food, and outright starvation, were found in many of these cases of deprivation. However, many of the children of this, and other studies, who were adequately fed, still suffered from growth failure. These children invariably lived in emotionally disturbing environments. Such environments can depress the secretion of digestive enzymes and the absorption of food, so that malnutrition may take place even if sufficient food is available (Parisi & de Martino, 1980). Thus, there is a possibility that many instances of psychological growth failure are actually cases of undernutrition caused only indirectly by a negative emotional environment.

What is more important, food or love?

Powell *et al.* (1967a,b) and Saenger *et al.* (1977) described a type of child with growth retardation simulating hypopituitary dwarfism, that is, short stature due to insufficient GH, that when removed from an emotionally stressful home environment has a total disappearance of symptoms. Usually, these children were admitted to hospital for the treatment of behavioral as well as physical symptoms. Before any treatment was administered, however, they experienced a spontaneous improvement in behavior, and increase in growth rate, and a rise in serum GH levels. Upon return to their home environments growth rate and GH levels both usually reverted to their previous low levels. According to Powell *et al.* (1967a), malnutrition and stress-induced intestinal malabsorption were not primary factors accounting for growth retardation in these patients. The authors concluded that the low GH levels and retarded growth of these children is due to emotional abuse rather than physical neglect.

Green *et al.* (1984) reviewed the literature relating to psychosocial dwarfism, with an aim towards evaluating the role that nutrition and endocrine factors play in the etiology of the disease. It was found that most cases of growth failure in children under three years old were due to malnutrition; these infants were usually denied food by their emotionally disturbed mothers. Children over three were usually not clinically malnourished. Moreover, it was commonly found that both GH and IGF-1 levels were significantly depressed in these older children. Since, as shown above, malnutrition is associated with low levels of serum IGF-1 and abnormally high secretion of GH, the endocrine profile of these children does not fit with starvation as the cause of their growth failure. Green *et al.* account for the growth disturbance in the children with a neuroendocrine hypothesis. Patton & Gardner (1963) proposed that emotional stress may affect some of the higher brain centers, particularly the amygdala and limbic cortex, which are known to control the emotions. Nerve impulses from these brain centers may pass to the hypothalamus where they are transduced into neuroendocrine messages that may affect the production and release of hypothalamic hormones. In this manner, psychological disturbances in the child might be translated into a cutoff of GHRF in the hypothalamus, a halt in GH secretion from the pituitary, and depressed levels of IGF-1 secretion from the body tissues.

An endocrine connection for the relationship between emotional stimulation, growth, and health has been established in both experimental studies with non-human animals and in clinical human studies. In a laboratory experiment, Meaney *et al.* (1988) compared infant rats who were licked by their mothers with infant rats who were not licked. The licked infants had lower levels of so called 'stress hormones', the glucocorticoids such as cortisol, high levels of GH, high growth rates, and even higher scores on tests of learning. In more recent research, Meaney and colleagues (Rosténe *et al.*, 1995) report direct pathways between physical stress, the release of glucocorticoid hormones, and several central nervous system neurotransmitters which regulate the activity of the hypothalamus and pituitary in humans as well as rats.

In clinical human studies, Skuse *et al.* (1996) report on a group of 29 children with psychosocial short stature and GH insufficiency accompanied with hyperphagia, that is, an excessive intake of food. The researchers found that when these children were removed from their stressful home circumstances the GH levels spontaneously returned to normal and the hyperphagia ended. Clearly, none of these children were denied food, but they were denied proper emotional care and their growth suffered. Studies with infants and children placed into foster care also find little evidence for a nutritional cause for growth retardation. These foster-care studies do find that one of the first physical changes that occurs with placement is an increased rate of growth in height and weight (Wyatt *et al.*, 1997). Wyatt and colleagues point out that few children placed into foster care show signs of clinical psychosocial short stature. However, in a study of 45 apparently healthy and well nourished infants and children aged 1.5 to 6.0 years placed into foster care, more than half experienced clinical catch-up growth following placement. The foster-care study shows that even when stature, weight, and food intake appear to be normal, a stressful home environment may be retarding growth. In these rat and human studies, emotional stress, and not starvation, is clearly the primary cause of growth retardation.

A case study, reported by Magner *et al.* (1984), provides a final example of the powerful interplay between emotions and growth. The study is of a 12-year-old boy who suffered growth retardation and delayed sexual maturation following an emotional trauma. The trauma was provoked by an argument between the boy and his stepfather, with whom the boy had a warm relationship. After the argument the boy verbalized a wish for his stepfather's death, and the next day the man seriously injured himself falling from a roof. The hospital where the man was recovering sent an erroneous 'notice of death' letter to the family's home which the boy received and read while at home alone. Though the man eventually recovered, the boy began a self-imposed period of food refusal and vomiting. He dropped from 34 kg to 25 kg in five months. At age 15 years, following periods of hospitalization and drug treatment, his eating behavior returned to normal, though at age 17 years he had the height of a normal child of 11.3 years, a bone age of 13.0 'years', and was, essentially, prepubertal in physical appearance. He was given treatment with growth hormone at age 19.3 years, and between ages 20 and 21 years experienced a growth spurt and sexual maturation. Growth in height continued until age 25 years, when the young man reached 171.0 cm. The authors of this report state that in this patient, an acute 'psychic trauma induced a deranged hormonal state that persisted for several years' (p. 741). Though malnutrition and drug treatments in the three years following the trauma may have also upset the hormonal balance, the boy was behaviorally normal and drug-free for about five years before treatment with GH returned his growth and maturation to normal.

This case, and the others previously discussed, exemplify the intimate and powerful influence that emotional factors, such as love, fear, and guilt, can have on the human endocrine system and the pattern of human growth. In the next, and final, chapter, I synthesize the emotional, endocrine, genetic, social, economic, and political influences on human development that are discussed throughout this book into a biocultural view of human growth.

8 A biocultural view of human growth

> As this whole volume is one long argument, it may be convenient to the
> reader to have the leading facts and inferences briefly recapitulated
>
> Charles Darwin, *The Origin of Species*, 1859.

Like Darwin, my intent in this final chapter is to recapitulate some of the
material from the preceding chapters. This review will take the form of a
synthesis of the many strands of my 'one long argument' for a biocultural
view of human growth. There are two essential messages of this synthesis.
First, human growth and development is best understood via a life history
perspective. Throughout this book the life history perspective is applied to
both the study of the stages of life of modern people, and to the evolution of
the human pattern of growth. The second essential message relates to a
newer way to conceptualize the biocultural nature of human development.
Since the late nineteenth century, anthropologists have used biocultural
models to explain human growth and development. Investigators such
Bowditch and Boas (Chapter 1) used biocultural models to dispel the
unscientific notions of the biological determinists and the racists, who
believed that human phenotypes were fixed and not amenable to environmental influences. Some years later, with the discovery of the nature of
DNA and other fundamentals of developmental biology, the biocultural
model considered human development as, basically, a biological process
which could be influenced to a greater or lesser extent by the social and
cultural environment. The newer, expanded biocultural view of the last
decade is that there is a recurring interaction between the biology of human
development and the sociocultural environment. Not only does the latter
influence the former, but human developmental biology modifies social
and cultural processes as well. It is now understood that environmental
forces, including the social, economic, and political environment, regulate
the expression of DNA as much, or more so, than DNA regulates the
growth process. This chapter reviews recent progress in both the application of life history theory to human growth and the expansion of the
biocultural perspective toward human development to a fully interactional
model.

387

Biocultural interactions in contemporary populations

Infant, child, juvenile, adolescent, adult, and post-menopause grandmother (and grandfather) are universal biological stages of human postnatal growth and development. There is much cultural variation, however, in the social and behavioral response to each of these stages. Moreover, the evolution of the new human stages of development, childhood and adolescence, bring about many biosocial benefits, but also incur new risks. The evolution of any new characteristic incurs both benefits and risks. Bipedalism, for example, is the method of human locomotion unique among the primates. Often considered to be one of the crucial feeding and reproductive adaptations of our species, bipedalism also brings about many physical ailments, including lower back pain, fallen arches, and inguinal hernias. In a similar fashion, the benefits of childhood need to be tempered against the hazards of this developmental stage. Dependency on older individuals for food and protection, small body size, slow rate of growth, and delayed reproductive maturation each entail liabilities to the child. The 'charms' of children and childhood do not provide for total security.

To illustrate this point one can examine traditional societies of both historic and prehistoric eras. In such societies, including hunter-gatherers and horticulturists, about 35 per cent of live born humans die by age seven, that is, by the end of childhood (Bogin, 1996). Even if two-thirds of these deaths occur during the infancy stage, the childhood period still has an appreciable risk of death. In hunter-gatherer societies, starvation, accidents, and predation account for most childhood deaths. !Kung parents, in an effort to minimizes these risks, protect their children by confining them to camp and the watchfulness of adults (Draper, 1976). Mbuti parents (hunter-gatherers of central African forests) allow children and juveniles to form age-graded play groups, also located near the camp. This method of grouping children with older juveniles and adolescents demands less parental supervision (Turnbull, 1983b). However, the risks of hunter-gather childhood can, perhaps, best be illustrated by Hadza society. Hadza children, aged three to eight years, form age-graded work/play groups that may wander far from camp, often out of sight of older individuals, and carry out costly and dangerous tasks including significant foraging for food (Blurton Jones, 1993). One study finds that of 301 Hadza infants born to 75 women aged 20 to 50 years old, 115 (39 per cent) of the offspring had died by the time of the survey (Blurton Jones *et al.* 1992). Demographic modeling indicates that most of the deaths occurred during infancy and childhood.

Today, hunter-gatherer and traditional horticultural societies account for less than one per cent of human cultures, so it may be more instructive to examine current risks to children in post-Colonial and industrialized societies. Mild-to-moderate energy undernutrition is, perhaps, the most common risk. World-wide, estimates are that between 60 to 75 per cent of all children are undernourished. Malnutrition is a serious threat to adequate growth, physical and cognitive development, health, and performance at school or work (see Chapter 6). Undernutrition may be due to food shortages alone, but equally likely it is due to work loads and infectious disease loads placed on children that compromise their energy balance (Worthman, 1993). Viewed in historical perspective, unreasonable workloads for children and many childhood diseases are products of the agricultural and industrial revolutions. Agriculture, for example, increased social stratification and reduced the variety and quality of foods consumed by people of the lower social classes (Bogin, 1997a). Industrialization further increased social disparities and resulted in the forced labor of children in mines and factories (Chapter 1).

Despite programs of legal protection and welfare for children in the twentieth century, there remain many risks for children in the contemporary world. Abuse and neglect are two such risks. One estimate of the world-wide mortality from abuse and neglect is between 13 and 20 infants and children per 1000 live births. The incidence of all suffering from abuse and neglect is probably higher, but very difficult to estimate since data are not reported by most nations. Some industrialized nations do maintain statistics for non-fatal abuse and neglect of children. In the United States, for example, '... in 1991 2.7 million abused or neglected children were reported to child protection agencies' (Kliegman, 1995). This is a rate of 38.6 children per 1000, but note that the term 'children' is used for any individual under age 18 years. A closer examination of the United States data by Putnam & Trickett (1997), indicates that rates of physical abuse for 'children' may be as high as 19/1000, and rates of sexual abuse may be 11/1000. Cases of neglect were not reported.

Some cases of child abuse and neglect may result from a severance between the biology of childhood and the rapid pace of technological, social, and ideological change relating to families and their children. It is now technologically possible to nourish infants without breast-feeding and this allows parents (mothers) an opportunity to pursue other economic activities, or have another baby. Among the poor populations of the developing countries, short birth intervals (less than 23 months) compromise the health of both the infant and the mother. A major negative effect on the infant is low birth weight, which is known to impair both physical growth and cognitive development during childhood and later life stages. In many Western cultures, such as the United States middle-class, fewer than 20% of infants are breast fed and the weaning process (from bottles) may begin by three months of age. This severely curtails the infant stage and prolongs childhood. These 'premature children' present a problem for care, often solved by sequestering them to restraining devices such as high chairs, playpens, and cribs (cots) and segregating them from the family by placement in crèches or pre-schools. When the infants react poorly to these arrangements, the frustrated parents or care givers may respond with abusive or neglectful behavior.

Older children, juveniles, and adolescents are age-segregated in schools. The evolutionarily derived learning needs of these stages of development are ignored by formal school curricula. With less than two offspring per woman, on average, juveniles do not participate in significant child care. Due to schooling laws, job market realities, and moral codes, adolescents are unable or strongly discouraged from participation in adult economic and sexual activities. Young adults increasingly require many years of post-secondary education before attaining the economic status required to marry and begin a family.

The Western model of behavior stands in sharp contrast to that of many non-Western societies (Whiting & Whiting, 1975; Whiting & Edwards, 1988). Thomas Weisner (1987) studied sibling child care as an institutionalized method to prepare juvenile girls for the demands of motherhood. Weisner worked with the Abaluyia tribe of Kenya, with Hawaiians, and has surveyed other published research on this topic, including non middle-class sub-cultures in the United States. He finds that 'In most non-Western societies, training for competent, culturally appropriate child care is an active apprenticeship experience for [juveniles], usually completed by adolescence and learned along with the performance of domestic tasks essential for family, and often community, survival' (p. 238). The importance of this apprenticeship in relation to human evolution was explained in Chapter 4. To reiterate, the juvenile girl as sibling caretaker allows the mother, and father, to do work that provides food, shelter, and other necessities for the whole family. One may also note that the mother is free to have another infant sooner than would be possible if she had to provide all the child care. Other research focuses on adolescence. Carol Worthman's biocultural research with the Kikuyu tribe of Kenya was highlighted in Chapter 4 (1986, 1993). Worthman found that adolescence is a well-defined and prolonged stage of development. As with most rural African cultures, the traditional Kikuyu do not use the concept of chronological age, rather they use the biological markers of sexual maturation to define adolescence. Kikuyu girls officially enter adolescence following circumcision, which entails removing 'the very tip of the clitoris'. This rite of passage is timed by the development of breasts, body hair, and growth in height to take place just before menarche. Following the circumcision ritual the girls begin receiving explicit teaching in reproductive function and sociosexual behavior. Girls also are expected to work with the older women of their household in domestic chores; learning about cooking, tending fields, and child care.

Adolescents in non-Western societies learn these skills in practical work settings, not in school settings. Thus the work that adolescents perform results in real productive contributions to the welfare of the family and the social group. The contributions of adolescents have an impact on the physical growth of children in these communities. Working in a rural agricultural village of Aymara people in highland Bolivia, Sara Stinson (1980) found that the number of productive adolescents living in a household is related to the growth of children in the same household. After accounting for differences in SES between families, the children living in households with more adolescents were taller than children in households with few or no adolescents.

Prior to Western acculturation, Copper Eskimos (Canadian Inuit), a hunting-based society, followed a course of adolescent social development that was closely tied to their biological development. Richard Condon (1990) reports that adolescent girls married at or just prior to menarche. The timing of such marriages was, most likely, predicted by the regular series of puberty events, such as breast development and the growth spurt, that always precede menarche. The new bride could not bear children for three to four years (adolescent sterility), so she had time to improve her economic and social skills, on-the-job as it were, prior to motherhood. Young men had to prove their abilities as hunters and providers prior to marriage. Until age 17 or 18 adolescent males were not strong enough to hunt large game or to build a snow house. At that age the late adolescent male might be accepted by a prospective father-in-law but a period of bride service prior to marriage was required. Thus, the groom finished growing and was a young adult, with years of experience at important adult economic and social activities, at the time of marriage.

These few examples (see the works cited here for additional cases) show that 'In many societies around the world training for parenthood is an apprenticeship experience, learned along with the performance of domestic and subsistence tasks within shared-function family systems. In such shared caretaking families, child-care skills are acquired first, followed only gradually by autonomy from parents and siblings, and then by full managerial control of a household' (Weisner, 1987 p. 265). This behavioral sequence is in harmony with the normal succession of stages in human biological development and promotes good physical growth.

Risks for grandparents

Finally, the evolution of a significant period of life after women experience menopause is associated with several risks for older women. The hormonal changes and bone loss that occur with the cessation of ovarian function, reviewed in detail in Chapter 4, are one type of risk. These biological changes may bring about several degenerative diseases, such as osteoporosis and heart disease. Men may also be at risk for these same diseases due to a decline in the production of several hormones, including GH and the gonadal hormones. Post-menopausal women, and older men, must often assume new social and economic roles for which they need adequate training and social support. However, in some 'modern' societies, grandmothers and grandfathers, just like their adolescent grandchildren, may no longer receive appropriate training for post-reproductive sexual, social, and economic expectations. In these societies the elderly may also be denied a productive social role, and may even be segregated away from productive society – in 'retirement communities' or 'old age homes' for those who can afford them, or poverty-level housing for those of limited means. The social isolation that these sequestered elderly people may experience is known to exacerbate the normal degenerative process of aging. Moreover, research shows that children living in households with little or no contact with grandparents suffer more abuse and neglect than children in multigenerational households – another testament to the value of grandmothers and grandfathers.

Social, economic and political disturbances to human growth

The influence of contemporary Western culture can introduce considerable discord into the normally harmonious relationship between human biology and behavior. Weisner (1987) reports that Kenyan mothers with more formal (European style) education believe that sibling care responsibilities teach juveniles to be passive and that domestic work, including child care, is menial. Condon (1990) finds that due to the loss of their hunting life style, the Inuit are forced to acculturate to settled life, the economics of wage labor, and the social values of television. The biocultural definition of Inuit life stages are also transformed. The biggest change is lengthening of the adolescent stage due to earlier sexual maturation, that is, menarche occurs at an earlier age than in the past. Condon reports that a longer adolescence affects girls especially, as they now marry several years later than in the past. Another modification is that during the juvenile and adolescent stages both sexes have much autonomy from parental control, and no longer participate in traditional social learning. A likely consequence of these shifts in values and behavior for future generations of Kenyans and Inuit will be a delay in the acquisition of parental knowledge of human growth and development until after the birth of the first infant. The consequences of these changes in social learning for the physical growth and development of the next generation are less clear.

A provocative hypothesis of possible consequences was proposed by Jay Belsky and colleagues (1991). In an attempt to develop an evolutionary theory of early socialization and later reproductive strategy, they describe two alternative developmental pathways. 'One is characterized, in childhood, by a stressful rearing environment and the development of insecure attachments to parents and subsequent behavior problems; in adolescence by early pubertal development and precocious sexuality; and in adulthood by unstable pair bonds and limited investment in child rearing' (p.647). These conditions could be the consequence of delayed and deficient learning of adult social, economic, and parenting skills. The other pathway '. . . is characterized by the opposite' (p. 647). Presumably, this means that childhood is non-stressful, with secure attachments and few behavior problems; pubertal development and adolescent sexual behavior occur at average or late ages, and in adulthood stable pair bonds and intensive investment in child-rearing occur.

This hypothesis is a good example of the power of the biocultural interaction model to integrate many seemingly disparate developmental events into a unified pattern. Belsky and colleagues (Moffitt *et al.*, 1992) tested their hypothesis using data from a longitudinal study of girls from New Zealand, when the girls were 16 years old. They found that the incidence of behavior problems in childhood does not predict either earlier or later menarche – the measure of sexual development used in this study. Family conflict and absence of the father from the family during a girl's childhood are associated with an earlier age of menarche. However, age of menarche is more parsimoniously predicted by a genetic inheritance model (i.e., mother's age at menarche) than by any of the behavioral data.

Even though Belsky's original hypothesis is not supported by these results, the biocultural interaction model may still offer the best interpretation of the results. Marcia Herman-Giddens and colleagues (1988) found a 1 in 15 prevalence of early pubertal development in a study of 105 girls, age

10 years old or younger, who were victims of confirmed or suspected sexual abuse. Although the 1 in 15 prevalence of early pubertal development and abuse is statistically weak, the authors propose two hypotheses to guide future research: (1) that a genetic tendency toward early expression of secondary sexual characteristics could lead to sexual abuse by stimulating the perpetrators, or (2) that sexual abuse is '. . . a stressor that in some way stimulates adrenal androgen secretion or early activation of the hypothalamic-pituitary-ovarian axis' (p. 433). Either hypothesis calls for an interactive biocultural model of sexual maturation and sexual abuse.

Stress, hormones, and human development

In a review of the relationship between the neuroendocrine system, stress, and growth, Stratakis *et al.* (1995, p. 162) offered this summary statement:

> A stressor above a threshold magnitude, or multiple stressors applied simultaneously, cause an organism to alter its behaviour and physiology, with the aim of maintaining homeostasis. The adaptive changes that occur are coordinated and mediated by the stress system in the central nervous system (which includes corticotrophin-releasing hormone and noradrenergic neurons in the hypothalamus and brainstem, respectively), and its peripheral limbs, the hypothalamic–pituitary–adrenal axis and the autonomic (sympathetic) system. Controlled or self-driven challenges to homeostasis and a normally functioning stress system are crucial for normal development and preservation of self and species. In childhood and adolescence, appropriately functioning neuroendocrine responses to stressors are necessary to allow growth and psychosexual maturation to progress normally. Maladaptive neuroendocrine responses, i.e. dysregulation of the stress system, may lead to disturbances in growth and development and cause psychiatric, endocrine/metabolic and/or autoimmune diseases or vulnerability to such diseases, not only during childhood and adolescence, but also in adulthood.

While the connection between sexual abuse and sexual maturation remains to be systematically investigated, there is considerable evidence for direct connections between other types of stress and growth. The last section of Chapter 7 detailed some of these connections, and here I will report on two longitudinal studies that exemplify the power of the newer biocultural model of human development.

Since 1988, Mark Flinn and his colleagues have been studying the interplay between socioeconomic conditions, psychosocial stress, and health of the people living in a rural village on the Caribbean island of Dominica. The study sample includes, '. . . 264 infants, children, adolescents, and young adults aged 2 months to 18 years' (Flinn & England, 1997, p. 33). This is a biocultural study, and the field research methods include ethnographic observation of people's behavior via participant observation, systematic behavioral observations, informal interviews, psychological questionnaires, clinical medical examinations, reviews of medical records, and repeated measures of salivary cortisol levels. Cortisol is one of the primary hormones released in response to psychosocial or physical stress. Flinn & England review the effects of cortisol on the human body by stating 'Cortisol modulates a wide range of somatic functions, including energy release (e.g. stimulation of hepatic gluconeogenesis in concert with glugagon and inhibition of the effects of insulin), immune activity (e.g., downregulation of inflammatory response and the cytokine cascade), mental activity (e.g., alertness, memorization, and learning), growth (e.g., inhibition of growth hormone and somatomedins), and reproductive function (e.g., inhibition of gonadal steroids, including testosterone)' (p. 35). By 1996, a total of 22 438 salivary cortisol samples were collected and analyzed by radioimmunoassay. Salivary samples are collected by having people spit into test tubes, or in the case of infants swabbing their mouths.

The hypothesis that directs the research is illustrated in Figure 8.1. 'Health' is the outcome of a complex relationship between the 'Family environment' and the 'Physical environment', both of which are influenced by socioeconomic conditions, or 'SECs', and both of which regulate levels of 'Stress' in family members. The research team evaluates health in terms of the immunological status and illnesses of the study subjects. Longitudinal measurements of weight, and perhaps other anthropometric variables, are also part of the database, but have not been published – hence the question mark for 'growth?'.

The most consistent finding of the study is that the quality of the family environment influences directly both the cortisol levels in the subjects and their health. High average levels of cortisol, or widely fluctuating levels of cortisol in individual infants, children, juveniles, or adolescents, are associated with significantly diminished immunity and a statistically higher frequency of illness. These high or unstable cortisol profiles in the subjects are more likely to be found when they are living in unstable families or households. Discord and separation of parents/caretakers and changes in the composition of the household are two stressors that elevate cortisol levels. Another important stressor is the degree of genetic relationship between caretakers and the study subjects. Mean cortisol levels and the percentage of days ill are lower for subjects living in households where they are the genetic offspring of the adult caretakers. Mean cortisol levels are

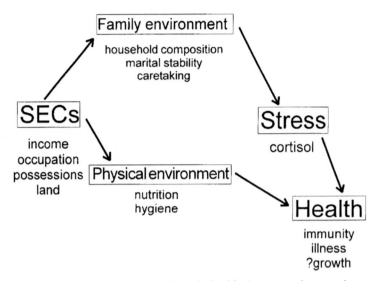

Figure 8.1. A model of the hypothesized relationships between socioeconomic
conditions (SECs), family environment, physical environment, stress, and health.
From Flinn & England (1997).

highest when the subjects live with a step-father and half-sibs, or when they
live with distant relatives. The next highest mean cortisol levels are found
in subjects living in single-mother households, where there was little or no
social or economic support from a mate or other kin.

The second longitudinal study is called 'The Psychobiological Effects of
Sexual Abuse' (Putnam & Trickett, 1997). This project is based in the
United States and draws its sample from the Washington, DC area. There
are 164 girls in the study, 6.0 to 15 years of age, divided between sexually
abused girls ($n = 77$) and non-abused girls matched for age, SES, ethnicity,
and family composition. The sexually abused sample were recruited into
the study via referrals from child protective service agencies. Accordingly,
the definition of 'sexual abuse' in this study is as defined by American social
service workers and the American legal system. The control sample was
recruited through newspaper advertisements and notices in social service
agencies. Putnam & Trickett state that, 'To avoid attracting comparison
families covertly interested in sexual abuse, the purpose of the study was
disguised under the title, "Normal female growth and development" (p.
152)'. After control families were selected the true nature of the study was
explained. The study began in 1987 and most girls have been evaluated
three or four times since then. The evaluations focus on the effects of sexual
abuse on the neuroendocrine system, especially cortisol, on immune system

function, and on psychological outcomes. Physical growth data are collected but their analysis has not been published so far.

Putnam & Trickett (1997) report that sexually abused girls differ from the control sample by having high morning, but lower afternoon levels of cortisol in blood samples. The abused girls also produce more cortisol in response to the stress of drawing the blood sample. The researchers state that this indicates a derangement in the response of the hypothalamic–pituitary–adrenal axis to normal levels of daily stress (e.g., waking-up and medical procedures). The abused girls show a tendency toward impaired immune systems, and report more days ill with colds and flu. The abused girls also show a complex variety of behavioral and emotional disorders. Despite their complexity, the higher number of clinical disorders is correlated with higher levels of stress hormones.

Some common patterns of interaction between childhood stress and human development emerge from these two longitudinal studies. Flinn & England conclude that 'Because SECs influence the family environment, they have consequences for child health that extend beyond direct material effects. And because health in turn may affect an individual's social and economic opportunities, a cycle of poor health and low SECs may be perpetuated generation after generation' (p. 49). Putnam & Trickett make the point that '. . . biological alterations are an important characteristic of responses to maltreatment. Biological alterations likely interact with development to influence the type and expression of pathological symptoms and behaviors in child and adult victims of maltreatment' (p. 158).

The 'recycling of poverty' and the transgenerational effect of the social, economic, emotional, and political environment on human growth and development are major themes of several chapters in this book. With the addition of these two longitudinal studies of stress, hormone regulation, and health, the multiple lines of research that are reviewed in earlier chapters may now be seen in sharper focus. These newer data combine an understanding of the biological and sociocultural correlates of growth. We now approach the point where the causes and consequences of prenatal growth, birth weight, and growth during the infancy stage of life may be synthesized into a comprehensive biocultural model of human development, both throughout the life of an individual and across the generations.

The future

It is my hope that this book will stimulate some readers to enter the field of human growth and development research. At times, readers may get the impression that all the major discoveries about the nature of human growth have been made. That is not the case. Human knowledge of how our bodies work, and how we relate to the biocultural world in which we live, builds incrementally, and in small steps. The need for a heavily revised second edition of this book is just one obvious example of such change in a ten-year period. It is my hope that in another decade this book will, again, be hopelessly out of date. Perhaps some of my readers will help to hasten that day.

Isaac Newton said 'If I have seen further, it is by standing on the shoulders of giants'. The research and ideas presented in this book stand on the shoulders of many 'giants'. Newton may have adapted his phrase from a line in Robert Burton's 1621 chrestomathy *The Anatomy of Melancholy* which states, 'Pygmies placed on the shoulders of giants see more than the giants themselves'. I can think of no better way to end this book than by referring to giants and pygmies, two extremes of the plasticity, and wondrous diversity, of the patterns of human growth.

Glossary

adolescent growth spurt The rapid and intense increase in the rate of growth in height and weight that occurs during the adolescent stage of the human life cycle.

adolescent phase A stage in the human life cycle, covering the five to eight years after the onset of puberty. The adolescent phase is characterized by a growth spurt in height and weight, virtual completion of permanent tooth eruption, development of secondary sexual characteristics, sociosexual maturation, intensification of interest and practice of adult social, economic, and sexual activities.

adolescent sterility A physiologic state of pubertal girls that begins after menarche and lasts until ovulatory cycles are established.

adrenarche The onset of secretion of androgen hormones from the adrenal gland, usually occurring at about six to eight years of age in most children.

adulthood The stage of the human life cycle that begins at about age 20 years. The prime of adulthood lasts until the end of child-bearing years (late forties) and is a time of homeostasis in physiology, behavior, and cognition. Old age and senescence mark the period of decline in the function of many body tissues or systems during adulthood. This later phase lasts from about the end of child-bearing years to death.

age-graded play group A social group of children and juveniles in which the older individuals provide basic caretaker behavior and enculturation for the younger individuals. The play group frees adults for subsistence activities and other adult behaviors.

allantosis A membranous sac that develops from the posterior part of the alimentary canal and which serves as a repository for wastes in the embryos of mammals, birds, and reptiles. It is important in the formation of the umbilical cord and placenta in mammals.

allometry Differential rates of growth of parts of the body relative to the growth of the body as a whole.

anthropometry The scientific measurement of the human body.

apoptosis Programmed cell death, a mechanism that allows cells of self-destruct when stimulated by the appropriate trigger. Cell deaths appear to be programmed by gene–environment interactions, and are initiated for various reasons, such as when a cell is no longer needed within the body or when it becomes a threat to the health of the organism.

attachment behavior The set of physical and psychological cues and responses that bond an infant with one of its caretakers.

Australopithecus Genus of early hominids from eastern and southern Africa, dating to between four to one million years before present.

autocrine action A process by which a hormone-like substance has its biological effect on the cells that produce the substance.

auxology The study of biological growth.

biacromial breadth The linear distance between the shoulders, measured between the most lateral points of the acromial processes of the scapula.

bicristal breadth The maximum linear distance between the iliac crests. Also called biiliac breadth or pelvic breadth.

biiliac width see *bicristal breadth*

biocultural Referring to a recurring interaction between the biology of human development and the sociocultural environment. Not only does the latter influence the former, but human developmental biology modifies social and cultural processes as well.

biological evolution. The continuous process of genetic adaption of organisms to their environments.

bitrochanteric breadth The linear distance between the most lateral extensions of the greater trochanters of the femurs.

body composition The make-up of the body in terms of the absolute and relative amounts of adipose tissue, muscle mass, skeletal mass, internal organs and other tissues.

body mass index (BMI) A mathematical indicator of the relative weight-for-height of a person. BMI is calculated as [weight in kilograms/(height in meters)^2]. Higher BMI scores indicate that an individual has relatively more weight-for-height than a person with a lower score. In the general population higher BMI scores usually indicate more body fatness.

calcification The process of mineral deposition, usually calcium, in tissues of the body. Regular sites of calcification are the skeleton and the dentition.

catch-up growth The rapid increase in growth velocity following recovery from disease or re-feeding after short-term starvation.

childhood phase A stage in the human life cycle that occurs between the end of infancy and the start of the juvenile growth period (about the ages two to ten years). Children are weaned from breast-feeding (or bottle-feeding) but must be provided specially prepared foods and require intensive care by older individuals. Childhood is characterized by relatively rapid neurological development and slow physical growth and development.

chorionic placenta A type of placenta in which there is a direct connection between the chorion and the fetus via the umbilical cord. The chorionic placenta is found in rodents, monkeys, apes and humans.

chorionic–allantoic placenta A type of placenta in which parts of the surface of the allantosis fuse with the chorion (a membrane surrounding the fetus composed of maternal and embryonic tissues). The chorionic–allantoic placenta is found in most orders of mammals, including Artiodactyla, Proboscidians, Carnivora, and others.

congenital States or conditions that exist in the individual at or prior to birth. The term is usually used to refer to hereditary or inborn medical conditions that are most often harmful.

cross-sectional growth study Measurement on a single occasion of individuals grouped by age and sex, and sometimes other characteristics.

cultural animal A phase describing the human species, meaning that humans possess all of the potentials and limitations of any living creature, but also add a cultural trilogy of: (1) dependence on technology, (2) codified social institutions, such as kinship and marriage, and (3) ideology.

deoxyribonucleic acid (DNA) One of the class of complex molecules called nucleic acids. DNA is found in the nucleus of virtually all living cells. DNA contains the genetic code needed for a cell to produce the proteins it requires to perform its function.

development Progression of changes from undifferentiated or immature state to a highly organized, specialized or mature state.

diaphysis The shaft of a long bone.

differential fertility One of the two fundamental mechanisms of natural selection (the other being differential mortality), referring to variation in the reproductive success of mature organisms.

differential mortality One of the two fundamental mechanisms of natural selection (the other being differential fertility), referring to the death of some individuals of a population prior to their reproductive maturation.

distance curve A graphic representation of the amount of growth achieved by an individual over time.

dizygotic twins Twins that result from two independent fertilizations. Such twins are no more alike genetically than ordinary brothers and sisters. Also referred to as 'fraternal twins'.

dual effector model A hypothesis for the regulation of cellular growth that posits that the role of growth hormone is to cause the differentiation of fibroblast cells into specific tissue types (bone, adipose, etc.) and that the role of insulin-like growth factors is to cause the multiplication (clonal expansion) of the newly differentiated cells.

ecology The relationship that an individual organism, or group of individuals of a species, has with its physical, biological, and social environment.

embryo Stage of prenatal development lasting from the second to tenth week following fertilization, characterized by the rapid differentiation of tissues and the formation of organs.

endocast A mold, or cast, of the interior of the skull which may be used to estimate brain size and shape.

endocrine action The secretion of a hormone from its tissue of origin and its distribution throughout the body via the bloodstream to its site of action.

endocrinology The study of hormones, their origins and actions.

endosteal The thin layer of cells lining the medullary cavity (inner surface) of a bone.

endotheliochorial placenta A sub-type of the chorionic placenta in which the uterine–epithelium barrier is eliminated. Found in rodents.

epidemiology The study of the causes and transmission of disease.

epiphysis An ossification center of a long bone, separated from the shaft of the bone by cartilage.

epitheliochorial placenta A sub-type of the chorionic placenta in which a separation between the tissues of the chorion and the epithelial lining of the uterus is maintained. Found in the lemurs and lorises (the strepsirhine primates).

essential nutrient A nutrient that cannot be manufactured by the human body from simpler elements and thus must be supplied from the diet.

expanding tissues Those tissues or cells that retain their mitotic potential even in the differentiated state. These tissues can increase in size and mass by cell division of all their cells. Examples include the liver, kidney and the endocrine glands.

feedback control A method used for the regulation of biological activity in an organism based on the flow of information between parts of the organism. An example is negative feedback control in the endocrine system, in which rising levels of a hormone in the bloodstream result in a lowering of the level of stimulation of cells that produce that hormone.

fetus Stage of prenatal development lasting from tenth week following fertilization to birth.

fibroblast cell A type of undifferentiated cell which may differentiate into bone, adipose or other types of tissue.

genetic potential A popular, but incorrect, concept that every human being has a genetically determined limit to many aspects of his or her phenotype.

genotype The genetic constitution of an individual.

germinative cells Undifferentiated cells, usually sequestered in well-defined regions within tissues, that give rise to the differentiated, specialized cells of mature tissues, organs, and subsystems in the body.

gonadarche Maturation of the gonads (testes or ovary) resulting in the secretion of gonadal hormones (androgens or estrogens).

growth Quantitative increase in size or mass.

growth factors Biochemical substances synthesized by specific cells within a wide variety of body tissues that have effects on growth, both independently and interactively within each other and with hormones.

growth plate region The site of formation of bone tissue in a growing long bone. The growth plate consists of highly ordered rows of cartilage cells; the row farthest removed from the bony shaft is a germinitive layer; it is responsible for cell replication and cartilage growth at the bone shaft. Over time the cartilage will be reformed into true bone tissue.

hemochorial placenta A sub-type of the chorionic placenta in which the cell walls of the maternal blood vessels break down and invade the chorion so that the chorion, and the fetal circulation, are directly bathed by maternal blood. Found in tarsiers,

New World monkeys, Old World monkeys, apes, and humans (the haplorhine primates).

heritability An estimate of the relative genetic contribution to the phenotypic expression of a physical or behavioral characteristic. The value of the heritability estimate is a function of genetic factors, environmental factors and the interaction of genetic and environmental factors.

heterochrony Referring to several processes in biology that bring about evolutionary change between ancestral and descendant species by modifying characters present in the ancestor via changes in developmental timing.

heterosis The mating of individuals born in different geographic regions, which may lead to greater genetic variability and 'genetic vitality' in their offspring. Human marriages of this type also may result in greater cultural variability in the generation of the offspring.

homeobox genes A highly conserved sequence of 180 DNA base pairs that codes for a 60 amino acid segment of a protein. The homeobox DNA sequence is found in all eukaryote organisms so far examined.

homeostasis Any self-regulating process by which biological systems tend to maintain stability while adjusting to conditions that are optimal for survival.

homology Likeness in the anatomy or behavior of different organisms due to an evolutionary differentiation from the same or a corresponding anatomical part or behavior of a remote ancestor.

hominid Living human beings and their extinct fossil ancestors. Primate species characterized by habitual bipedal locomotion.

hominoid The group of the Primates that includes human beings, the apes and their extinct ancestors.

homoiothermy Self-regulation of a relatively constant body temperature.

hormone Chemical substance secreted from a specific tissue, usually into the general blood circulation, where it travels to its site of action.

Hox **genes** A category of homeobox genes that encode transcription, that is, proteins that initiate and regulate the conversion of the DNA code to the RNA sequence that is used to make amino acid polypeptide chains.

hypermorphosis A type of heterochrony that extends one or more stages of growth and development of the descendant species beyond that of the ancestral species.

hyperplasia Cellular growth by cell division (mitosis).

hypertrophy Cellular growth by an increase of material within each cell.

hypothalamus An evolutionarily ancient midbrain structure, which in mammals provides a connection between the nervous system and the endocrine system. It is a neuroendocrine transducer. The human hypothalamus secretes hormones that stimulate or inhibit the production of other hormones by the pituitary which regulate growth and development.

hypoxia The lack of sufficient oxygen supplied to the tissues of the body. May be the result of disease or may be due to residence at high altitude (3000 meters or more above sea level).

infant phase or infancy (of growth and development) A stage in the life cycle of all mammals. For human being it lasts from the second month after birth to end of lactation, usually by age 36. Human infancy is characterized by: (1) rapid growth velocity with a steep deceleration in velocity with time, (2) feeding by lactation, (3) deciduous tooth eruption, and (4) many developmental milestones in physiology, behavior, and cognition.

intracrine action The process whereby a hormone may act on the nucleus of the cell which manufactures the hormone.

juvenile phase (of growth and development) A stage in the life cycle of most social mammals, including all of the higher primates. The juvenile stage is defined as the time of life when an individual is no longer dependent on adults (parents) for survival, and prior to that individual's sexual maturation.

karyotype A description of the chromosomes of an organism in terms of their number, shape and size.

kwashiorkor A disease of protein deficiency, characterized by failure to grow, wasting of the muscles, loss of appetite, irritability, changes in the hair and skin and anemia.

lactation The secretion or formation of milk by the mammary glands, as well as the period during which the milk is the major food provided to the infant.

Ladino One of the major ethnic groups of Guatemala, defined as the cultural descendants of the Spanish conquerors of Guatemala, who speak Spanish as their primary language, wear western-style clothing, and deny any Maya ancestry or heritage.

life cycle The stages of growth, development, and maturation from conception to dath of any organism.

life history theory A field of biology concerned with the strategy an organism uses to allocate its energy toward growth, maintenance, reproduction, raising offspring to independence, and avoiding death. For a mammal, it is the strategy of when to be born, when to be weaned, how many and what type of pre-reproductive stages of development to pass through, when to reproduce, and when to die.

longitudinal growth study Measurement of the same individual or group of individuals, repeated at regular intervals.

low birth weight A weight at birth of 2500 grams or less for a neonate of normal gestation length.

marasmus A disease of severe undernutrition, especially energy and protein malnutrition, characterized by failure to grow, wasting of muscles, edema, apathy and a ravenous appetite.

maturation The process and the state of reaching functional capacity in terms of biological, behavioral, and cognitive capacities.

Maya One of the major ethnic groups of Guatemala, defined as the cultural descendants of the pre-Conquest Maya, speak a Maya language and Spanish, wear traditional non-Western clothing (especially during social and religious rituals), and claim Maya ancestry or heritage.

menarche The first menstrual period.

menopause The sudden or gradual cessation of the menstrual cycle subsequent to the loss of ovarian function.

mid-growth spurt A relatively small increase in the rate of growth in height that occurs in many children between the ages of six to eight years.

mid-parent height The average of the stature of the mother and the father. Used in formulas to predict the adult stature of offspring.

migration The movement of people from place to place. In recent years, much of this migration has been from rural to urban areas, requiring substantial changes in the environment and lifestyle of the migrants.

mitosis Cell division resulting in two 'daughter' cells with the same genetic constitution as the original 'parent' cell.

model A representation that displays the pattern, mode of structure, or formation of an object or organism.

monochorionic placenta A type of placenta usually associated with twin pregnancies in which a single chorion (one of the placental tissues) is shared by both twins.

monozygotic twins Twins that result from a single fertilization and, therefore, share the same genotype. Also referred to as 'identical twins'.

natural selection The process by which environmental constraints lead to varying degrees of reproductive success among individuals of a population of organisms. The individuals must vary in terms of genetically inherited characteristics. Natural selection determines the course of evolutionary change by maintaining favorable genotypes and phenotypes in a constant environment (stabilizing selection) or improving adaptation in a direction appropriate to environmental changes (directional selection). Charles Darwin and Alfred Wallace fist proposed this concept in 1858.

negative secular trend A decrease in the mean size (height, weight, leg length, etc.) of the individuals of a population from one generation to the next. Negative secular trends usually indicate a deterioration in the quality of the biocultural environment for human development.

neocortex Region of the mammalian brain associated with 'higher' level motor and sensory activities and the integration of these activities into complex patterns of behavior.

neonate and **neonatal period** A stage in the human life cycle lasting from birth to 28 days after birth.

neoteny A type of heterochrony that results in the retention of infantile or juvenile traits into adulthood. This is achieved by having sexual maturation take place while the individual is still in a pre-adult stage of phenotypic development.

Ontogeny The process of growth and development of individuals of a species.

organogenesis The formation of body organs and systems during the first trimester of prenatal life.

ossification The process of bone formation in skeletal tissue.

paedomorphism Having features in the adult like those of the child. Unlike neoteny, paedomorphism results in only the appearance of child-like features, which are, in fact, unlike the childhood condition in actual form and function.

paleoecology The study of extinct forms of life and their relations with the environment (e.g., types of foods eaten, requirements for reproduction).

paleontology The study of extinct forms of life, usually as represented by fossilized remains.

paracrine action The process by which a hormone-like substance has its biological effect on nearby cells within the same tissue as the cells that produce the substance.

parental investment The allocation of resources, such as time or energy, to offspring that occurs at some cost to the parents.

peak growth velocity The maximum rate of growth in height, weight, etc. achieved during the adolescent growth spurt.

pelvic inlet The bony opening of the birth canal through which the fetus must pass during parturition.

percentile of growth Method of ranking growth status for height, weight, etc. of an individual relative to other members of a sample or population of people. Example: a child at the 75th percentile for height is taller than 75 percent of the other children in the group under consideration.

periosteal The dense fibrous membrane covering the outer surface of bones, except at the joints, and serving as an attachment for muscles and tendons.

phenotype The physical or behavioral appearance of an individual, resulting from the interaction of the genotype and the environment during growth and development.

photogammetry A method of measuring the human body by taking one or more photographs of an individual posed in particular positions. Photogammetry is often used in somatotyping.

phylogeny The evolutionary history of a species.

pituitary An endocrine gland of vertebrate animals located at the base of the brain, below the hypothalamus to which it is connected via blood vessels and nerves. The hormones secreted by the pituitary stimulate and control the functioning of almost all the other endocrine glands in the body.

placenta An organ of some mammals composed of fetal and naternal tissues that transfers nutrients and oxygen from the mother's blood circulation to the fetus and fetal wastes to the mother's blood circulation for disposal.

plastic or plasticity The concept that the development of the phenotype is responsive to variations in the quality and quantity of environmental factors required for life. Such variations produce many of the differences in growth observed between individuals or groups of people.

polygenic Referring to characteristics of the phenotype that are influenced by any group of nonallelic genes. Height, for example, is influenced by many genes located on several different chromosomes.

preformation Erroneous notion that the prenatal human body is essentially adult-like in form.

prematurity A state at birth for human neonates who are born prior to 37 weeks of gestation.

psychosocial short stature A type of growth retardation produced by a negative physical and emotional environment for growth.

puberty An event of short duration (days or a few weeks) that marks the reactivation of the central nervous system regulation of sexual development. The onset of puberty is accompanied by a dramatic increase in secretion of sex hormones. In social mammals, including humans, puberty occurs at the end of the juvenile stage.

regulatory genes Genes that initiate and terminate the action of structural genes. Regulatory genes are important in the control of growth in terms of size, shape, and the timing and duration of each stage of the life cycle.

remodeling of bone The process that maintains the characteristic shape and function of a bone as that bone grows in size. Remodeling is achieved by destroying or creating new bone cells as a bone grows.

renewing tissues Those tissues, or cells, of the body that are replaced by a two-step process: (1) the mitotic division of germanitive cells, and (2) the differentiation of some of these newly divided cells into mature tissues. Examples of renewing tissues are the blood and the skin cells of the epidermis.

risk focusing A concept used by epidemiologists that refers to the process by which economic, political, and cultural systems act to differentially allocate the benefits and risks for health and disease between groups of people. Risk focusing is often used to study the sociocultural process of allocating exposure to toxic or infectious materials to groups whose members enter the group partly because of previous exposure to those materials (Schell, 1997, p. 70).

secondary sexual characteristics Physical traits associated with the onset of sexual maturation, including the development of facial hair and muscularity in boys and the development of the breast and adult fat distribution in girls.

secular trend The process that results in a change in the mean size or shape of individuals of a population from one generation to the next. Such trends can be positive (increasing size) or negative (decreasing size).

sedente A person living in his or her geographic region of birth; a non-migrating individual.

senescence The period of the adult phase of the life cycle characterized by a decline in the function of many body tissues or systems. Senescence usually begins after the end of child-bearing years and lasts until death.

sexual dimorphism Differences between the sexes in physical appearance, behavioral performance, and psychological characteristics.

skeletal age and skeletal maturation A measure of biological maturation (as distinguished from chronological age) based on stages of formation of the bones.

socioeconomic status (SES) An indicator, often defined by measures of occupation and education of the parents or heads of household, used as a proxy for the general quality of the environment for growth and development of an individual.

somatotyping A method of classifying the human physique based on external, usually visual, assessment of body shape. The goal of somatotyping is to find associations between types of physiques and physiological function, habitual behavior, or risk for disease.

static tissues Those tissues or cells that are incapable of growth by hyperplasia once they have differentiated from precursor germinative cells. Examples are nerve cells and striated muscle.

structural genes Genes that code for specific proteins of the body.

subcutaneous fat The layer, or compartment, of adipose tissue that lies just under the skin.

synergistic action The interaction of two or more agents or forces so that their combined effect is greater than the sum of their individual effects.

tempo of growth Based on a metaphor from classical music, this phrase refers to the pace at which individuals pass through the stages of growth and development. Some humans grow rapidly and/or mature early, while others grow slowly or are late maturers. The tempo of growth is, generally, unrelated to amount of growth that an individual will achieve.

transformation grids A method developed by D'Arcy Thompson (based on drawings of the artist Albrecht Duerer) to describe two-dimensional changes in growth and form, both within and between species of organisms.

trimesters (of pregnancy) The division of the nine calander months of human pregnancy into three, three-month periods. Usually called first, second and third trimester.

velocity curve A graphic representation of the rate of growth of an individual over time.

viviparity Giving birth to living offspring that develop within the mother's body.

weaning The termination of breast-feeding.

yolk sac A membranous sac attached to an embryo, providing early nourishment in the form of yolk in bony fishes, sharks, reptiles, birds, and primitive mammals and functioning as the circulatory system of the human embryo before internal circulation begins.

yolk sac placenta A type of placenta found in some marsupials and rabbits, in which blood vessels connect the yolk sac with the uterine wall.

References

Able, E. L. (1982). Consumption of alcohol during pregnancy: a review of effects on growth and development of offspring. *Human Biology*, **54**, 421–53.

Acheson, R. M. (1954). The Oxford method of assessing skeletal maturity. *Journal of Anatomy* (London), **88**, 498–508.

Adair, L. S. & Guilkey, D. K. (1997). Age-specific determinants of stunting in Filipino children. *Journal of Nutrition*, **127**, 314–20.

Adair, L. S., Vanderslice, J. & Zohoori, N. (1993). Urban–rural differences in growth and diarrhoeal morbidity of Filipino infants. In *Urban Ecology and Health in the Third World*, ed. L. M. Schell, M. T. Smith & A. Bilsborough, pp. 75–98. Cambridge: Cambridge University Press.

Aiello, L. C. & Wheeler, P. (1995). The expensive-tissue hypothesis. *Current Anthropology*, **36**, 199–221.

Alexander, RD (1990). *How Did Humans Evolve? Reflections on the Uniquely Unique Species*. Special Publication No. 1. Ann Arbor: University of Michigan Museum of Zoology.

Alexander, R. D., Hoogland, J. L., Howard, R. D., Noonan, K. M. & Sherman, P. W. (1979). Sexual dimorphisms and breeding systems in pinnipeds, ungulates, primates, and humans. In *Evolutionary Biology and Human Social Behavior: An Anthropological Perspective*, ed. N. A. Chagnon & W. Irars, pp. 402–35. North Scituate, Massachusetts: Duxbury.

Alley, T. R. (1983). Growth-produced changes in body shape and size as determinants of perceived age and adult caregiving. *Child Development*, **54**, 241–8.

Altmann, J. (1980). *Baboon Mothers and Infants*. Cambridge: Harvard University Press.

Alvesalo, L. (1997). Sex chromosomes and human growth: a dental approach. *Human Genetics*, **101**, 1–5.

Ama, P. F. M. & Ambassa, S. (1997). Bouyancy of African and European white males. *American Journal of Human Biology*, **9**, 87–92.

Ammon, O. (1899). Zur Anthropologie der Badener. Jena, cited in Boas, F. (1922).

Amoroso, E. C. (1961). Histology of the placenta. *British Medical Bulletin*, **17**, 81–8.

Anderson, D. C. (1980). The adrenal androgen-stimulating hormone does not exist. *Lancet*, **2**, 454–6.

Anderssen, L., Haley, C. S., Ellegren, H. *et al.* (1994). Genetic mapping of quantitative trait loci for growth and fatness in pigs. *Science*, **262**, 1771–4.

Anonymous (1995). *The Economist*, December 23, 19–22.

Argente, J., Chowen, J. A., Zeitler, P., Clifton, D. K. & Steiner, R. A. (1992). Sex steroid and developmental regulation of somatostatin and growth hormone-releasing hormone gene expression. In *Human Growth: Basic and Clinical Aspects*, ed. M. Hernandez & J. Argente, pp. 295–306. Amsterdam: Elsevier.

Ariès, P. (1965). *Centuries of Childhood: A Social History of Family Life* (trans. R. Baldich). New York: Vintage Books.

Armstrong, E. Schleicher, A. Omran, H. Curtis, M & Zilles, K. (1995). The ontogeny of human gyrification. *Cerebral Cortex*, **5**, 56–63.

Arslanian, S., Suprasongsin, C., Kalhan, S. C., Drash, A. L., Brna, R. & Janosky, J. E. (1998). Plasma leptin in children: relationship to puberty, gender, body composition, insulin sensitivity, and energy expenditure. *Metabolism*, **47**, 309–12.

Ashcroft, M. T. & Lovell, H. A. (1964). Heights and weights of Jamaican children of various racial origins. *Tropical and Geographical Medicine*, **4**, 346–53.

Ashcroft, M. T., Heneage, P. & Lovell, H. A. (1966). Heights and weights of Jamaican schoolchildren of various ethnic groups. *American Journal of Physical Anthropology*, **24**, 35–44.

Ashizawa, K. & Kawabata, M. (1990). Daily measurements of the heights of two children from June 1984 to May 1985. *Annals of Human Biology*, **17**, 437–43.

Austad S. N. (1994) Menopause: an evolutionary perspective. *Experimental Gerontology*, **29**, 255–63.

Austad, S. (1997). *Why We Age: What Science Is Discovering About the Body's Journey Through Life*. New York: John Wiley & Sons.

Backman, G. (1934). Das Wachstum der Korperlange des Menchen. *Kunglicke Svenska Verenskapsakademiens Handlingar*, **14**, 145.

Baer, K. E. von (1886). *Autobiography of Karl Ernst von Baer*. English translation by J. Oppenheimer. Canton, Massachusetts: Science History Publications.

Bailey, R. C. (1991). The comparative growth of Efe pygmies and African farmers from birth to age 5 years. *Annals of Human Biology*, **18**, 72–9.

Bailey, S. M., Gershoff, S. N., McGandy, R. B., Nondasuta, A., Tantiwongse, P., Suttapreyasri, D., Miller, J. & McCree, P. (1984). A longitudinal study of growth and maturation in rural Thailand. *Human Biology*, **56**, 539–57.

Baker, P. ed. (1977). *The Biology of High Altitude Peoples*. Cambridge: Cambridge University Press.

Bakwin, H, (1942). Loneliness in infants. *American Journal of Disease of Children*, **63**, 30–40.

Baten, J. (1996). Der Einfluß von regionalen Wirtschaftsstrukuren auf den biologischen Lebensstandard. Eine anthropometrische Studie zur bayerischen Wirtschaftsgeschichte im frühen 19. Jahrhundert. *Vierteljahresschrift für Sozial- und Wirtschaftsgeschichte*, **83**(2), 180–213.

Bayley, N. & Pinneau, S. R. (1952). Tables for predicting adult height from skeletal age: revised standards for use with the Greulich–Pyle hand standards. *Journal of Pediatrics*, **40**, 423–41.

Barinaga, M., Bilezikjian, L. M., Vale, W. W., Rosenfeld, M. G. & Evans, R. M. (1985). Independent effects of growth hormone releasing factor on growth hormone release and gene transcription. *Nature*, **314**, 279–81.

Barnicot, N. A. (1977). Biological variation in modern populations. In *Human Biology*, 2nd edn, ed. G. A. Harrison, J. S. Weiner, J. M. Tanner & N. A. Barnicot. pp. 181–298. Oxford: Oxford University Press.

Baughan, B., Brault-Dubuc, M., Demirjian, A. & Gagnon, G. (1980). Sexual dimorphism in body composition changes during the pubertal period: as shown by French-Canadian children. *American Journal of Physical Anthropology*, **52**, 85–94.

Baumann, G., Shaw, M. A. & Merimee, T. J. (1989). Low level of high affinity growth hormone-binding protein in African pygmies. *New England Journal of Medicine*, **320**, 1705–9.

Baumann, G., Shaw, M. A., Brumbaugh, R. C. & Schwartz, J. (1991). Short stature and decreased serum growth hormone-binding protein in the Mountain Ok people of Papua New Guinea. *Journal of Clinical Endocrinology and Metabolism*, **72**, 1346–9.

Baumgartner, R. M. (1997). Body-composition studies. In *History of Physical Anthropology: An Encyclopedia*, ed. F. Spencer, pp. 190–5. New York: Garland Press.

Bayley, N. & Davis, F. C. (1935). Growth changes in bodily size and proportions during the first three years: a developmental study of 61 children by repeated measurements. *Biometrika*, **27**, 26–87.

Beall, C. M. & Reichsman, A. B. (1984). Hemoglobin levels in a Himalayan high altitude population. *American Journal of Physical Anthropology*, **63**, 301–6.

Beall, C. M., Baker, P. T., Baker, T. S. & Haas, J. D. (1977). The effects of high altitude on adolescent growth in southern Peruvian Amerindians. *Human Biology*, Vol. 49, pp. 109–24.

Beck, B. B. (1980). *Animal Tool Behavior*. New York: Garland.

Behar, M. (1977). Protein-caloric deficits in developing countries. *Annals of the New York Academy of Sciences*, **300**, 176.

Bekoff, M. & Byers, J. A. (1985). The development of behavior from evolutionary and ecological perspectives in mammals and birds. *Evolutionary Biology*, **19**, 215–86.

Belgorosky, A. & Rivarola, M. A. (1998). Irreversible increase of serum IGF-1 and IGFBP-3 levels in GnRH-dependent precocious puberty of different etiologies: implications for the onset of puberty. *Hormone Research*, **49**, 226–32.

Bellamy, C. (1996). *The State of the World's Children*. Oxford: Oxford University Press.

Belsky, J., Steinberg, L. & Draper, P. (1991). Childhood experience, interpersonal development, and reproductive strategy: an evolutionary theory of socialization. *Child Development*, **62**, 647–70.

Bemprad, J. R. (1991). Dementia praecox as a

failure of neoteny. *Theoretical Medicine*, **12**, 45–53.

Benfer, R. (1984). The challenges and rewards of sedentism: the preceramic village of Paloma, Peru. In *Paleopathology at the Origins of Agriculture*, ed. M. N. Cohen & G. J. Armelagos, pp. 531–58. New York: Academic Press.

Benfer, R. (1990). The preceramic period site of Paloma, Peru: bioindications of improving adaptation to sedentism. *Latin American Antiquity*, **1**, 284–318.

Bercu, B. B., Lee, B. C., Spiliotis, B. E., Pineda, J. L., Denman, D. W., Hoffman, H. J., Brown, T. J. & Sachs, H. C. (1983). Male sexual development in the monkey. I. Cross-sectional analysis of pulsatile hypothalmic–pituitary–testicular function. *Journal of Clinical Endocrinology and Metabolism*, **56**, 1214–26.

Berge, C. (1998). Heterochronic processes in human evolution: an ontogenetic analysis of the hominid pelvis. *American Journal of Physical Anthropology*, **105**, 441–59.

Berkey, C. S., Reed, R. B. & Valadian, I. (1983). Midgrowth spurt in height of Boston children. *Annals of Human of Biology*, **10**, 25–30.

Bertalanffy, L. von (1960). Principles and theory of growth. In *Fundamental Aspects of Normal and Malignant Growth*, ed. W. N. Nowinski, pp. 137–259. Amsterdam: Elsevier.

Bielicki, T. (1975). Interrelationships between various measures of maturation rate in girls during adolescence. *Studies in Physical Anthropology*, **1**, 51–64.

Bielicki, T. & Charzewski, J. (1983). Body height and upward social mobility. *Annals of Human Biology*, **10**, 403–8.

Bielicki, T. & Welon, Z. (1982). Growth data as indicators of social inequalities: the case of Poland. *Yearbook of Physical Anthropology*, **25**, 153–67.

Bielicki, T., Szczotke, H. & Charzewski, J. (1981). The influence of three socioeconomic factors on body height in Polish military conscripts. *Human Biology*, **53**, 543–55.

Bielicki, T., Koniarek, J. & Malina, R. M. (1984). Interrelationships among certain measures of growth and maturation rate in boys during adolescence. *Annals of Human Biology*, **11**, 201–10.

Bierman, J. M., Siegel, E., French, F. E. & Simonian, K. (1965). Analysis of the outcome of all pregnancies in a community. *American Journal of Obstetrics and Gynecology*, **91**, 37–45.

Billewicz, W. Z. (1967). A note on body weight and seasonal variation. *Human Biology*, **39**, 241–50.

Billewicz, W. Z. & McGregor, I. A. (1982). A birth-to-maturity longitudinal study of heights and weights in two West African (Gambian) villages. *Annals of Human Biology*, **9**, 309–20.

Billewicz, W. Z., Fellows, H. M. & Thompson, A. M. (1981). Pubertal changes in boys and girls in Newcastle upon Tyne. *Annals of Human Biology*, **8**, 211–19.

Bindon, J. R. & Baker, P. T. (1985). Modernization, migration and obesity among Samoan adults. *Annals of Human Biology*, **12**, 67–76.

Birdsell, J. B. (1979). Ecological influences on Australian Aboriginal social organization. In *Primate Ecology and Human Origins*, ed. I. S. Bernstein & E. O. Smith, pp. 117–51. New York: Garland.

Bjorntorp, P. (1997). Body fat distribution, insulin resistance, and metabolic diseases. *Nutrition*, **13**, 795–803.

Blank, M. S., Panerai, S. & Griesan, H. (1979). Opioid peptides modulate luteinizing hormone secretion during sexual maturation. *Science*, **203**, 1129–31.

Bloom, B. (1964). *Stability and Change in Human Characteristics*. New York: Wiley.

Blurton-Jones N. G. (1993). The lives of hunter-gather children: effects of parental behavior and parental reproductive strategy. In *Juvenile Primates*, ed. M. E. Pereira & L. A. Fairbanks, pp. 309–26. Oxford: Oxford University Press.

Blurton-Jones, N. G., Smith, L. C., O'Connel, J. F. & Handler, J. S. (1992). Demography of the Hadza, an increasing and high density population of savanna foragers. *American Journal of Physical Anthropology*, **89**, 159–81.

Boas, F. (1892). The growth of children, II. *Science*, **19**, 281–82.

Boas, F. (1910). *Changes in the Bodily Form of Descendants of Immigrants* (abstract). (Report submitted to the Congress of the United States of America.) New York: Columbia University.

Boas, F. (1911). *Changes in the Bodily Form of Descendants of Immigrants*. 61st Congress, 2nd session. S. Doc. 208. Washington, DC: US Government Printing Office.

Boas, F. (1912). Changes in the bodily form of descendents of immigrants. *American Anthroplolgist*, **14**, 530–63.

Boas, F. (1922). Report on an anthropometric investigation of the population of the United States. *Journal of the American Statistical Association*, **18**, 181–209.

Boas, F. (1930). Observations on the growth of children. *Science*, **72**, 44–8.

Boas, F. (1940). *Race, Language, & Culture*. New York: Free Press.

Boaz, N.T. & Almquist, A. J. (1997). *Biological Anthropology: A Synthetic Approach to Human Evolution*. Upper Saddle River, New Jersey: Prentice Hall.

Bock, R. D. (1986). Unusual growth patterns in the Fels data. In *Human Growth: A Multidisciplinary Review*, ed. A. Demirjian, pp. 69–84. London: Taylor & Francis.

Bock, R. D. & Thissen, D. M. (1976). Fitting multi-component models for growth in stature. *Proceedings of the Ninth International Biometric Conference*, **1**, 431–42.

Bock, R. D. & Thissen, D. (1980). Statistical problems of fitting individual growth curves. In *Human Physical Growth and Maturation, Methodologies and Factors*, ed. F. E. Johnston, A. F. Roche & C. Susanne, pp. 265–90. New York: Plenum Bock, R. D., Wainer, H., Peterson, A., Thissen, J. M. & Roche, A. (1973). A parameterization for individual human growth curves. *Human Biology*, **45**, 63–80.

Bogin, B. A. (1977). Periodic Rhythm in the Rates of Growth in Height and Weight of Children and its Relation to Season of the Year. Ph.D. Dissertation. Ann Arbor: University Microfilms.

Bogin, B. A. (1978). Seasonal pattern in the rate of growth in height of children living in Guatemala. *American Journal of Physical Anthropology*, **49**, 205–10.

Bogin, B. (1979). Monthly changes in the gain and loss of growth in weight of children living in Guatemala. *American Journal of Physical Anthropology*, **51**, 287–92.

Bogin, B. (1980). Catastrophe theory model for the regulation of human growth. *Human Biology*, **52**, 215–27.

Bogin, B. (1986). Auxology and anthropology. *Reviews in Anthropology*, **13**, 7–13.

Bogin, B. (1988a). Rural-to-urban migration. In *Biological Aspects of Human Migration*, ed. C. G. N. Mascie-Taylor & G. W. Lasker, pp. 90–129. Cambridge: Cambridge University Press.

Bogin, B. (1988b). *Patterns of Human Growth*. Cambridge: Cambridge University Press.

Bogin, B. (1989). Biological effects of urban migration on Hispanic populations (abstract). *American Journal of Physical Anthropology*, **78**, 194.

Bogin, B. (1991). The evolution of human childhood. *BioScience*, **40**, 16–25.

Bogin, B. (1993). Why must I be a teenager at all? *New Scientist*, **137**, 34–8.

Bogin, B. (1994a) Adolescence in evolutionary perspective. *Acta Paediatrica*, Supplement 406, 29–35.

Bogin, B. (1994b). The evolution of learning. In *The International Encyclopedia of Education*, 2nd edn, ed. T. Husen & T. N. Postlethwaite, pp. 2681–5. Oxford: Pergamon Press.

Bogin, B. (1995). Growth and development: recent evolutionary and biocultural research. In *Biological Anthropology: The State of the Science*, ed. N. Boaz & L. D. Wolfe, pp. 49–70. Bend, Oregon: International Institute for Human Evolutionary Research.

Bogin, B. (1996). Childhood in evolutionary and biocultural perspective. In *Long Term Consequence of Early Environment*, ed. C. J. K. Henry & S. J. Ulijaszek, pp. 7–22. Cambridge: Cambridge University Press.

Bogin, B. (1997a). The evolution of human nutrition. In *The Anthropology of Medicine*, 3rd edn, ed. L. Romanucci-Ross, D. Moerman & L. R. Tancredi, pp. 96–142. South Hadley, Massachusetts: Bergen & Garvey.

Bogin, B. (1997b). Evolutionary hypotheses for human childhood. *Yearbook of Physical Anthropology*, **40**, 63–89.

Bogin, B. (1998). The tall and the short of it. *Discover*, **19**, 40–4.

Bogin, B. & Kapell, M. (1997). Growth studies. In *History of Physical Anthropology: An Encyclopedia*, ed. F. Spencer, pp. 461–6. New York: Garland Press.

Bogin, B. & Keep, R. (1998). Eight thousand years of human growth in Latin America: economic and political history revealed by anthropometry. In *The Biological Standard of Living and Economic Development: Nutrition, Health, and Well Being in Historical Perspective*, ed. J. Komlos & J. Baten, pp. 277–308. Munich: Fritz Steiner.

Bogin, B. & Loucky, J. (1997). Plasticity, political economy, and physical growth status of Guatemala Maya children living in the United States. *American Journal of Physical Anthropology*, **102**, 17–32.

Bogin, B. & MacVean, R. B. (1978). Growth in height and weight of urban Guatemalan primary school children of high and low socioeconomic class. *Human Biology*, **50**, 477–88.

Bogin, B. & MacVean, R. B. (1981a). Body composition and nutritional status of urban Guatemalan children of high and low socioeconomic class. *American Journal of Physical Anthropology*, **55**, 543–51.

Bogin, B. & MacVean, R. B. (1981b). Nutritional and biological determinants of body fat patterning in urban Guatemalan children. *Human Biology*, **53**, 259–68.

Bogin, B. & MacVean, R. B. (1981c). Biosocial effects of migration on the development of families and children in Guatemala. *The American Journal of Public Health*, **71**, 1373–7.

Bogin, B. & MacVean, R. B. (1982). Ethnic and secular influences on the size and maturity of seven year old children living in Guatemala City. *American Journal of Physical Anthropology*, **59**, 393–8.

Bogin, B. & MacVean, R. B. (1983). The relationship of socioeconomic status and sex to body size, skeletal maturation, and cognitive status of Guatemala City schoolchildren. *Child Development*, **54**, 115–28.

Bogin, B. & MacVean, R. B. (1984). Growth status of non-agrarian, semi-urban living Indians in Guatemala. *Human Biology*, **56**, 527–38.

Bogin, B. & Smith, B. H. (1996) Evolution of the human life cycle. *American Journal of Human Biology*, **8**, 703–16.

Bogin, B. & Sullivan, T. (1986). Socioeconomic status, sex, age, and ethnicity as determinants of body fat distribution for Guatemalan children. *American Journal of Physical Anthropology*, **69**, 527–35.

Bogin B., Wall, M. & MacVean, R. B. (1992). Longitudinal analysis of adolescent growth of ladino and Mayan school children in Guatemala: effects of environment and sex. *American Journal of Physical Anthropology*, **89**, 447–57.

Bolk, L. (1926). *Das Problem der Menschwerdung*. Jena: Gustav Fischer.

Bolzan, A. G., Guimarey, L. M. & Pucciarelli, H. M. (1993). Crecimiento y dismorfismo sexual de escolares segun la ocupacion laboral paterna. *Archivos Latinoamericanos de Nutricion*, **43**, 132–8.

Bonaventure, J., Rousseau, F., Legeai-Mallet, L., LeMerrer, M., Munnich, A. & Marteaux, P. (1996). Common mutations in the gene encoding fibroblast growth factor receptor 3 account for achondroplasia, hypochondro-

plasia and thanatophoric dysplasia. *Acta Paediatrica*, Supplement, **417**, 33–8.

BonJour, J. P., Carrie, A. L. Ferrari, S., Clavien, H., Slosman, D., Theintz, G. & Rizzoli, R. (1997). Calcium-enriched foods and bone mass growth in prepubertal girls: a randomized 'double-blind' placebo-controlled trial. *Journal of Clinical Investigation*, **99**, 1287–94.

Bonner, J. T. (1965). *Size and Cycle*. Princeton, NJ: Princeton University Press.

Bonner, J. T. (1993). *Life Cycles: Reflections of an Evolutionary Biologist*. Princeton, NJ: Princeton University Press.

Bookstein, F. C. (1978). *The Measurement of Biological Shape and Shape Change*. New York: Springer-Verlag.

Borkan, G. A., Hults, D. E., Cardarelli, J. & Burrows, B. A. (1982). Comparison of ultrasound and skinfold measurements in assessment of subcutaneous and total fatness. *American Journal of Physical Anthropology*, **58**, 307–13.

Borms, J. (1984). Preface. In *Human Growth and Development*, ed. J. Borms, R. Hauspie, C. Sand, C. Susanne & M. Hebbelinck, pp. v–vii. New York: Plenum.

Borms, J., Hauspie, R., Sand, C., Susanne, C. & Hebbelinck, M. (eds.) (1984). *Human Growth and Development*. New York: Plenum.

Bowlby, R. (1969). *Attachment and Loss*. New York: Basic Books.

Bowman, J. E. & Lee, P. C. (1995) Growth and threshold weaning weights among captive rhesus macaques. *American Journal of Physical Anthropology*, **96**, 159–75.

Boyce, A. J., ed. (1984). *Migration and Mobility*. London: Taylor & Francis.

Boyd, E. (1980). *Origins of the Study of Human Growth*, ed. B. S. Savara and J. F. Schilke. Eugene: University of Oregon Press.

Boyer, R. M., Rosenfeld, R. S., Kapan, S., Finkelstein, J. W., Roffwarg, H. P., Weitzman, E. D. & Hellman, L. (1974). Simultaneous augmented secretion of luteinizing hormone and testosterone during sleep. *Journal of Clinic Investigation*, **54**, 609–18.

Boyer, R. M., Wu, R. H. K., Roffwatg, H., Kapen, S. Weitzman, E. D., Hellman, L. & Finkelstein, J. W. (1976). Human puberty – 24 hour estradiol patterns in prepubertal girls. *Journal of Clinical Endocrinology and Metabolism*, **43**, 1418–21.

Boym, C. (1994). Cover story: By design. *Metropolis* **13**, 38.

Britten, R. J. & Davidson, E. H. (1969). Gene regulation for higher cells: a theory. *Science*, **165**, 349–57.

Brody, S. (1945). *Bioenergetics and Growth*. New York: Reinhold Publishing Co.

Bromage, T. G. & Dean, M. C. (1985). Re-evaluation of the age at death of immature fossil hominids. *Nature*, **317**, 523–7.

Brook, C. G. D., Gasser, T., Werder, E. A., Prader, A. & Vanderschuren-Ludewyks, M. A. (1977). Height correlations between parents and mature offspring in normal subjects and in subjects with Turner's and Klinefelter's and other syndromes. *Annals of Human Biology*, **4**, 17–22.

Brown, F., Harris, J., Leakey, R. & Walker, A. (1985). Early *Homo erectus* skeleton from west Lake Turkana, Kenya. *Nature*, **316**, 788–92.

Brown, S. (1995). Through the lens of play. *ReVision* **17**, 4–8.

Brown, T. (1983). The Preece–Baines growth function demonstrated by personal computer: a teaching and research aid. *Annals of Human Biology*, **10**, 487–9.

Brozek, J. (1960). The measurement of body composition. In *A Handbook of Anthropometry*, ed. M. F. Ashley Montagu, pp. 78–120. Springfield: Illinois, Charles C. Thomas.

Brundtland, G. H., Liestøl, K. & Walløe, L. (1980). Height, weight, and menarcheal age of Oslo schoolchildren during the last 60 years. *Annals of Human Biology*, **7**, 307–22.

Brundtland, G. H. & Walløe, L. (1973). Menarcheal age in Norway: halt in trend towards earlier maturation. *Nature*, **241**, 478–9.

Buckler, J. (1990). *A Longitudinal Study of Adolescent Growth*. London: Springer-Verlag.

Buffon, G. (1777). *Histoire Naturelle, générale et particulière avec la description du cabinet de Roi*. 44 Vols. Fourth Supplement. Paris: Imprimerie Royale.

Burt, C. (1937). *The Backward Child*. New York: Appleton-Century.

Butler, N. R. & Bonham, D. G. (1963). *Perinatal Mortality: The First Report of the 1958 British Perinatal Mortality Survey*. Edinburgh: E & S Livingstone.

Butler, G. E., McKie, M. & Ratcliffe, S. G. (1990). The cyclical nature of prepubertal growth. *Annals of Human Biology*, **17**, 177–98.

Byard, P. J., Siervogel, R. M. & Roche, A. F. (1983). Familial correlations for serial measurements of recumbent length and stature. *Annals of Human Biology*, **10**, 281–93.

Cabana, T., Jolicoeur, P. & Michaud, J. (1993). Prenatal and postnatal growth and allometry of stature, head circumference, and brain weight in Québec children. *American Journal of Human Biology*, **5**, 93–9.

Cameron, N. (1984). *The Measurement of Human Growth*. London: Croom Helm.

Cameron, N. (1997). Growth and health in a developing country: the South African Experience 1984–1994. In *Growth and Development in a Changing World*, ed. D. F. Roberts, P. Rudan & T. Skaric-Juric, pp. 131–56. Zagreb: Croatian Anthropological Society.

Cameron, N., Tanner, J. M. & Whitehouse, R. H. (1982). A longitudinal analysis of the growth of limb segments in adolescence. *Annals of Human Biology*, **9**, 211–20.

Cameron, N., Mitchell, J., Meyer, D., Moodie, A., Bowie, M. D., Mann, M. D., Hansen J. D. L. (1988) Secondary sexual development of 'Cape Coloured' girls following kwashiorkor. *Annals of Human Biology*, **15**, 65–76.

Cameron, N., Mitchell, J., Meyer, D., Moodie, A., Bowie, M. D., Mann, M. D. & Hansen J. D. L. (1990) Secondary sexual development of 'Cape Coloured' boys following kwashiorkor. *Annals of Human Biology*, **17**, 217–28.

Cameron, N., Grieve, C. A., Kruger, A. & Leschner, K. F. (1993). Secondary sexual development in rural and urban South African black children. *Annals of Human Biology*, **20**, 583–93.

Carlson, D. S., Armelagos, G. J. & Van Gerven, D. P. (1976). Patterns of age-related cortical bone loss (osteoporosis) within the femoral diaphysis. *Human Biology*, **48**, 295–314.

Carroll, S. B. (1995). Homeotic genes and the evolution of arthropods and chordates. *Nature*, **376**, 479–85.

Case, T. J. (1978). On the evolution and adaptive significance of postnatal growth rates in terrestrial vertebrates. *Quarterly Review of Biology*, **53**, 243–82.

Chang, K. S. F., Ng, P. H., Lee, M. M. C. & Chan, S. J. (1966). Sexual maturation of Chinese boys in Hong Kong. *Pediatrics*, **37**, 804–11.

Charlesworth B. (1980). *Evolution in Age-Structured Populations*. Cambridge: Cambridge University Press.

Cheek, D. B. (1968). Muscle cell growth in normal children. In *Human Growth*, ed. D. B. Cheek, pp. 337–51. Philadelphia: Lea & Febiger.

Cheverud, J. M., Wilson, P. & Dittus, W. P. J. (1992). Primate population studies at Polannaruwa. III. Somatometric growth in a natural population of toque macaques (*Macaca sinica*). *Journal of Human Evolution*, **23**, 51–77.

Chow, B. (1974). Effect of maternal dietary protein on anthropometric and behavioral development of the offspring. In *Nutrition and Malnutrition: Identification and Measurement*, ed. A. F. Roche & F. Falkner, pp. 189–219. New York: Plenum.

Clegg, E. J. (1982). The influence of social, geographical and demographic factors on the size of 11–13-year-old children from the Isle of Lewis, Scotland. *Human Biology*, **54**, 93–109.

Clegg, E. J., Pawson, I. G., Ashton, E. H. & Flinn, R. M. (1972). The growth of children at different altitudes in Ethiopia. *Philosophical Transactions of the Royal Society of London*, **264B**, 403–37.

Clemmons, D.R., Klibanski, A., Underwood, L.E., McArthur, J.W., Ridgway, E.C., Beitins, I.Z. & Van Wyk, J.J. (1981). Reduction of plasma immunoreactive somatomedin-C during fasting in humans. *Journal of Clinical Endocrinology and Metabolism*, **53**, 1247–50.

Clemmons, D.R. & Van Wyk, J.J. (1984). Factors controlling blood concentration of somatomedin-C. *Journal of Clinical Endocrinology and Metabolism*, **13**, 113–43.

Coelho, A. M. Jr. (1985). Baboon dimorphism: growth in weight, length and adiposity from birth to 8 years of age. In *Nonhuman Primate Models for Human Growth*, ed. E. S. Watts, pp. 125–59. New York: Alan R. Liss.

Cohen, M.N. and Armelagos, G.J. (eds.) (1984). *Paleopathology at the Origins of Agriculture*. New York: Academic Press.

Cohen, M.N., O'Connor, K., Danforth, M. Jacobi, K. & Armstrong, C. (1994). Health and death at Tipu. In *In the Wake of Conquest*, ed. C. S. Larsen & G. R. Milner, pp. 121–33. New York: Wiley-Liss.

Collins, J. W. Jr. & Butler, A. G.(1997). Racial differences in the prevalence of small-for-dates infants among college educated women. *Epidemiology*, **8**, 315–17.

Collins, J. W. Jr. & David, R. J. (1993). Race and birthweight in biracial infants. *American Journal of Public Health*, **83**, 1125–9.

Collins, J. W. Jr., Derrick, M., Hilder, L. & Kempley, S. (1997). Relation of maternal ethnicity to infant birthweight in east London, England. *Ethnicity and Disease*, **7**, 1–4.

Condon, R. G. (1990). The rise of adolescence: social change and life stage dilemmas in the Central Canadian Arctic. *Human Organization*, **49**, 266–79.

Conroy, G. C. & Vannier, M. W. (1991). Dental development in South African australopithecines. Part I: problems of pattern and chronology. *American Journal of Physical Anthropology*, **86**, 121–36.

Cook, S.F. and Borah, W.W. (1979). *Essays in Population History: Mexico and California*. Berkeley: University of California.

Copeland, K. C., Eichberg, J. W., Parker, C. R. Jr. & Bartke, A. (1985). Puberty in the chimpanzee: somatomedin-C and its relationship to somatic growth and steroid hormone concentrations. *Journal of Clinical Endocrinology and Metabolism*, **60**, 1154–60.

Count, E. W. (1943). Growth patterns of the human physique: an approach to kinetic anthropometry. *Human Biology*, **15**, 1–32.

Cousins, R. J. & Deluca, H. F. (1972). Vitamin D and bone. In *The Biochemistry and Physiology of Bone*, ed. G. H. Bourne, pp. 281–335. New York: Academic Press.

Crawford, B. A., Harewood, W. J. & Handelsman, D. J. (1997). Growth and hormone characteristics of pubertal development in the hamadryas baboon. *Journal of Medical Primatology*, **26**, 153–63.

Crognier, E. (1981). Climate and anthropometric variations in Europe and the Mediterranean area. *Annals of Human Biology*, **8**, 99–107.

Crooks, D. L.(1995). American children at risk: poverty and its consequences for children's health, growth, and school achievement. *Yearbook of Physical Anthropology*, **38**, 57–86.

Cunningham, A. S. (1995). Breastfeeding: Adaptive behavior for child health and longevity. In *Breastfeeding: Biocultural Perspectives*, ed. P. Stuart-Macadam & K. A. Detwyller, pp. 243–64. New York: Aldine de Gruyter.

Cutler, G. B. Jr., Glenn, M., Bush, M., Hodgen, G. D., Graham, C. E. & Loriaux, D. L. (1978). Adrenarche: a survey of rodents, domestic animals, and primates. *Endocrinology*, **103**, 2112–18.

Dasgupta, I., Dasgupta, P. & Daschaudhuri, A. B. (1997). Familial resemblence in height and weight in an endogamous Mahisya caste population of rural West Bengal. *American Journal of Human Biology*, **9**, 7–9.

Dasgupta, P. & Das, S. R. (1997). A cross-sectional growth study of trunk and limb segments of the Bengali boys of Calcutta. *Annals of Human Biology*, **24**, 363–9.

David, R. J. & Collins, J. W. Jr. (1991). Bad outcomes in black babies: race or racism? *Ethnicity and Disease*, **1**, 236–44.

Davies, P. S. W., Jones, P. R. M. & Norgan, N. G. (1986). The distribution of subcutaneous and internal fat in man. *Annals of Human Biology*, **13**, 189–92.

Davis, T. A. & Reeds, P. J. (1998). The roles of nutrition, development and hormone sensitivity in the regulation of protein metabolism: an overview. *Journal of Nutrition*, **128**, 340S–41S.

Dean, M. C. & Wood, B. A. (1984). Phylogeny, neoteny and growth of the cranial base in hominoids. *Folia Primatologia*, **43**, 157–80.

Dechant, J. J., Mooney, M. P., Burrows, A. M., Smith, T. D. & Siegel, M. I. (1997). 'Month of birth effect' does not alter longitudinal growth in an experimental animal model. *American Journal of Human Biology*, **9**, 481–6.

Dietz, W. H., Marino, B., Peacock, N. R. & Bailey, R. C. (1989). Nutritional status of Efe pygmies and Lese horticulturalists. *American Journal of Physical Anthropology*, **78**, 509–18.

D'Ercole, A. J. & Underwood, L. E. (1986). Regulation of fetal growth by hormones and growth factors. In *Human Growth*, Vol. 1, 2nd edn, ed. F. Falkner & J. M. Tanner, pp. 327–38. New York: Plenum.

D'Esposito, M., Detre, J. A., Alsop, D. C., Shin, R. K., Atlas, S. & Grossman, M. (1995). The neural basis of the central executive system of working memory. *Nature*, **378**, 279–81.

Demirjian, A. (1986). Dentition. In *Human Growth, Volume 2, Postnatal Growth*, ed. F. Falkner & J. M. Tanner, pp. 269, 298. New York: Plenum.

Dettwyler, K. A. (1995) A time to wean: the hominid blueprint for the natural age of weaning in modern human populations. In *Breastfeeding: Biocultural Perspectives*, ed. P. Stuart-Macadam & K. A. Detwyller, pp. 39–74. New York: Aldine de Gruyter.

DeVilliers, H. (1971). A study of morphological variables in urban and rural Venda male populations. In *Human Biology of Environmental Change*, ed. D. J. M. Vorster. London: International Biological Program.

Dickinson, F., Cervera, M., Murguia, R. &

Uc, L. (1990). Growth, nutritional status and environmental change in Yucatan, Mexico. *Studies in Human Ecology*, **9**, 135–49.

Dittus, W. P. J. (1977). The social regulation of population density and age–sex distribution in the Toque Monkey. *Behaviour*, **63**, 281–322.

Dobzanshky, T. (1962). *Mankind Evolving*. New Haven: Yale University Press.

Dobzhansky, T. (1973). Nothing in biology makes sense except in the light of evolution. *American Biology Teacher*, **35**, 125–9.

Donaldson, H. H. (1895). *The Growth of the Brain*. New York: Charles Scribner's Sons.

Draper, P. (1976). Social and economic constraints on child life among the !Kung. In *Kalahari Hunter-Gatherers*, ed. R. B. Lee & I. DeVore, pp. 199–217. Cambridge: Harvard University Press.

Dubos, R. (1965). *Man Adapting*. New Haven: Yale University Press.

Eccles, J. C. (1979). *The Understanding of the Brain*. New York: McGraw-Hill.

Eigenmann, J. E. Patterson, D. F. & Froesch, E. R. (1984). Body size parallels insulin-like growth factor I levels but not growth hormone secretory capacity. *Acta Endocrinologica*, **106**, 448–53.

Ekwo, E., Gosselink, C., Roizen, N. & Brazdziunas, D. (1991). The effect of height on family income. *American Journal of Human Biology*, **3**, 181–8.

Elkin, A. P. (1964). *The Australian Aboriginies*. New York: Doubleday/Anchor.

Ellis, L. (1994). *Social Stratification and Socioeconomic Inequality. Volume 2: Reproductive and Interpersonal Aspects of Dominance and Status*. London: Praeger.

Ellison, P. T. (1982). Skeletal growth, fatness, and menarcheal age: a comparison of two hypotheses. *Human Biology*, **54**, 269–81.

Ellison, P. T. (1990) Human ovarian function and reproductive ecology: new hypotheses. *American Anthropologist* **92**, 933–52.

Eltis, D. (1982). Nutritional trends in Africa and the Americas: heights of Africans, 1819–1839. *Journal of Interdisciplinary History*, **12**, 453–75.

Emanuel, I. (1993). Intergenerational factors in pregnancy outcome: implications for teratology? *Issues and Reviews in Teratology*, **6**, 47–84.

Emanuel, I., Hale, C. B. & Berg, C. J. (1989). Poor birth outcomes of American black women: an alternative hypothesis. *Journal of Public Health Policy*, **10**, 299–308.

Emanuel, I., Filakti, H., Alberman, E. & Evans, S. J. W. (1992). Intergenerational studies of human birthweight from the 1958 birth cohort. 1. Evidence for a multigenerational effect. *British Journal of Obstectrics and Gynecology*, **99**, 67–74.

Enlow, D. H. (1963). *Principles of Bone Remodeling*. Springfield: C. C. Thomas.

Enlow, D. H. (1976). The remodeling of bone. *Yearbook of Physical Anthropology*, **20**, 19–34.

Ericksen, J. A., Ericksen, E. P. Hostetler, J. A. & Huntington, G. E. (1979). Fertility patterns and trends among the Old Order Amish. *Population Studies*, **33**, 255–76.

Eveleth, P. B. & Tanner, J. M. (1976). *World-Wide Variation in Human Growth*. Cambridge: Cambridge University Press.

Eveleth, P. B. & Tanner, J. M. (1990). *World-Wide Variation in Human Growth*, 2nd edn. Cambridge: Cambridge University Press.

Ewer, R. F. (1973). *The Carnivores*. Ithaca, NY: Cornell University Press.

Falkner, F. (1966). General considerations in human development. In *Human Development*, ed. F. Falkner, pp. 10–39. Philadelphia: Saunders.

Falkner, F. (1978). Implications for growth in human twins. In *Human Growth, Vol. 1*, ed. F. Falkner & J.M. Tanner, pp. 397–413. New York: Plenum.

Fedigan L. M. & Pavelka, M. S. M. (1994). The physical anthropology of menopause. In *Strength in Diversity*, ed. A. Herring & M. S. M. Pavelka, pp. 103–26. Toronto: Canadian Scholars Press.

Field, T. (1988). Stimulation of preterm infants. *Pediatric Reviews*, **10**, 149–53.

Finch, C. E. & Rose, M. R. (1995). Hormones and the physiological architecture of life history evolution. *The Quarterly Review of Biology*, **70**, 1–52.

Fischbein, S. (1977). Onset of puberty in MZ and DZ twins. *Acta Geneticae and Medicae Gemellologiae*, **26**, 151–58.

Flinn, M. V. & England, B. G. (1997). Social economics of childhood glucocorticoid stress response and health. *American Journal of Physical Anthropology*, **102**, 33–53.

Fogel, R. W. (1986). Physical growth as a measure of the economic wellbeing of populations: The eighteenth and nineteenth centuries. In *Human Growth*, 2nd edn, Vol. 3, ed. F. Falkner & J. M. Tanner, pp. 263–81. New York: Plenum.

Fomon, S. J., Owen, G. M., Filer, L. J. &

Maresh, M. (1966). Body composition of the infant, parts I and II. In *Human Development*, ed. F. Falkner, pp. 239–53. Philadelphia: Saunders.

Forbes, G. (1986). Body composition during adolescence. In *Human Growth*, 2nd edn, Vol. 2, ed. F. Falkner & J. M. Tanner, pp. 119–46. New York: Plenum.

Foster, J. W., Dominguez-Steglich, M. A., Guioli, S., Kwok, C., Weller, P. A., Stevanovic, Weissenbach, J., Mansour, S., Young, I. D., Goodfellow, P. N., Brook, J. D. & Schafer, A. J. (1994). Campomelic dysplasia and autosomal sex reversal caused by mutations in *SRY*-related gene. *Nature*, **372**, 525–30.

Frayer, D. W., Horton, W. A., Macchiarelli, R. & Mussi, M. (1987). Dwarfism in an adolescent from the Italian late Upper Palaeolithic. *Nature*, **330**, 60–2.

Freeman, H. E., Klein, R. E., Townsend, J. W. & Lechtig, A. L. (1980). Nutrition and cognitive development among rural Guatemalan children. *American Journal of Public Health*, **70**, 1277–85.

Frisancho, A. R. (1977). Human growth and development among high-altitude populations. In *The Biology of High Altitude Peoples*, ed. P. Baker, pp. 117–71. Cambridge: Cambridge University Press.

Frisancho, A. R. (1981). New norms of upper limb fat and muscle areas for assessment of nutritional status. *American Journal of Clinical Nutrition*, **34**, 2540–5.

Frisancho, A. R. (1990). *Anthropometric Standards for the Assessment of Growth and Nutritional Status*. Ann Arbor: University of Michigan Press.

Frisancho, A. R. & Baker, P. T. (1970). Altitude and growth: a study of the patterns of physical growth of a high altitude Peruvian Quechua population. *American Journal of Physical Anthropology*, **32**, 279–92.

Frisancho, A. R., Garn, S. M. & Ascoli, W. (1970). Childhood retardation resulting in reduction of adult body size due to lesser adolescent skeletal delay. *American Journal of Physical Anthropology*, **33**, 325–36.

Frisancho, A. R., Sanchez, J., Pallardel, D. & Yanez, L. (1973). Adaptive significance of small body size under poor socio-economic conditions in Southern Peru. *American Journal of Physical Anthropology*, **39**, 255–62.

Frisancho, A. R., Borkan, G. A. & Klayman, J. F. (1975). Pattern of growth of Lowland and Highland Peruvian Quechua of similar genetic compostiion. *Human Biology*, **47**, 233–43.

Frisancho, A. R., Guire, K., Babler, W., Borkan, G, & Way, A. (1980). Nutritional influence of childhood development and genetic control of adolescent growth of Quechuas and Mestizos from the Peruvian Lowlands. *American Journal of Physical Anthropology*, **52**, 367–75.

Frisancho, R. A., Matos, J., Leonard, W. R. & Yaroch, L. A. (1985). Developmental and nutritional determinants of pregnancy outcome among teenagers. *American Journal of Physical Anthropology*, **66**, 247–61.

Frisch, R. E. (1974). Critical weight at menarche, initiation of the adolescent growth spurt and control of puberty. In *Control of the Onset of Puberty*, eds. M. M. Grumbach, G. D. Grace & F. F. Mayer, pp. 403–23. New York: Wiley.

Frisch, R. E. & Revelle, R. (1970). Height and weight at menarche and a hypothesis of critical body weights and adolescent events. *Science*, **169**, 397–8.

Frisch, R. E., Wyshak, G. & Vincent, L. (1980). Delayed menarche and amenorrhea in ballet dancers. *New England Medical Journal*, **303**, 17–19.

Froment A. & Hiernaux, J. (1984). Climate-associated anthropometric variation between populations of the Niger bend. *Annals of Human Biology*, **11**, 189–200.

Fulwood, R., Abraham, S. & Johnson, C. (1981). *Height and Weight of Adults Ages 18–74 Years by Socioeconomic and Geographic Variables*. Vital and Health Statistics, Series 11, No. 224, DHEW Pub. No. (PHS) 81–1674. Washington DC: US Government Printing Office.

Galdikas, B. M. & Wood, J. W. (1990). Birth spacing patterns in humans and apes. *American Journal of Physical Anthropology*, **83**, 185–91.

Garber, P. A. & Leigh, S. R. (1997). Ontogenetic variation in small-bodied New World primates: implications for patterns of reproduction and infant care. *Folia Primatologica*, **68**, 1–22.

Garn, S. M. (1958). Fat, body size, and growth in the newborn. *Human Biology*, **29**, 337–53.

Garn, S.M. (1970). *The Earlier Gain and Later Loss of Cortical Bone*. Springfield, Illinois: Charles C. Thomas.

Garn, S. M. (1985). Smoking and human biology. *Human Biology*, **57**, 505–23.

Garn, S. M. (1987). The secular trend in size and maturational timing and its implication for nutritional assessment. *Journal of Nutrition*, **117**, 817–23.

Garn, S. M. & Bailey, S. M. (1978). Genetics of the maturational processes. In *Human Growth*, Vol. 1, ed. F. Falkner & J. M. Tanner, pp. 307–30. New York: Plenum.

Garn, S. M. & Clark, D. C. (1975). Nutrition, growth, development, and maturation: findings from the Ten-State Nutrition Survey of 1968–1970. *Pediatrics*, **56**, 300–19.

Garn, S. M. & Petzold, A. S. (1983). Characteristics of the mother and child in teenage pregnancy. *American Journal of Diseases of Children*, **137**, 365–8.

Garn, S. M. & Rohmann, C. G. (1962). X-linked inheritance of developmental timing in man. *Nature*, **196**, 695–6.

Garn, S. M. & Rohmann, C. G. (1966). Interaction of nutrition and genetics in the timing of growth and development. *Pediatric Clinics of North America*, **13**, 353–79.

Garn, S. M., Owen, G. M. & Clark, D. C. (1974). Ascorbic acid: the vitamin of affluence. *Ecology of Food and Nutrition*, **3**, 151–3.

Garn, S. M., Bailey, S. M. & Higgins, I.T.T. (1976). Fatness similarities in adopted pairs. *American Journal of Clinical Nutrition*, **29**, 1067–8.

Garn, S. M., Shaw, H. A. & McCabe, K. D. (1977). Effect of socioeconomic status and race on weight defined and gestational prematurity in this United States. In *The Epidemiology of Prematurity*, ed. D. W. Reed & F. J. Stanley, pp. 127–43. Baltimore: Urban & Scharzenberg.

Garn, S. M., Cole, P.E. & Bailey, S. M. (1979). Living together as a factor in family-line resemblances. *Human Biology*, **51**, 565–87.

Garn, S. M., Pesick, S. D. & Pilkington, J. J. (1984). The interaction between prenatal and socioeconomic effects on growth and development in childhood. In *Human Growth and Development*, ed. J. Borms *et al.*, pp. 59–70. New York: Plenum.

Gavan, J. A. (1953). Growth and development of the chimpanzee, a longitudinal and comparative study. *Human Biology*, **25**, 93–143.

Gavan, J. A. (1971). Longitudinal postnatal growth in the chimpanzee. In *The Chimpanzee*, Vol. 4, ed. G. Bourne, pp. 46–102. Basel: Karger.

Gavan, J. A. (1982). Adolescent growth in non-human primates: an introduction. *Human Biology*, **54**, 1–5.

Geffner, M. E., Bersch, N., Scott, M., Bailey, R. C. & Golde, D. W. (1996). IGF-1 does not mediate T-lymphoblast colony formation in response to estradiol, testosterone, 1,25(OH)$_2$ vitamin D$_3$, and triiodothyronine: studies in control and pygmy T-cell lines. *Biochemical and Molecular Medicine*, **59**, 72–9.

German, J., Simpson, J. & McLemore, Y. (1973). Abnormalities of human sex chromosomes. I. A ring Y without mosaicism. *Annals de Génétique*, **16**, 225–31.

Gillett, R. M. (1998). Permanent tooth emergence among Zambian schoolchildren: a standard for the assignment of ages. *American Journal of Human Biology*, **10**, 45–51.

Godfrey, L. & Sutherland, M. R. (1996). Paradox of peramorphic paedomorphosis: heterochrony and human evolution. *American Journal of Physical Anthropology*, **99**, 17–42.

Goff, D. J. & Tabin, C. J. (1997). Analysis of Hoxd-13 and Hoxd-11 misexpression in chick limb buds reveals that Hox genes affect both bone condensation and growth. *Development*, **124**, 627–36.

Goldizen, A. W. (1987) Tamarins and marmosets: comunal care of offspring. In *Primate Societies*, ed. B. B. Smuts, D. L. Cheney, R. M. Seyfarth, R. W. Wrangham & T. T. Struhsaker, pp. 34–43. Chicago: University of Chicago Press.

Goldsmith, J. (1993). Language and learning as properties of our species. *Modern Philosophy*, **90**, 34–8.

Goldstein, M. S. (1943). *Demographic and Bodily Changes in Descendants of Mexican Immigrants*. Austin: Institute of Latin American Studies.

Goodall, J. (1983). Population dynamics during a 15-year period in one community of free-living chimpanzees in the Gombe National Park, Tanzania. *Zietschrift fur Tierpsychologie*, **61**, 1–60.

Goodman, F. R., Mundlos, S. Muragaki, Y., Donnai, D., Giovannucci-Uzielli, M. L., Lapi, E., Majewski, F., McGaughran, J., McKeown, C., Reardon, W., Upton, J., Winter, R. M., Olsn, B. R. & Scambler, P. J. (1997). Synpolydactyly phenotypes correlate with size of expansion in HOXD13 polyalanine tract. *Proceedings of the the National Academy of Science, USA*, **94**, 7458–63.

Gordon-Larsen, P., Zemel, B. S. & Johnston, F. E. (1997). Secular changes in stature, weight, fatness, overweight, and obesity in urban African American adolescents from the mid-1950's to the mid-1990's. *American Journal of Human Biology*, **9**, 675–88.

Goss, R. (1964). *Adaptive Growth*. New York: Academic Press.

Goss, R. (1978). *The Physiology of Growth*. New York: Academic Press.

Goss, R. (1986). Modes of growth and regeneration. In *Human Growth*, 2nd edn, Vol. 1, ed. F. Falkner & J. M. Tanner, pp. 3–26. New York: Plenum.

Gould, J. B. (1986). The low birth weight infant. In *Human Growth*, 2nd edn, vol.1, ed. F. Falkner & J. M. Tanner, pp. 391–413. New York: Plenum.

Gould, K. G., Flint, M. & Graham, C. E. (1981). Chimpanzee reproductive senescence: a possible model for the evolution of the menopause. *Maturitas* **3**, 157–66.

Gould, S. J. (1977). *Ontogeny and Phylogeny*. Cambridge: Belknap.

Gould, S. J. (1979). Mickey Mouse meets Konrad Lorenz. *Natural History*, **88**(4), 30–6.

Gould, S. J. (1981). *The Mismeasure of Man*. New York: Norton.

Graham, C. E., Kling, O. R. & Steiner, R. A. (1979). Reproductive senescence in female nonhuman primates. In *Aging in Nonhuman Primates*, ed. D. J. Bowden, pp. 183–209. New York: Van Nostrand Reinhold.

Grant, D. B., Hambley, J., Becker, D. & Pimstone, B. L. (1973). Reduced sulphation factor in undernourished children. *Archives of Disease in Childhood*, **48**, 596–600.

Green, H., Morikawa, M. & Nixon, T. (1985). A duel effector theory of growth-hormone action. *Differentiation*, **29**, 195–8.

Green, W. H., Campbell, M. & David, R. (1984). Psychosocial dwarfism: a critical review of the evidence. *Journal of the American Academy of Child Psychiatry*, 23, 39–48.

Greene, L. S. (1973). Physical growth and development, neurological maturation and behavioral functioning in two Ecuadorian Andean communities in which goiter is endemic. *American Journal of Physical Anthropology*, **38**, 119–34.

Greil, H. (1991). Urbanization and heavy physical work as influencing factors on physiques. *Collegium Anthropologicum*, **1**, 123–9.

Greil, H. (1997). Sex, body type and timing in bodily development – trend statements based on a cross-sectional anthropometric study. In *Growth and Development in a Changing World*, ed. D. F. Roberts, P. Rudan & T.

Škaric-Juric, pp. 57–88. Zagreb: Croatian Anthropological Society, pp. 59–88.

Gregor, T. (1979). Short people. *Natural History*, February, 14–23.

Greska, L. (1986). Growth patterns of European and Amerindian high altitude natives. *Current Anthropology*, **27**, 72–4.

Greulich, W. W. (1976). Some secular changes in the growth of American-born and native Japanese children. *American Journal of Physical Anthropology*, **45**, 553–68.

Greulich, W. W. & Pyle, S. I. (1959). *Radiographic Atlas of Skeletal Development of the Hand and Wrist*, 2nd edn. Stanford: Stanford University Press.

Grumbach, M. M., Roth, J. C., Kaplan, S. L. & Kelch, R. P. (1974). Hypothalmic–pituitary regulation of puberty in man: evidence and concepts derived from clinical research. In *Control of the Onset of Puberty*, ed. M. M. Grumbach, G. D. Grave & F. E. Mayer, pp. 115–66. New York: Wiley.

Guillemant, J., Cabrol, S. Allemandou, A. Peres, G. & Guillemant, S. (1995). Vitamin D-dependent seasonal variation of PTH in growing male adolescents. *Bone*, **17**, 513–16.

Gupta, R. & Basu, A. (1981). Variations in body dimensions in relation to altitude among the Sherpas of the eastern Himalayas. *Annals of Human Biology*, **8**, 145–51.

Gurney, J. M. & Jelliffe, D. B. (1973). Arm anthropometry in nutritional assessment: nomogram for the rapid calculation of muscle circumference and cross-sectional muscle and fat areas. *American Journal of Clinical Nutrition*, **26**, 912–15.

Gurri, F. D. & Dickinson F. (1990). Effects of socioeconomic, ecological, and demographic conditions on the development of the extremities and the trunk: a case study with adult females from Chiapas. *Journal of Human Ecology*, **1**, 125–38.

Guthrie, H. & Picciano, M. F. (1995). *Human Nutrition*. St. Louis: Mosby.

Habicht, J.-P., Yarbrough, C., Martorell, R. , Malina, R. M. & Klein, R. E. (1974). Height and weight standards for preschool children. *The Lancet*, **1**, 611–15.

Haddad, J. G. & Hahn, T. J. (1973). Natural and synthetic sources of circulating 25-hydroxy vitamin D in man. *Nature*, **224**, 515–16.

Haeckel, E. (1874). *Anthropogenie oder Entwicklungsgeschichte des Menschen*. Leipzig: Englemann.

Halder, G., Callaerts, P. & Gehring, W. J. (1995). New perspectives on eye evolution. *Current Opinion in Genetics and Development*, **5**, 602–9.

Halpern, C. T., Udry, R.J. Campbell, B. & Suchinddran, C. (1993). Testosterone and pubertal development as predictors of sexual activity: a panel analysis of adolescent males. *Psychosomatic Medicine*, **55**, 436–47.

Halpern, C. T., Udry, R. J. & Suchinddran, C. (1997). Testosterone predicts initiation of coitus in adolescent females. *Psychosomatic Medicine*, **59**, 161–71.

Hamada, Y. (1994). Standard growth patterns and variations in growth patterns of Japanese monkeys (*Macaca fuscata*) based on an analysis by the spline function method. *Anthropological Science*, **102** (Suppl.), 57–76.

Hamada, Y., Udono, T., Teramoto, M. & Sugawara (1996). The growth pattern of chimpanzees: somatic growth and reproductive maturation in *Pan troglodytes*. *Primates*, **37**, 279–95.

Hamill, P. V. V., Johnston, F. E. & Lemshow, S. (1973). *Body weight, stature, and sitting height: white and Negro youths 12–17 years, United States*. DHEW Publication No. (HRA) 74–1608. Washington, DC: US Government Printing Office.

Hamill, P. V. V., Johnson, C. L., Reed, R. B. & Roche, A. F. (1977). *NCHS Growth Curves for Children Births – 18 years, United States*. DHEW Publications, (PHS) 78–1650. Washington, DC: US Government Printing Office.

Hamilton, W. (1966). The moulding of senescence by natural selection. *Journal of Theoretical Biology*, **12**, 12–45.

Hamilton, W. J. & Mossman, H. (1972). *Human Embryology: Prenatal Development of Form and Function*, 4th edn. Cambridge: Heffer & Sons.

Hansman, C (1970). Anthropometry and related data: anthropometry, skinfold thickness measurements. In *Human Growth and Development*, ed. R. W. McCammon, pp. 101–54. C. C. Thomas: Springfield, Illinois.

Harlow, H. F. & Zimmermann, R. R. (1959). Affectional responses in the infant monkey. *Science*, **130**, 421–32.

Harsha, D. W., Voors, A. W. & Berenson, G. S. (1980). Racial differences in subcutaneous fat patterns in children aged 7–15 years. *American Journal of Physical Anthropology*, **53**, 333–7.

Harvey, P. H., Martin, R. D. & Clutton-Brock, T. H. (1986). Life histories in comparative perspective. In *Primate Societies*, ed. B. B. Smuts, D. L. Cheney, R. M. Seyfarth, R. W. Wrangham & T. T. Struhsaker, pp. 181–96. Chicago: University of Chicago Press.

Harvey, S. Hull, K. L. & Fraser, R. A. (1993). Mini review: growth hormone: neurocrine and neuroendocrine perspectives. *Growth Regulation*, 3, 161–71.

Hass, J. D., Frongillo, E. A., Jr., Stepick, C. D., Beard, J. L. & Hurtado, G. (1980). Altitude, ethnic and sex difference in birth weight and length in Bolivia. *Human Biology*, 52, 459–77.

Hattori, Y., Vera, J. C., Rivas, C. I., Bersch, N. Bailey, R. C., Geffner, M. E. & Golde, D. W. (1996). Decreased insulin-like growth factor I receptor expression and function in immortalized African Pygmy T cells. *Journal of Clinical Endocrinology and Metabolism*, 81, 2257–63.

Hauspie, R. C. Vercauteren, M. & Susanne, C. (1996). Secular changes in growth. *Hormone Research*, 45, supplement 2, 8–17.

Haviland, W. A. & Moholy-Nagy, H. (1992). Distinguishing the high and mighty from the hoi polloi at Tikal, Guatemala. In *Mesoamerican Elites: An Archaeological Assessment*, ed. A. F. Chase & D. Z. Chase, pp. 50–60. Norman, Oklahoma: University of Oklahoma Press.

Hayflick, L. (1980). The cell biology of human aging. *Scientific American*, 242, 58–65.

Hayes, F. J. & Crowley, W. F., Jr (1998). Gonadotropin pulsations across development. *Hormone Research*, 49, 163–8.

Hediger, M. L. & Katz, S. H. (1986). Fat patterning, overweight, and adrenal androgen interactions in black adolescent females. *Human Biology*, 58, 585–600.

Heinrichs, C., Munson, P. J., Counts, D. R., Cutler Jr, G. B. & Baron, J. (1995). Patterns of human growth. *Science*, 268, 442–5.

Henneberg, M. & Louw, G. J. (1990). Height and weight differences amoung South African Urban schoolchildren born in various months of the year. *American Journal of Human Biology*, 2, 227–33.

Henneberg, M. & Louw, G. J. (1993). Further studies on the month-of-birth effect on body size: rural schoolchildren and an animal model. *American Journal of Physical Anthropology*, 91, 235–44.

Henneberg, M. & Van Den Berg, E. R. (1990). Test of socioeconomic causation of secular trend: stature changes among favored and oppressed South Africans are parallel. *American Journal of Physical Anthropology*, 83, 459–65.

Hensley, W. E. (1993). Height as a measure of success in academe. *Psychology: A Journal of Human Behavior*, 30, 40–6.

Hensley, W. E. & Cooper, R. (1987). Height and occupational success: a review and critique. *Psychological Reports*, 60, 843–49.

Herman-Giddens, M. E., Sandler, A. D. & Friedman, N. E. (1988). Sexual precocity in girls. An association with sexual abuse? *American Journal of Diseases of Childhood*, 142, 431–33.

Herman-Giddens, M. E., Slora, E. J., Wasserman, R. C., Bourdony, C. J., Bhapkar, M. V., Koch, G. C. & Hasemeier, C. M. (1997). Secondary sexual characteristics and menses in young girls seen in office practice: a study from the National Pediatric Research in Office Settings Network. *Pediatrics*, 99, 505–12.

Hermanussen, M., Geiger-Benoit, K., Burmeister, J. & Sippel, W. G. (1988). Periodical changes of short term growth velocity ('mini growth spurts') in human growth. *Annals of Human Biology*, 15, 103–9.

Hiernaux, J. (1974). *The People of Africa*. London: Weidenfeld & Nicolson.

Higham, E. (1980). Variations in adolescent psychohormonal development. In *Handbook of Adolescent Psychology*, ed. J. Adelson, pp. 472–94. New York: Wiley.

Higham, P. A. & Carmet, D. W. (1992). The rise and fall of politicians: the judged height of Broadbent, Mulroney and Turner before and after the 1988 Canadian federal election. *Canadian Journal of Behavioral Science*, 24, 404–9.

Hill, J. P. & Lynch, M. E. (1983). The intensification of gender-related role expectations during early adolescence. In *Girls at Puberty*, ed. J. Brooks-Gunn & A. C. Petersen, pp. 201–28. New York: Plenum Press.

Hill, K. & Hurtado, A. R. (1991). The evolution of premature reproductive senescence and menopause in human females: an evaluation of the 'Grandmother Hypothesis'. *Human Nature*, 2, 313–50.

Hill, K. & Kaplan, H. (1988). Trade offs in male and female reproductive strategies among the Ache. Parts 1 and 2. In *Human Reproductive Behavior: A Darwininian Perspective*, ed. L. Betzig, M. Borgerhoff-Mulder & P. Turke, pp. 277–89, 291–305. Cambridge: Cambridge University Press.

Hirsch, M., Shemesh, J., Modan, M. & Lunenfeld, B. (1979). Emission of spermatoza. Age of onset. *International Journal of Andrology*, **2**, 289–98.

Hojo, T., Takemoto, R. & Shinoda, K. (1981). The secular unchangeability in relative sitting height of female Kyushuites. *Journal of the University of Occupational and Environmental Health, Japan*, **3**, 203–5.

Holland, P. W. H. & Garcia-Fernàndez, J. (1996). *Hox* genes and chordate evolution. *Developmental Biology*, **173**, 382–95.

Holliday, M. A. (1986). Body composition and energy needs during growth. In *Human Growth*, 2nd edn., vol. 2, ed. F. Falkner & J. M. Tanner, pp. 101–17. New York: Plenum.

Horton, W. A. & Machado, M. A. (1992). Molecular structure of the growth plate. In *Human Growth: Basic and Clinical Aspects*, ed. M. Hernández and J. Argente, pp. 75–80. Amsterdam: Elsvier.

Hoshi, H. & Kouchi, M. (1981). Secular trend of the age at menarche of Japanese girls with special regard to the secular acceleration of the age at peak height velocity. *Human Biology*, **53**, 593–8.

Howe, P. E. & Schiller, M. (1952). Growth responses of the school child to changes in diet and environmental factors *Journal of Applied Physiology*, **5**, 51–61.

Howell, N. (1979). *Demography of the Dobe !Kung*. New York: Academic Press.

Hulanicka, B. & Kotlarz, K. (1983). The final phase of growth in height. *Annals of Human Biology*, **10**, 429–34.

Hulanicka, B. & Waliszko, A. (1991). Deceleration of age at menarche in Poland. *Annals of Human Biology*, **18**, 507–13.

Hulse, F. (1969). Migration and cultural selection in human genetics. In *The Anthropologist*, ed. P. C. Biswas, pp. 1–21. Delhi, India: University of Delhi.

Hunt, E. E. (1966). The developmental genetics of man. In *Human Development*, ed. F. Falkner, pp. 76–122. Philadelphia: W. B. Saunders.

Hunt, E. E. Jr. & Heald, F. P. (1963). Physique, body composition, and sexual maturation in adolescent boys. *Annals of the New York Academy of Sciences*, **110**, 532–44.

Huxley, J. S. (1932). *Problems of Relative Growth*. London: Methuen, 2nd edition. Reprint 1972, New York: Dover.

Huxley, T. H. (1863). *Evidence as to Man's Place in Nature*. London: Williams & Norgate.

IBNMRR (1995). Third report on nutrition monitoring in the United States: executive summary. Washington, DC: US Government Printing Office (report of the Interagency Board for Nutrition Monitoring and Related Research).

Illsley, R., Finlayson, A. & Thompson, B. (1963). The motivation and characteristics of internal migrants: a socio-medical study of young migrants in Scotland. *Milbank Memorial Fund Quarterly*, **41**, 115–44 & **41**, 217–48.

Isaksson, O. G. P., Jansson, J-O. & Gause, I. A. M. (1982). Growth hormone stimulates longitudinal bone growth directly. *Science*, **216**, 1237–8.

Isley, W. L., Underwood, L. E. & Clemmons, D. R. (1983). Dietary components that regulate serum somatomedin-C concentrations in humans. *Journal of Clinical Investigation*, **71**, 175–82.

Jakacki, R. I., Kelch, R. P., Sander, S. E., Lloyd J. S., Hopwood, N. J. & Marshall, J. C. (1982). Pulsatile secretion of luteinizing hormone in children. *Journal of Endocrinology and Metabolism*, **55**, 453–8.

Janson, C. H. & Van Schaik, C. P. (1993) Ecological risk aversion in juvenile primates: slow and steady wins the race. In *Juvenile Primates: Life History, Development, and Behavior*, ed. M. E. Perieira & L. A. Fairbanks, pp. 57–74. New York: Oxford University Press.

Jaswal, S. (1983). Age and sequence of permanent tooth emergence among Khasis. *American Journal of Physical Anthropology*, **62**, 177–86.

Jelliffe, D. B. (1966). *The Assessment of the Nutritional Status of the Community*. WHO Monograph No. 53. Geneva: World Health Organization.

Jenkins, C. L. (1981). Patterns of growth and malnutrition among preschoolers in Belize. *American Journal of Physical Anthropology*, **56**, 169–78.

Jenkins, C. L., Orr-Ewing, A. K. & Heywood, P. F. (1984). Cultural aspects of early childhood growth and nutrition among the Amele of lowland Papua, New Guinea. *Ecology of Food and Nutrition*, **14**, 261–75.

Jerison, H. S. (1973). *Evolution of the Brain and Intelligence*. New York: Academic Press.

Jerison, H. S. (1976). Paleoneurology and the evolution of mind. *Scientific American*, **234**(1), 90–101.

Johnson, C. L., Fulwood, R., Abraham, S. &

Bryner, J. D. (1981). *Basic Data on Anthropometric Measurements and Angular Measurements of the Hip and Knee Joints for Selected Age Groups 1–74 Years of Age.* DHHS Publication, no. (PHS) 81-1669. Washington, DC: US Government Printing Office.

Johnson, T. D. (1982). Selective costs and benefits in the evolution of learning. *Advances in the Study of Animal Behavior*, **12**, 65–106.

Johnston, F. E. (1974). Control of age at menarche. *Human Biology*, **46**, 159–71.

Johnston, F. E. (1980). Nutrition and growth. In *Human Physical Growth and Maturation*, ed. F. E. Johnston, A. Roche & C. Sussane, pp. 291–300. New York: Plenum.

Johnston, F. E. (1986). Somatic growth of the infant and preschool child. In *Human Growth*, 2nd edn, Vol. 2, ed. F. Falkner & J. M. Tanner, pp. 3–24. New York: Plenum.

Johnston, F. E. & Beller, A. (1976). Anthropometric evaluation of the body composition of black, white, and Puerto Rican newborns. *American Journal of Clinical Nutrition*, **29**, 61.

Johnston, F. E. & Low, S. M. (1995). *Children of the Urban Poor: The Sociocultural Environment of Growth, Development, and Malnutrition in Guatemala City.* Boulder, Colorado: Westview.

Johnston, F. E., Borden, M. & MacVean, R. B. (1973). Height, weight and their growth velocities in Guatemalan private schoolchildren of high socio-economic class. *Human Biology*, **45**, 627–41.

Johnston, F. E., Hamill, P. V. V. & Lemeshow, S. (1974). Skinfold thicknesses in a national probability sample of U.S. males and females 6 through 17 years. *American Journal of Physical Anthropology*, **40**, 321–4.

Johnston, F. E., Dechow, P. C. & MacVean, R. B. (1975). Age changes in skinfold thickness among upper class school children of differing ethnic backgrounds residing in Guatemala. *Human Biology*, **47**, 251–62.

Johnston, F. E., Wainer, H., Thissen, D. & MacVean, R. B. (1976). Hereditary and environmental determinants of growth in height in a longitudinal sample of children and youth of Guatemalan and European ancestry. *American Journal of Physical Anthropology*, **44**, 469–76.

Johnston, F. E., Scholl, T. O., Newman, B. C., Cravioto, J. & De Licardie, E. R. (1980). An analysis of environmental variables and factors associated with growth failure in a Mexican village. *Human Biology*, **52**, 627–37.

Johnston, F. E., Bogin, B., MacVean, R. B. & Newman, B. C. (1984). A comparison of international standards versus local reference data for the triceps and subscapular skinfolds of Guatemalan children and youth. *Human Biology*, **56**, 157–71.

Johnston, F. E., Low, S. M., de Baessa, Y. & MacVean, R.B. (1985). Growth status of disadvantaged urban Guatemalan children of a resettled community. *American Journal of Physical Anthropology*, **68**, 215–24.

Jolicoeur, P., Pontier, J. & Abidi, H. (1992). Asymptotic models for the longitudinal growth of human stature. *American Journal of Human Biology*, **68**, 461–8.

Jolly, A. (1985). *The Evolution of Primate Behavior*, 2nd edn. New York: Macmillan.

Jones, D. (1995). Sexual selection, physical attractiveness, and facial neoteny. *Current Anthropology*, **36**, 723–48.

Jones, J. I. & Clemmons, D. R. (1995). Insulin-like growth factors and their binding proteins: biological actions. *Endocrinology Reviews*, **16**, 3–34.

Jones, P. R. M. (1995). From tape to technology. In *Essays on Asuxology*, ed. R. Hauspie, G. Lindgren & F, Falkner, pp. 3–17. Welwyn Garden City, England: Castlemead.

Jones, P. R. M. & Dean, R. F. A. (1956). The effects of Kwashiorkor on the development of the bones of the hand. *Journal of Tropical Pediatrics*, **2**, 51.

Judd, H. L., Parker, D. C. & Yen, S.S.C. (1977). Sleep-wake patterns of LH and testosterone release in propubertal boys. *Journal of Clinical Endocrinology and Metabolism*, **44**, 865–9.

Kaplan, B. (1954). Environment and human plasticity. *American Anthropologist*, **56**, 780–99.

Kaplan, L. (1984). *Adolescence: The Farewell to Childhood.* New York: Simon & Schuster.

Kaplowitz, H., Martorell, R. & Engle, P. L. (1993). Selection for rural-to-urban migrants in Guatemala. In *Urban Ecology and Health in the Third World*, ed. L. M. Schell, M. T. Smith & A. Bilsborough, pp. 144–62. Cambridge: Cambridge University Press.

Karlberg, J. (1987). On the modelling of human growth. *Statistics in Medicine*, **6**, 185–92.

Katch, V. L., Campaigne, B., Freedson, P., Sady, S., Katch, F. I. & Behnke, A. R. (1980). Contribution of breast volume and weight to body fat distribution in females. *American Journal of Physical Anthropology*, **53**, 93–100.

Katz, S. H., Hediger, M. L., Zemel, B. S. & Parks, J. S. (1985). Adrenal androgens, body fat and advanced skeletal age in puberty: new evidence for the relations of adrenarche and gonadarche in males. *Human Biology*, **57**, 401–13.

Kelch, R., Jenner, M., Weinstein, R., Kaplan, S. & Grumbach, M. (1972). Estradiol and testosterone secretion by human, simian and canine testes, in males with hypogonadism and in male pseudo hermaphrodites with the feminizing testes syndrome. *Journal of Clinical Investigation*, **51**, 824–30.

Kember, N. F. (1992) The physiology of the growth plate. In *Human Growth: Basic and Clinical Aspects*, ed. M. Hernández & J. Argente, pp. 81–6. Amsterdam: Elsevier.

Kerr, G. R., Chamove, A. S. & Harlow, H. F. (1969). Environmental deprivation: its effect on the growth of infant monkeys. *Journal of Pediatrics*, **75**, 833–7.

Keyes, R. (1979). The height report. *Esquire*, November, 31–43.

Kimura, K. (1984). Studies on growth and development in Japan. *Yearbook of Physical Anthropology*, **27**, 179–214.

Kimura, K. & Kitano, S. (1959). Growth of the Japanese physiques in four successive decades before World War II. *Zinruigaku Zassi*, **67**, 37–46.

King, M. C. & Wilson, A. C. (1975). Evolution at two levels: molecular similarities and differences between humans and chimpanzees. *Science*, **188**, 107–16.

Kirkwood, T. B. L. (1977). Evolution of aging. *Nature*, 270, 301–4.

Kirkwood, T. B. L. & Holliday, R. (1986). Selection for optimal accuracy and the evolution of aging. In *Accuracy in Molecular Processes*, ed. T. B. L. Kirkwood, R. F. Rosenberger & D. J. Galas, pp. 363–79. New York: Chapman & Hall.

Klein, R.G. (1989) *The Human Career: Human Biological and Cultural Origins*. Chicago: University of Chicago Press.

Kliegman, R. M. (1995). Neonatal technology, perinatal survival, social consequences, and the perinatal paradox. *American Journal of Public Health*, **85**, 909–13.

Kobyliansky, E. & Arensburg, B. (1974). Changes in morphology of human populations due to migrations and selection. *Annals of Human Biology*, **4**, 57–71.

Kondo, S. & Eto, M. (1975). Physical growth studies on Japanese–American children in comparison with native Japanese. In *Comparative Studies of Human Adaptability of Japanese, Caucasians, and Japanese–Americans*, ed. S. M. Horvath, S. Kondo, H. Matsui, & H. Yoshimena. pp. 13–45. Tokyo: Japanese International Biological Program.

Konner, M. (1976). Maternal care, infant behavior and development among the !Kung. In *Kalahari Hunter-Gatherers*, ed. R. B. Lee & I. DeVore, pp. 218–45. Cambridge: Harvard University Press.

Kramer, P. (1998). The costs of human locomotion: maternal investment in infant transport. *American Journal of Physical Anthropology*, **107**, 71–85.

Krogman, W. M. (1970). Growth of the head, face, trunk, and limbs in Phildelphia white and Negro children of elementary and high school age. *Monographs of the Society for Research in Child Development*, **20**, 1–91.

Krogman, W. M. (1972). *Child Growth*. Ann Arbor: University of Michigan Press.

Kummer, B. (1953). Untersuchungen uberdie ontogenetische Entwicklung des menschlichen Schadelbasiswinkels. *Zietschrift fur Morphologie und Anthropologie*, **43**, 331–60.

La Barre, W. (1991). *Shadow of Childhood: Neoteny and the Biology of Religion*. Norman, Oklahoma: University of Oklahoma Press.

Laird, A. K. (1967). Evolution of the human growth curve. *Growth*, **31**, 345–55.

Laitman, J. T. & Heimbuch, R. C. (1982). The basicranium of Plio-Pleistocene hominids as an indicator of their upper respiratory systems. *American Journal Physical Anthropology*, **59**, 323–43.

Lampl, M., Johnston, F. E. & Malcolm, L. A. (1978). The effects of protein supplementation on growth and skeletal maturation of New Guinean school children. *Annals of Human Biology*, **5**, 219–27.

Lampl, M., Veldhuis, J. D. & Johnson, M. L. (1992). Saltation and stasis: a model of human growth. *Science*, **258**, 801–3.

Lancaster, J. B. (1985). Evolutionary perspectives on sex differences in the higher primates. In *Gender and the Life Course*, ed. A. S. Rossi, pp. 3–28. New York: Aldine.

Lancaster, J. B. & Lancaster, C. S. (1983). Parental investment: The hominid adaptation. In *How Humans Adapt*, ed. D. J. Ortner, pp. 33–65. Washington: Smithsonian Institution Press.

Largo, R. H., Gasser, Th., Prader, A., Stutzle, W. & Huber, P. J. (1978). Analysis of the adolescent growth spurt using smoothing spline functions. *Annals of Human Biology*, 5, 421–34.

Laron, Z., Arad, J., Gurewitz, R., Grunebaum, M. & Dickerman, Z. (1980). Age at first conscious ejaculation – a milestone in male puberty. *Helvatica Paediatrica Acta*, 35, 13–20.

Lasker, G. W. (1952). Environmental growth factors and selective migration. *Human Biology*, 24, 262–89.

Lasker, G. W. (1954). The question of physical selection of Mexican migrants to the United States of America. *Human Biology*, 26, 52–8.

Lasker, G. W. (1969). Human biological adaptability. *Science*, 166, 1480–6.

Lasker, G. W. & Mascie-Taylor, C. G. N. (1989). Effects of social class differences and social mobility on growth in height, weight and body mass index in a British cohort. *Annals of Human Biology*, 16, 1–8.

Lasker, G. W. & Mascie-Taylor, C. G. N. (1996). Influence of social class on the correlation of stature of adult children with that of their mothers and fathers. *Journal of Biosocial Science*, 28, 117–22.

Lee, J. W. Y. (1996). An XXYY male with schizophrenia. *Australian & New Zealand Journal of Psychiatry*, 30, 553–6.

Lee, M. M. C., Chang, K. S. F. & Chan, M. M. C. (1963). Sexual maturation of Chinese girls in Hong Kong. *Pediatrics*, 32, 389–98.

Lee, P. C., Majluf, P. & Gordon, I. J. (1991). Growth, weaning, and maternal investment from a comparative perspective. *Journal of the Zoological Society of London*, 225, 99–114.

Lee, P. C. & Moss, C. J. (1995). Statural growth in known-age African elephants (*Loxodonta africana*). *Journal of the Zoological Society of London*, 236, 29–41.

Lee, W. T., Leung, S. S., Leung, D. M., Tsang, H. S., Lau, J. & Cheng, J. C. (1995). A randomized double-blind controlled calcium supplementation trial and bone and height acquisition in children. *British Journal of Nutrition*, 74, 125–39.

Leigh, S. R. (1992). Patterns of variation in the ontogeny of primate body size dimorphism. *Journal of Human Evolution*, 23, 27–50.

Leigh, S. R. (1994a). Ontogenetic correlates of diet in anthropoid primates. *American Journal of Physical Anthropology*, 94, 499–522.

Leigh, S. R. (1994b). Growing up to be a primate (book review). *Evolutionary Anthropol-*
ogy, 3, 106–8.

Leigh, S. R. (1996). Evolution of human growth spurts. *American Journal of Physical Anthropology*, 101, 455–74.

Leigh, S. R. & Shea, B. T. (1996). The ontogeny of body size variation in African apes. *American Journal of Physical Anthropology*, 99, 43–65.

Leighton, G. & Clark, M. L. (1929). Milk consumption and the growth of schoolchildren. *Lancet*, 1, 40–3.

Leonard W. R. & Robertson M. L. (1992). Nutritional requirements and human evolution: a bioenergetics model. *American Journal of Human Biology*, 4, 179–95.

Leonard, W. R. & Robertson, M. L. (1994). Evolutionary perspectives on human nutrition: the influence of brain and body size on diet and metabolism. *American Journal of Human Biology*, 6, 77–88.

Leonard, W. R., Leatherman, T. L., Carey, J. W. & Thomas, R. B. (1990). Contributions of nutrition versus hypoxia to growth in rural Andean populations. *American Journal of Human Biology*, 2, 613–26.

Lewin, R. (1993). *Human Evolution: An Illustrated Introduction*. Oxford: Blackwell.

LeVine, R. (1977). Child rearing as a cultural adaptation. In *Culture and Infancy: Variations in the Human Experience*, ed. P. H. Leiderman, S. Tulkin & A. Rosenfeld, pp. 15–27. Academic Press: New York.

Lichty, J. A., Ting, R. Y., Bruns, P. & Dyar, E. (1957). Studies of babies born at high altitude. I. Relation of altitude to birth-weights. *American Journal of Disease in Childhood*, 93, 666–9.

Lieberman, P., Crelin, E. S. & Klatt, D. H. (1972). Phonetic ability and related anatomy of the newborn and adult human, Neanderthal man, and the chimpanzee. *American Anthropologist*, 74, 287–307.

Lindeman, M. & Sundvik, L. (1994). Impact of height assessment on Finnish female job applicants' managerial abilities. *Journal of Social Psychology*, 134, 169–74.

Lindgren, G. (1976). Height, weight, and menarche in Swedish urban school children in relation to socio-economic and regional factors. *Annals of Human Biology*, 3, 501–28.

Lindgren, G. (1978). Growth of schoolchildren with early, average, and late ages of peak height velocity. *Annals of Human Biology*, 5, 253–67.

Lindgren, G. (1995). Socio-economic background, growth, educational outcome and health. In *Essays on Auxology*, ed. R. Hauspie, G. Lindgren & F. Falkner, pp. 408–24. Welwyn Garden City, England: Castlemead.

Little, M. A., Galvin, K. & Mugambi, M. (1983). Cross-sectional growth of nomadic Turkana pastoralists. *Human Biology*, 55, 811–30.

Livi, R. (1896). *Antropometica Militare*. Rome. Cited in Boas, F. (1922).

Loh, E. S. (1993). The economic effects of physical appearance. *Social Science Quarterly*, 74, 420–38.

Lohman, T. G., Roche, A. F. & Martorell, R. (1988). *Anthropometric Standardization Reference Manual*. Champaign, Illinois: Human Kinetics Publishers.

Lopez-Blanco, M. (1995). Growth as a mirror of conditions of a developing society: the case of Venezuela. In *Essays on Auxology*, ed. R. Hauspie, G. Lindgren & F. Falkner, pp. 312–21. Welwyn Garden City, UK: Castlemead.

Lorenz, K. (1971). Part and parcel in animal and human societies: a methodological discussion. In *Studies in Animal and Human Behavior*, Vol. II (translated by Robert Martin), pp. 115–95. Cambridge: Harvard University Press.

Lourie, J. A., Taufa, T., Cattani, J. & Anderson, W. (1986). The Ok Tedi Health and Nutrition Project, Papua New Guinea: physique, growth and nutritional status of the Wopkaimin of the Star Mountains. *Annals of Human Biology*, 13, 517–36.

Lovejoy, A. O. (1936). *The Great Chain of Being*. Cambridge: Harvard University Press.

Lovejoy, O. (1981). The origin of man. *Science*, 211, 341–50.

Low, W. D., Kung, L. S. & Leong, J. C. Y. (1982). Secular trend in the sexual maturation of Chinese girls. *Human Biology*, 54, 539–51.

Lowe, C. U., Forbes, G., Garn, S., Owen, G. M., Smith, N. J., Weil, W. B. Jr & Nichaman, M. Z. (1975). Reflections of dietary studies with children in the Ten-State Nutrition Survey of 1968–1970. *Pediatrics*, 56(2), 320–6.

Lowery, G. H. (1986). *Growth and Development of Children*, 8th edn. Chicago: Yearbook Medical Publishers.

Luft, U. C. (1972). Principles of adaptation to altitude. In *Physiological Adaptations: Desert and Mountain*, ed. M. K. Yousef, S. M. Horvath & R. W. Ballard, pp. 143–56. New York: Academic Press.

MacBeth, H. M. (1984). The study of biological selectivity in migrants. In *Migration and Mobility*, ed. A. J. Boyce, pp. 195–207. London: Taylor & Francis.

MacGillivray, M. H., Morishima, A., Conte, F., Grumbach, M. & Smith, E. P. (1998). Pediatric endocrinology update: an overview. The essential role of estrogens in pubertal growth, epiphyseal fusion and bone turnover: lessons from mutations in the genes for aromatase and the estrogen receptor. *Hormone Research*, 49, Supplement 1, 2–8.

MacKay, D. H. & Martin, W. J. (1952). Dentition and physique of Bantu children. *Journal of Tropical Medicine and Hygiene*, 55, 265–75.

Magner, J. A., Rogol, A. D. & Gorden, P. (1984). Reversible growth hormone deficiency and delayed puberty triggered by a stressful experience in a young adult. *The American Journal of Medicine*, 76, 737–42.

Malcolm, L. A. (1970). Growth and development in the Bundi children of the New Guinea highlands. *Human Biology*, 42, 293–328.

Malina, R. M. (1966). Patterns of development in skinfolds of Negro and white Philadelphia children. *Human Biology*, 38, 89–103.

Malina, R. M. (1979). Secular changes in size and maturity: causes and effects. In Secular Trends in Human Growth, maturation, and development, ed. A. F. Roche, *Monographs of the Society for Research in Child Development*, 44, 59–102.

Malina, R. M. (1986). Growth of muscle tissue and muscle mass. In *Human Growth, Volume 2, Postnatal Growth*, ed. F. Falkner & J. M. Tanner, pp. 77–99. New York: Plenum.

Malina, R. M. (1990). Research on secular trends in auxology. *Anthropologie Anzeiger*, 48, 209–27.

Malina, R. M., Harper, A. B., Avent, H. H. & Campbell (1973). Age at menarche in athletes and non-athletes. *Medicine and Science in Sports*, 5, 11–13.

Malina, R. M., Himes, J. H., Stepick, C. D., Lopez, F. G. & Buschang, P. H. (1981). Growth of rural and urban children in the Valley of Oaxaca, Mexico. *American Journal of Physical Anthropology*, 55, 269–80.

Malina, R. M., Bushang, P. H., Aronson, W. L. & Selby, H. (1982a). Childhood growth status of eventual migrants and sedentes in a rural Zapotec community in the valley of

Oaxaca, Mexico. *Human Biology*, **54**, 709–16.

Malina, R. M., Mueller, W. H., Bouchard, C., Shoup, R. F. & Lariviere, G. (1982b). Fatness and fat patterning among athletes at the Montreal Olympic Games, 1976. *Medicine and Science in Sports and Exercise*, **14**, 445–52.

Malina, R. M., Little, B. B., Stern, M. P., Gaskill, S. P. & Hazuda, H. P. (1983). Ethnic and social class differences in selected anthropometric characteristics of Mexican American and Anglo adults: the San Antonio heart study. *Human Biology*, **55**, 867–83.

Malinowski, A. & Wolanski, N. (1985). Anthropology in Poland. *Teoria I Emperia W Polskiej Szkole Antropologicznej*, **11**, 35–69.

Mann, A. E., Lampl, M. & Monge, J. (1990). Patterns of ontogeny in human evolution: evidence from dental development. *Yearbook of Physical Anthropology*, **33**, 111–50.

Mark, M., Rijli, F. M. & Chambon, P. (1997). Homeobox genes in embryogenesis and pathogenesis. *Pediatric Research*, **42**, 421–9.

Markowitz, S. D. (1955). Retardation in growth of children in Europe and Asia during World War II. *Human Biology*, **27**, 258–73.

Marshall, J. C. & Kelch, R. P. (1986). Neuroendocrine regualtion of reproduction. The critical role of pulsatile GnRH secretion and implications for therapy. *New England Journal of Medicine*, **315**, 1459–67.

Marshall, W. A. (1975). The relationship of variations in children's growth rates to season climatic variation. *Annals of Human Biology*, **2**, 243–50.

Marshall, W. A. (1978). Puberty. In *Human Growth, Vol. 2*, Postnatal Growth, ed. F. Falkner & J. M. Tanner, pp. 141–81. New York: Plenum.

Marshall, W. A. & Swan, A. V. N. (1971). Seasonal variation in growth rates of normal and blind children. *Human Biology*, **43**, 502–16.

Marshall, W. A. & Tanner, J. M. (1969). Variation in the pattern of pubertal changes in girls. *Archives of the Diseases of Childhood*, **44**, 291–303.

Marshall, W. A. & Tanner, J. M. (1970). Variation in the pattern of pubertal changes in boys. *Archives of the Diseases of Childhood*, **45**, 13–23.

Marsiglio, W. (1987). Adolescent fathers in the United States: their initial living arrangements, marital experience and educational outcomes. *Family Planning Perspectives*, **19**, 240–51.

Martin, D. E., Swenson, R. B. & Colins, D. Cl. (1977). Correlation of serum testosterone levels with age in male chimpanzees. *Steroids*, **29**, 471–81.

Martin, R. D. (1968). Reproduction and ontogeney in tree shrews (*Tupaia belangeri*) with reference to their general behavior and taxonomic relationships. *Zeitschrift fur Tierpsychologie*, **25**, 409–95 & **25**, 505–32.

Martin, R.D. (1983). *Human Brain Evolution in an Ecological Context*. Fifty-second James Arthur Lecture. New York: American Museum of Natural History.

Martin, R. D. (1990). *Primate Origins and Evolution: A Phylogenetic Reconstruction*. Princeton: Princeton University Press.

Martin, W. J. (1949). The physique of young adult males. *Medical Research Council Memorandum*, No. 20. London, HMSO.

Martorell, R. (1989). Body size, adaptation, and function. *Human Organization*, **48**, 15–20.

Martorell, R. (1995). Results and implications of the INCAP follow-up study. *Journal of Nutrition*, **125**, 1127S–38S.

Martorell, R., Yarbrough, C., Klein, R.E. & Lechtig, A (1975). Malnutrition, body size, and skeletal maturation: interrelationships and implications for catch-up growth. *Human Biology*, **51**, 371–89.

Martorell, R., Yarbrough, C., Lechtig, A., Delgado, H. & Klein (1976). Upper arm anthropometric indicators of nutritional status. *American Journal of Clinical Nutrition*, **29**, 46–53.

Martorell, R., Yarbrough, C., Lechtig, A., Delgado, H. & Klein, R.E. (1977). Genetic-environmental interactions in physical growth. *Acta Paediatrica Scandinavia*, **66**, 579–84.

Martorell, R., Mendoza, F. S. & Castillo, R. O. (1984). Genetic and environmental determinants of growth in Mexican-Americans. *Pediatrics*, **84**, 864–71.

Marubini, E. & Milani, S. (1986). Approaches to the analysis of longitudinal data. In *Human Growth*, ed. F. Falkner & J. M. Tanner, pp. 79–109. New York: Plenum.

Mascie-Taylor, C. G. N. (1984). The interaction between geographical and social mobility. In *Migration and Mobility*, ed. A. J. Boyce, pp. 161–78. London: Taylor and Francis.

Mascie-Taylor, C. G. N. & Boldsen, J. L. (1985). Regional and social analysis of height variation in a contemporary British sample. *Annals of Human Biology*, **12**, 315–24.

Mascie-Taylor, C. G. N. & Bogin, B. (eds.) (1995). *Human Variability and Plasticity*. Cambridge Studies in Biological Anthropology. Cambridge: Cambridge University Press.

Mascie-Taylor, C. G. N. & Lasker, G. W., eds. (1988). *Biological Aspects of Human Migration*. Cambridge: Cambridge University Press.

Mathers, K. & Henneberg, M. (1995). Were we ever that big? Gradual increase in hominid body size over time. *Homo*, **46**, 141–73.

Matsumoto, K. (1982). Secular acceleration of growth in height in Japanese and its social background. *Annals of Human Biology*, **9**, 399–410.

McCabe, V. (1988). Facial proportions, perceived age, and caregiving. In *Social and Applied Aspects of Perceiving Faces*, ed. T. R. Alley, pp. 89–95. Hillsdale, New Jersey: Lawerence Earlbaum Associates.

McGarvey, S. T. (1984). Subcutaneous fat distribution and blood pressure of Samoans. (Abstract). *American Journal of Physical Anthropology*, **63**, 192.

McKinney, M. L. (1998). The juvenilized ape myth: our 'overdeveloped' brain. *BioScience*, **48**, 109–16.

McKinney, M. L. & McNamara, K. J. (1991). *Heterochrony: The Evolution of Ontogeny*. New York: Plenum Press.

McKnight, S. L. (1991). Molecular zippers in gene regulation. *Scientific American*, **264**, 54–64.

Meaney, M. J., Aitken, D. H., Bhatnagar, S., Van Berkel, C. & Sapolsky, R. M. (1988). Postnatal handling attenutates neuroendocrine, anatomical and cognitive impairments related to the aged hippocampus. *Science*, **239**, 766–8.

Medawar, P. B. (1945). Size, shape and age. In *Essays on Growth and Form*, ed. W. E. LeGros Clark & P. B. Medawar, pp. 157–87. Oxford: Clarendon Press.

Medwar, P. B. (1952). *An Unsolved Problem in Biology*. London: H. K. Lewis.

Meire, H. B. (1986). Ultrasound measurement of fetal growth. In *Human Growth*, 2nd edn., vol.1, ed. F. Falkner & J. M. Tanner, pp. 275–90. New York: Plenum.

Meredith, H. V. (1935). The rhythm of physical growth. *University of Iowa Studies in Child Welfare*, **11**, 124.

Meredith, H. V. (1979). Comparative findings on body size of children and youths living at urban centers and in rural areas. *Growth*, **43**, 95–104.

Meredith, H. V. (1981). An addendum on presence and absence of a mid-childhood spurt in somatic dimensions. *Annals of Human Biology*, **8**, 473–6.

Merimee, T. J. (1993). Why are pygmies short? *Growth Matters*, **12** (February), 4–6 (published by Kabi Pharmacia Peptide Hormones, Stockholm, Sweden).

Merimee, T. J., Rimoin, D. L., Rabinowitz, D., Cavalli-Sforza, L. L. & McKusick, V. A. (1968). Metabolic studies in the African Pygmy. *Transactions of the Association of American Physicians*, **81**, 221–320.

Merimee, T. J., Zapf, J. & Froesch, E. R. (1981). Dwarfism in the pygmy, an isolated deficiency of insulin-like growth factor I. *New England Journal of Medicine*, **305**, 965–8.

Merimee, T. J., Zapf, J., Hewlett, B. & Cavalli-Sforza, L. L. (1987). Insulin-like growth factors in pygmies: the role of puberty in determining final stature. *New England Journal of Medicine*, **316**, 906–11.

Migone, A., Emanuel, I., Mueller, B., Daling, J. & Little, R. E. (1991). Gestational duration and birthweight in White, Black, and mixed-race babies. *Pediatric and Perinatal Epidemiology*, **5**, 378–91.

Miklashevskaya, N. M., Solovyeva, V. S. & Godina, E. Z. (1973). Process of human growth in high altitudes. *Proceedings of the IX International Congress of Anthropological and Ethnological Sciences, Chicago*. Moscow: Moscow State University, pp. 243–54.

Mills, M. G. L. (1990). *Kalahari Hyenas*. London: Unwin Hyman.

Milton, K, (1983). Morphometric features as tribal predictors in North-Western Amazonia. *Annals of Human Biology*, **10**, 435–40.

Ministerio de Salud Publica (1989). *Encuesta Nacional de Salud Materno Infantil 1987*. Guatemala City: Ministerio de Salud Publica.

Moerman, M. L. (1982). Growth of the birth canal in adolescent girls. *American Journal of Obstetrics and Gynecology*, **143**, 528–32.

Moffitt, T. E., Caspi, A., Belsky, J. & Silva, P. A. (1992). Childhood experience and the onset of menarche: a test of a sociobiological model. *Child Development*, **63**, 47–58.

Molinari, L., Largo, R. H. & Prader, A. (1980). Analysis of the growth spurt at age seven (mid-growth spurt). *Helvetica Paediatrica*

Acta, **35,** 325–34.

Montagu, A. (1989). *Growing Young,* 2nd edition. Massachusetts: Bergin & Garvey.

Morikawa, M., Nixon, T. & Green, H. (1982). Growth hormone and the adipose conversion of 3T3 cells. *Cell,* **29,** 783–9.

Mossman, H. W. (1937). Comparative morphogenesis of the fetal membranes and accessory uterine structures. *Contributions to Embryology,* **26,** 129–246.

Mueller, W. H. (1977). Sibling correlations in growth and morphology in a rural Columbian population. *Annals of Human Biology,* **4,** 133–42.

Mueller, W. H. (1995). Can thought influence body build? In *Essays on Auxology,* ed. R. Hauspie, G. Lindgren & F. Falkner, pp. 387–95. Welwyn Garden City, England: Castlemead.

Mueller, W. H. & Pollitt, E. (1983). The Bacon Chow Study: genetic analysis of physical growth in assessment of energy–protein malnutrition. *American Journal of Physical Anthropology,* **62,** 11–17.

Mueller, W. H., Schull, V. N. & Schull, W. J. (1978). A multinational Andean genetic and health program: growth and development in an hypoxic environment. *Annals of Human Biology,* **5,** 329–52.

Mueller, W. H., Murillo, F., Palamino, H., Badzioch, M., Chakraborty, R., Fuerst, P. & Schull, W. J. (1980). The Aymara of Western Bolivia: V. Growth and development in an hypoxic environment. *Human Biology,* **52,** 529–46.

Mueller, W. H., Shoup, R. F. & Malina, R. M. (1982). Fat patterning in athletes in relation to ethnic origin and sport. *Annals of Human Biology,* **9,** 371–6.

Mueller, W. H., Joos, S. K., Janis, C. L., Zavalita, A. N., Eicher, J. & Schull, W. J. (1984). Diabetes alert study: growth fatness, and fat patterning, adolescence through adulthood, in Mexican-Americans. *American Journal of Physical Anthropology,* **64,** 389–99.

Muirhead, C. R. (1995). Childhood leukemia in metropolitan regions in the United States: a possible relation to population density? *Cancer Causes Control,* **6,** 383–88.

Muller, J., Nielsen, C. T. & Skakkebaek, N.E. (1989). Testicular maturation and pubertal growth and development in normal boys. In *The Physiology of Human Growth,* ed. J. M. Tanner & M. A. Preece, pp. 201–7. Cambridge: Cambridge University Press.

Mullis, K., Faloona, F., Scharf, S., Saiki, R., Horn, G. & Erlich, H. (1986). Specific enzymatic amplification of DNA in vitro: the polymerase chain reaction. *Cold Spring Harbor Symposium in Quantitative Biology,* **51,** Pt 1, 263–73.

Muragaki, Y., Mundlos, S., Upton, J. & Olsen, B. R. (1996). Altered growth and branching in synpolydactyly caused by mutations in HOXD13. *Science,* **272,** 548–51.

Murguia, R., Dickinson, F., Cervera, M. & Uc, L. (1990). Socio-economic activities, ecology and somatic differences in Yucatan, Mexico. *Studies in Human Ecology,* **9,** 111–34.

Nakano, Y. & Kimura, T. (1992) Development of bipedal walking in *Macaca fuscata* and *Pan troglodytes.* In *Topics in Primatology,* Vol. 3, ed. S. Matano, R. H. Tuttle, H. Ishida & M. Goodman, pp. 177–90. Tokyo: University of Tokyo.

Napier, J. R. & Napier, P. H. (1967). *A Handbook of Living Primates.* London: Academic Press.

National Center for Health Statistics (1994). *Vital Statistics of the United States, Vol. 1, Natality.* Washington, DC: Public Health Service.

Neel, J. V. & Weiss, K. (1975). The genetic structure of a tribal population, the Yanomamö Indians. Biodemographic studies XII. *American Journal of Physical Anthropology,* **42,** 25–52.

Nellhaus, G. (1968). Head circumference from birth to eighteen years. *Pediatrics,* **41,** 106.

Newell-Morris, L. & Fahrenbach, C. F. (1985). Practical and evolutionary considerations for use of the non-human primate model in pre-natal research. In *Non-human Primate Models for Human Growth and Development,* ed. E. Watts, pp. 9–40. New York: Alan R. Liss.

Newman, M. T. (1953). The application of ecological rules to racial anthropology of the aboriginal new-world. *American Anthropologist,* **55,** 311–27.

Newman, M. T. (1962). Evolutionary changes in body size and head form in American Indians. *American Anthropologist,* **64,** 237–57.

Newth, D. R. (1970). *Animal Growth and Development.* London: Edward Arnold.

Neyzi, O., Alp, H. & Orhon, A. (1975a). Sexual maturation in Turkish girls. *Annals of Human Biology,* **2,** 49–59.

Neyzi, O., Alp, H., Yalcindag, A., Yakacikli, S.

& Orphon. A. (1975b). Sexual maturation in Turkish boys. *Annals of Human Biology*, **2**, 251–9.

Nicolson, A. B. & Hanley, C. (1953). Indices of physiological maturity: derivation and interrelationships. *Child Development*, **24**, 3–38.

Nijhout, H. F. & Emlen, D. J. (1998). Competition between body parts in the development and evolution of insect morphology. *Proceedings of the National Academy of Sciences*, **95**, 3685–9.

Nilsson, A., Isgaard, J., Lindahl, A., Dahlstrom, A., Skottner, A. & Isaksson, O. G. P. (1986). Regulation by growth hormone of number of chondrocytes containing IGF-I in rat growth plate. *Science*, **233**, 571–4.

Nishida, T., Takasaki, H., & Takahata, Y. (1990). Demography and reproductive profiles. In *The Chimpanzees of the Mahale Mountains: Sexual and Life History Strategies*, ed. T. Nishida, pp. 63–97. Tokyo: University of Tokyo Press.

Nixon, T. & Green, H. (1984). Contribution of growth hormone to the adipogenic activity of serum. *Endocrinology*, **114**, 527–32.

Nylin, G. (1929). Periodical variation in growth, standard metabolism and oxygen capacity of the blood in children. *Acta Medica Scandinavica*, **31**, 1–207.

Nystroem, M. & Lundberg, O. (1995). Short stature as an effect of economic and social conditions in childhood. *Social Science & Medicine*, **41**, 733–8.

Onat, T. & Ertem, B. (1974). Adolescent female height velocity: relationships to body measurements, sexual and skeletal maturity. *Human Biology*, **46**, 199–217.

O'Rahilliy, R. & Muller, F. (1986). Human growth during the embryonic period proper. In *Human Growth*, 2nd edn., Vol. 1, ed. F. Falkner & J. M. Tanner, pp. 245–53. New York: Plenum.

Orlosky, F. J. (1982). Adolescent midfacial growth in *Macaca nemestrina* and *Papio cynocephalus*. *Human Biology*, **54**, 23–9.

Orr, J. B. (1928). Milk consumption and the growth of schoolchildren. *Lancet*, **1**, 202–3.

Palmert, M. R., Radovick, S. & Boepple, P. A. (1998). The impact of reversible gonadal sex steroid suppression on serum leptin concentrations in children with central precocious puberty. *Journal of Clinical Endocrinology and Metabolism*, **83**, 1091–6.

Panek, S. & Piasecki, M. (1971). Nowa Huta: intergration of the population in the light of anthropometric data. *Materialyi I Prace Anthropologiczne*, **80**, 1–249 (in Polish with English summary).

Parisi, P. & de Martino, V. (1980). Psychosocial factors in human growth. In *Human Physical Growth and Maturation: Methodologies and Factors*, ed. F. E. Johnston, A. F. Roche & C. Susanne, pp. 339–56. New York: Plenum.

Parker, L. N. (1991). Adrenarche. *Endocrinology and Metabolism Clinics of North America*, **20**, 71–83.

Parker, S. T. (1996). Using cladistic analysis of comparative data to reconstruct the evolution of cognitive development in hominids. In *Phylogenies and the Comparative Method in Animal Behavior*, ed. E. Martins, pp. 433–48. Oxford: Oxford University Press.

Parnell, R. W. (1954). The physique of Oxford undergraduates. *Journal of Hygiene*, **52**, 396–78.

Patton, R. G. & Gardner, L. I. (1963). *Growth Failure in Maternal Deprivation*. Springfield, Illinois: Charles C. Thomas.

Pavelka, M. S. & Fedigan, L. M. (1991). Menopause: A comparative life history perspective. *Yearbook of Physical Anthropology*, **34**, 13–38.

Pawson, I. G. & Janes, C. (1981). Massive obesity in a migrant Samoan population. *American Journal of Public Health*, **71**, 508–13.

Pelto, G. H. & Pelto, P. J. (1989). Small but healthy? An anthropological perspective. *Human Organization*, **48**, 11–15.

Perera, A. D. & Plant, T. M. (1992). The neurobiology of primate puberty. In *Proceedings of the Ciba Foundation Symposium No. 168: Functional Anatomy of the Neuroendocrine Hypothalamus*, ed. D. J. Chadwick & J. Marsh, pp. 252–67. Chichester: Wiley.

Perera, A. D. & Plant, T. M. (1997). Ultrastructural studies of neuronal correlates of the pubertal reaugmentation of hypothalamic gonadotropin-releasing hormone (GnRH) release in the rhesus monkey (*Macaca mulatta*). *Journal of Comparative Neurology*, **385**, 71–82.

Pereira, M. E. & Altman, J. (1985). Development of social behavior in free-living nonhuman primates. In *Nonhuman Primate Models for Human Growth and Development*, ed. E. S. Watts, pp. 217–309. New York: Alan R. Liss.

Perieira, M. E. & Fairbanks, L. A. (eds.) (1993). *Juvenile Primates: Life History, De-*

velopment, and Behavior. New York: Oxford University Press.

Peschel, R. E. & Peschel, E. R. (1987). Medical insights into the castrati of opera. *American Scientist*, 75, 578–83.

Petersen, A. C. & Taylor, B. (1980). The biological approach to adolescence: biological change and psychological adaptation. In *Handbook of Adolescent Psychology*, ed. J. Adelson, pp. 117–55. New York: Wiley.

Petty, C. (1989). Primary research and public health: the prioritization of nutrition research in inter-war Britain. In: *Historical Perspectives on the Role of the MRC*, ed. J. Austoker & L. Bryder, pp. 83–108. Oxford: Oxford University Press.

Piaget, J. (1954). *The Construction of Reality in the Child.* New York: Basic Books.

Piaget, J. & Inhelder, B. (1969). *The Psychology of the Child.* New York: Basic Books.

Pimstone, B. L., Barbezat, G., Hansen, J. D. & Murray, P. (1968). Studies on growth hormone secretion in protein-calorie malutrition. *American Journal of Clinical Nutrition*, 21, 482–7.

Piscopo, J. (1962). Skinfold and other anthropometrical measures of pre-adolescent boys from three ethnic groups. *Research Quarterly*, 33, 255–62.

Plant, T. M. (1994). Puberty in primates. In *The Physiology of Reproduction*, 2nd edn, ed. E. Knobil & J. D. Neill, pp. 453–85. New York: Raven Press.

Plant, T. M. & Durrant, A. R. (1997). Circulating leptin does not appear to provide a signal for triggering the initiation of puberty in the male rhesus monkey (*Macaca mulatta*). *Endocrinology*, 138, 4505–8.

Pollitt, E. & Lewis, N. (1980) Nutrition and educational achievement. Part 1, Malnutrition and behavioral test indicators. *Food and Nutrition Bulletin*, 2, 32–4.

Pond, C. M. (1977) The significance of lactation in the evolution of the mammals. *Evolution*, 31, 177–99.

Post, G. B., Kemper, H. C. G., Welten, D. C. & Coudert, J. (1997) Dietary pattern and growth of 10–12-year-old Bolivian girls and boys: relationship between altitude and socioeconomic status. *American Journal of Human Biology*, 9, 51–62.

Potts, R. (1988). *Early Hominid Activities at Olduvai.* New York: Aldine de Gruyter.

Powell, G. F., Brasel, J. A. & Blizzard, R. M. (1967a). Emotional deprivation and growth retardation simulating idiopathic hypo-pituitarism. I. Clinical evaluation of the syndrome. *New England Journal of Medicine*, 276, 1271–8.

Powell, G. F., Raiti, S. & Blizzard, R. M. (1967b). Emotional deprivation and growth retardation simulating idiopathic hypopituitarism. II. Endocrine evaluation of the syndrome. *New England Journal of Medicine*, 276, 1279–83.

Prader, A. (1984). Biomedical and endocrinological aspects of normal growth and development. In *Human Growth and Development*, ed. J. Borms, R. R. Hauspie, A. Sand, C. Susanne & M. Hebbelinck, pp. 1–22. New York: Plenum.

Prader, A., Tanner, J.M. & Von Harnack, G. A. (1963). Catch-up growth following illness or starvation. *Journal of Paediatrics*, 62, 646–59.

Preece, M. A. (1986). Prepubertal and pubertal endocrinology. In *Human Growth*, Vol. 2, 2nd edition, ed. F. Falkner & J. M. Tanner, pp. 211–24. New York: Plenum.

Preece, M. A. & Baines, M. J. (1978). A new family of mathematical models describing the human growth curve. *Annals of Human Biology*, 5, 1–24.

Preece, M. A. & Holder, A. T. (1982). The somatomedins: A family of serum growth factors. In *Recent Advances in Endocrinology and Metabolism*, Vol. 2, ed. L. H. O'Riordan, pp. 47–72. Edinburgh: Churchill Livingstone.

Preece, M. A., Cameron, N., Donnall, M. C., Dunger, D. B., Holder, A. T., Baines-Preece, J., Seth, J., Sharp, G. & Taylor, A. M. (1984). The endocrinology of male puberty. In *Human Growth and Development*, ed. J. Borms *et al.*, pp. 23–37. New York: Plenum.

Price, B., Cameron, N. & Tobias, P. V. (1987). A further search for a secular trend of adult body size in South African Blacks: evidence from the femur and tibia. *Human Biology*, 59, 467–75.

Prokopec, M. & Lhotská, L. (1989). Growth analysis of marginal cases of normal variation. *Anthrop. Közl.*, 32, 65–79.

Putnam, F. W. & Trickett, P. K. (1997). Psychological effects of sexual abuse. *Annals of the New York Academy of Sciences*, 821, 150–9.

Ramirez, M. E. & Mueller, W. H. (1980). The development of obesity and fat patterning in Tokelau children. *Human Biology*, 52, 675–88.

Rappaport, R. (1984). Growth hormone secretion in children of short stature. In *Human Growth*, ed. J. Borms *et al.*, pp. 39 48. New York: Plenum.

Rasmusen, H. (1974). Parathyroid hormone, calcitonin and the calciferols. In *Texbook of Endocrinology*, ed. B. H. Williams, pp. 660 773. Philadelphia: Saunders.

Ratcliffe, S. G. (1976). The development of children with sex chromosome abnormalities. *Proceedings of the Royal Society of Medicine*, **69**, 189 91.

Ratcliffe, S. G. (1995). The ontogenesis of sex chromosomal effects on human growth. In *Essays on Auxology*, ed. R. Hauspie, G. Lindgren & F. Falkner, pp. 480 8. Welwyn Garden City, UK: Castlemead

Reed, T., Spitz, E. Vacher-Lavenu, M.-C. & Carlier, M. (1997). Evaluation of dermatoglyphic index to detect placental type variation in MZ twins. *American Journal of Human Biology*, **9**, 609 15.

Reiter, R. J. (1998). Melatonin and human reproduction. *Annals of Medicine*, **30**, 103 8.

Relethford, J. H. (1997). *The Human Species*. London: Mayfield.

Reynolds, E. L. & Wines, J. V. (1948). Individual differences in physical changes associated with adolescence in girls. *American Journal of Disease of Children*, **75**, 329 50.

Richardson, D. W. & Short, R. V. (1978). Time of onset of sperm production in boys. *Journal of Biosocial Science* (Supplement) 5, 15 24.

Rimoin, D. L., Merimee, T. J., Rabinowitz, D., Cavalli-Sforza, L. L. & McKusick, V. A. (1968). Genetic aspects of clinical endocrinology. *Recent Progress in Hormone Research*, **24**, 365 467.

Roberts, D. F. (1953). Body weight, race and climate. *American Journal of Physical Anthropology*, **11**, 533 58.

Roberts, D. F., Billewicz, W. Z. & McGregor, I. A. (1978). Heritability of stature in a west African population. *Annals of Human Genetics*, **42**, 15 24.

Robertson, T. B. (1908). On the normal rate of growth of an individual, and its biochemical significance. *Archiv fur Entiwicklungs Mechanik den Organismen*, **25**, 581 614.

Robson, E. B. (1978). The genetics of birth weight. In *Human Growth*, Vol. 1, ed. F. Falkner & J.M. Tanner, pp. 285 97. New York: Plenum.

Robson, J. R. K., Bazin, M. & Soderstrom, B. S. (1971). Ethnic differences in skin-fold thickness. *The American Journal of Clinical Nutrition*, **24**, 864 8.

Roche, A. F., ed. (1979). Secular trends in human growth, maturation, and development. *Monographs of the Society for Research in Child Development*, **44**, 1 120.

Roche, A. F. (1992). *Growth, maturation, and body composition : the Fels Longitudinal Study, 1929–1991*. Cambridge: Cambridge University Press.

Roche, A. F. & Davila, G. H. (1972). Late adolescent growth in stature. *Pediatrics*, **50**, 874 80.

Roche, A. F., Wainer, H. & Thissen, D. (1975a). *Skeletal Maturity. The Knee Joint as a Biological Indicator*. New York: Plenum.

Roche, A. F., Wainer, H. & Thissen, D. (1975b). Predicting adult stature for individuals. *Monographica Paediatrica*, **3**, 41 96.

Roede, M. J. (1985). The privilege of growing. *Acta Medica Auxologica*, **17**, 217 26.

Roede, M. J. & van Wieringen, J. C. (1985). Growth diagrams 1980. *Tijdschrift voor Sociale Gezondheidszorg*, Supplement 1985, 1 34.

Rogers, A. & Williamson, J. C. (1982). Migration, urbanization, and third world development: an overview. *Economic Development and Cultural Change*, **30**, 463 82.

Rogoff, B., Seller, M. J., Pirrotta, S., Fox, N. & White, S. H. (1975) Age assignment of roles and responsibilities of children: a cross cultural survey. *Human Development*, **18**, 353 69.

Romer, A. S. (1966). *Vertebrate Paleontology*. Chicago: University of Chicago Press.

Rosenbloom, A. L., Savage, M. O., Blum, W. F., Guevara-Aguirre, J. & Rosenfeld, R. G. (1992). Clinical and biochemical characteristics of growth hormone receptor deficiency (Laron syndrome). *Acta Paediatrica*, Supplement 383, 121 4.

Rosenfeld, R. L., Furlanetto, R. & Bock, D. (1983). Relationship of somatomedin-C concentrations to pubertal changes. *Journal of Pediatrics*, **103**, 723 8.

Rosténe, W., Sarrieau, A., Nicot, A., Scarceriaux, V., Betancur, C., Gully, D., Meaney, M., Rowe, W., De Kloet, R., Pelaprat, D. *et al.* (1995). Steriod effects on brain functions: an example of the action of glucocorticoids on central dopaminergic and neurotensinergic systems. *Journal of Psychiatry and Neuroscience*, **20**, 349 56.

Rousseau, F., Bonaventure, J., Legeai-Mallet, L., Pelet, A., Rozet, J-M., Marteaux, P., LeMerrer, M. & Munnich, A. (1994). Mutations in the gene encoding fibroblast growth factor receptor 3 in achondroplasia. *Nature*,

371, 252–4.

Ruble, D. N. & Brooks-Gunn, J. (1982). The experience of menarche. *Child Development*, 1557–66.

Ruff, C. B. (1991). *Aging and Osteoporosis in Native Americans from Pecos Pueblo, New Mexico: Behavioral and Biomechanical Effects*. New York: Garland.

Ruff, C. B. (1994). Morphological adaptation to climate in modern and fossil hominids. *Yearbook of Physical Anthropology*, **37**, 65–107.

Ruff, C. B., Trinkhaus, E., Walker, A. & Larsen C. S. (1993). Postcranial robusticity in *Homo*. I: Temporal trends and mechanical interpretation. *American Journal of Physical Anthropology*, **91**, 21–53.

Ruff, C. B., Trinkaus, E. & Holliday, T. W. (1997). Body mass and encephalization in Pleistocene *Homo*. *Nature*, **387**, 126–7.

Russell, M. (1976). Parent–child and sibling–sibling correlations of height and weight in a rural Guatemalan population of preschool children. *Human Biology*, **48**, 501–15.

Ryan, A. S. (1997a). The resurgence of breast-feeding in the United States. *Pediatrics*, **99**(4): e12 [http://www.pediatrics.org/cgi/content/full/99/4/e12/].

Ryan, A. S. (1997b). Iron-deficiency anemia in infant development: implications for growth, cognitive development, resistance to infection, and iron supplementation. *Yearbook of Physical Anthropology*, **40**, 25–62.

Sacher, G. A. (1975). Maturation and longevity in relation to cranial capacity in hominid evolution. In *Primate Functional Morphology and Evolution*, ed. R. Tuttle, pp. 417–41. The Hague: Mouton.

Saenger, P., Levine, L. S., Wiedemann, E., Schwartz, E., Korth- Schutz, S., Paretra, J. Heinig, B. & New, M.I. (1977). Somatomedin and growth hormone in psychosocial dwarfism. *Padiatrie und Padologie Supplementum*, **5**, 1–12.

Sanderson, M., Emanuel, I. & Holt, V. (1995). The intergenerational relationship between mother's birthweight, infant birthweight and infant mortality in black and white mothers. *Paediatric and Perinatal Epidemiology*, **9**, 391–405.

Sanger, R. Tippett, P. & Gavin, J. (1971). Xg groups and sex abnormalities in people of Northern European ancestry. *Journal of Medical Genetics*, **8**, 417–26.

Satake, T. (1994). Individual variation in seasonal growth of Japanese children 3–6 years of age. *Humanbiologie Budapestensis*, **25**, 381–6.

Satake, T., Kirutka, F. & Ozaki, T. (1993). Ages at peak velocity and peak velocities for seven body dimensions in Japenese children. *Annals of Human Biology*, **20**, 67–70.

Satake, T., Malina, R. M., Tanaka, S. & Kirutka, F. (1994) Individual variation in the sequence of ages at peak velocity in seven body dimensions. *American Journal of Human Biology*, **6**, 359–67.

Satyanarayana, K., Naidu, A. N. & Rao, B. S. N. (1980). Adolescent growth spurt among rural Indian boys in relation to their nutritional status in early childhood. *Annals of Human Biology*, **7**, 359–65.

Satyanarayana, K., Radhaiah, G., Murali Mohan, K. R., Thimmayamma, B. V. S., Pralhad Rao, N. & Narasinga Rao, B. S. (1989). The adolescent growth spurt of height among rural Indian boys in relation to childhood nutritional background. *Annals of Human Biology*, **16**, 289–300.

Scammon, R. E. (1927). The first seriation study of human growth. *American Journal of Physical Anthropology*, **10**, 329–36.

Scammon, R. E. (1930). The measurement of the body in childhood. In *The Measurement of Man*. ed. J. A. Harris *et al.*, pp. 173–215. Minneapolis: University of Minnesota Press.

Scammon, R. E. & Calkins, L. A. (1929). *The Development and Growth of the External Dimensions of the Human Body in the Fetal Period*. Minneapolis: University of Minnesota Press.

Scariati, P. D., Grummer-Strawn, L. M. & Fein, S. B. (1997). A longitudinal analysis of infant morbidity and the extent of breast-feeding in the United States. *Pediatrics*, (6): e5 [http://www.pediatrics.org/cgi/content/full/99/6/e5/].

Schally, A. V., Kastin, A. J. & Arimura, A. (1977). Hypothalmic hormones: the link between brain and body. *American Scientist*, **65**, 712–19.

Schell, L. M. (1991a). Human growth and urban pollution. *Collegium Anthropologicum*, **15**, 59–71.

Schell, L. M. (1991b). Effects of pollutants on human prenatal and postnatal growth: noise, lead, polychlorinated compounds and toxic wastes. *Yearbook of Physical Anthropology*, **34**, 157–88.

Schell, L. M. (1997). Culture as a stressor: a revised model of biocultural interaction.

American Journal of Physical Anthropology, **102**, 67–77.

Schell, L. M. & Hodges, D. C. (1985). Variation in size at birth and cigarette smoking during pregnancy. *American Journal of Physical Anthropology*, **68**, 549–54.

Schell, L. M., Madan, M. & Davidson, G. (1992). Auxological epidemiology and methods for the study of effects of pollution. *Acta Medica Auxologica*, **24**, 181–7.

Schlegel, A., (ed.) (1995). Special issue on adolescence. *Ethos*, **23**, 3–103.

Schlegel, A. & Barry, H. (1991). *Adolescence: An Anthropological Inquiry*. New York: Free Press

Schreider, E. (1964a). Ecological rules, body-heat regulation, and human evolution. *Evolution*, **18**, 1–9.

Schreider, E. (1964b). Recherches sur la stratification sociale des caracteres biologiques. *Biotypologie*, **26**, 105–35.

Schultz, A. H. (1924). Growth studies on primates bearing upon man's evolution. *American Journal of Physical Anthropology*, **7**, 149–64.

Schultz, A. H. (1960). Age changes in primates and their modification in Man. In *Human Growth*, ed. J. M. Tanner, pp. 1–20. Oxford: Pergamon.

Schwartz, J., Brumbaugh, R. C. & Chiu, M. (1987). Short stature, growth hormone, insulin-like growth factors and serum proteins in the Mountain Ok people of Papua New Guinea. *Journal of Clinical Endocrinology and Metabolism*, **65**, 901–5.

Scott, E. C. & Johnston, F. E. (1982). Critical fat, menarche, and the maintenance of menstrual cycles. *Journal of Adolescent Health Care*, **2**, 249–60.

Scott, E. M., Illsley, I. P. & Thomson, A. M. (1956). A psychological investigation of primigravidae. Maternal social class, age, physique and intelligence. *Journal of Obstetrics and Gynaecology of the British Empire*, **63**, 338–43.

Scott, J. P. (1967). Comparative psychology and ethnology. *Annual Review of Psychology*, **18**, 65–86.

Scrimshaw, N. S. (1968). *Interactions of Nutrition and Infection*. Geneva: World Health Organization.

Seckler, D. (1980). Malnutrition: an intellectual odyssey. *Western Journal of Agricultural Economics*, **5**, 219–27.

Seckler, D. (1982). 'Small but healthy': a basic hypothesis in the theory, measurement, and policy of malnutrition. In *Newer Concepts of Nutrition and their Implication for Policy*, ed. P. V. Sukhatme, pp. 127–37. Pune, India: Maharashta Association for the Cultivation of Science Research Institute.

Service, E. R. (1978). The Arunta of Australia. In *Profiles in Ethnology*, pp. 13–34. New York: Harper & Row.

Shapiro, H. L. (1939). *Migration and Environment*. Oxford: Oxford University Press.

Shapiro, S. & Unger, J. (1965). *Relation of weight at birth to cause of death and age at death in the neonatal period: United States, early 1950*. Public Health Service Pub. no. 1000–Series 21–No. 6, Washington DC: US Government Printing Office.

Sharma, J. C. & Sharma, K. (1984). Estimates of genetic variance for some selected morphometric characters: a twin study. *Acta Genetica Gemellologica* (Roma), **33**, 509–14.

Shea, B. T. (1989). Heterochrony in human evolution: the case for neoteny reconsidered. *Yearbook of Physical Anthropology*, **32**, 69–101.

Shea, B. T. & Bailey, R. C. (1996). Allometry and adaptation of body proportions and stature in African pygmies. *American Journal of Physical Anthropology*, **100**, 311–40.

Shein, M. (1992). *The Precolumbian Child*. Culver City, California: Labyrinthos.

Sheldon, W. H. (1940). *The Varieties of Human Physique*. New York: Harper's.

Shock, N. W. (1966). Physiological growth. In *Human Development*, ed. F. Falkner, pp. 150–77. Saunders: Philadelphia, PA.

Shohoji, T. & Saski, H. (1984). The growth process of the stature of Japanese: growth from early childhood. *Acta Medica Auxologica*, **16**, 101–11.

Short, R. V. (1976). The evolution of human reproduction. *Proceedings, Royal Society, Series B*, **195**, 3–24.

Shuttleworth, F. K. (1937). Sexual maturation and the physical growth of girls age six to nineteen. *Monographs of the Society for Research in Child Development*, **2**, No. 5.

Shuttleworth, F. K. (1939). The physical and mental growth of girls and boys age six to nineteen in relation to age at maximum growth. *Monographs of the Society for Research in Child Development*, **4**, No. 3.

Sills, R. H. (1978). Failure to thrive: the role of clinical and laboratory evaluation. *American Journal of Disease of Children*, **132**, 967–9.

Silman, R. E. (1993). Melatonin: a contraceptive for the nineties. *European Journal of*

Obstetrics, Gynecology, and Reproductive Biology, **49**, 3–9.

Silman, R. E., Leone, R. M., Hooper, R. J. L. & Preece, M. A. (1979). Melatonin, the pineal gland and human puberty. *Nature*, **282**, 301–2.

Simmons, K. & Greulich, W. W. (1943). Menarcheal age and the height, weight and skeletal age of girls, age 7 to 17 years. *Journal of Pediatrics*, **22**, 518–48.

Simons, E. L. (1989). Human origins. *Science* **245**, 1343–50.

Simpson, S. W., Russell, K. F. & Lovejoy, C. O. (1996). Comparison of diaphyseal growth between the Libben Population and the Hamann-Todd chimpanzee sample. *American Journal of Physical Anthropology*, **99**, 66–78.

Singer, C. (1959). *A Short History of Scientific Ideas to 1900*. London: Oxford University Press.

Siniarska, A. (1995). Family environment and body build in adults of Yucatan (Mexico). *American Journal of Physical Anthropology*, Supplement 20, 196.

Sirianni, J. E., VanNess, A. L. & Swindler, D. R. (1982). Growth of the mandible in adolescent pigtailed macaques (*Macaca nemestrina*). *Human Biology*, **54**, 31–44.

Sizonenko, P. C. & Aubert, M. L. (1986). Pre- and perinatal endocrinology. In *Human Growth*, Vol. 1, 2nd edn, ed. F. Falkner & J. M. Tanner, pp. 339–76. New York: Plenum.

Sizonenko, P. C. & Paunier, L. (1982). Failure of DHEA-oenenthate to promote growth. *Pediatric Research*, **16**, 888 (abstract).

Skjaerven, R., Wilcox, A. J., Oyen, N. & Magnus, P. (1997). Mothers' birth weight and survival of their offspring: population based study. *British Medical Journal*, **314**, 1376–80.

Skuse, D., Albanese, A., Stanhope, R., Gilmour, J. & Voss, L. (1996). A new stress-related syndrome of growth failure and hyperphagia in children, associated with reversibility of growth-hormone insufficiency. *Lancet*, **348**, 353–58.

Smail, P. J., Faiman, C., Hobson, W. C., Fuller, G. B., & Winter, J. S. (1982). Further studies on adrenarche in nonhuman primates. *Endocrinology*, **111**, 844–8.

Smith, B. H. (1991a). Age at weaning approximates age of emergence of the first permanent molar in non-human primates. *American Journal of Physical Anthropology*, Suppl. 12, 163–4 (abstract).

Smith B. H. (1991b). Dental development and the evolution of life history in Hominidae. *American Journal of Physical Anthropology*, **86**, 157–74.

Smith, B. H. (1992). Life history and the evolution of human maturation. *Evolutionary Anthropology*, **1**, 134–42.

Smith, B. H. (1993). Physiological age of KMN-WT 15000 and its significance for growth and development of early *Homo*. In *The Nariokotome* Homo erectus *Skeleton*, ed. A. C. Walker & R. F. Leakey, pp. 195–220. Cambridge, Massachusetts: Belknap Press.

Smith, B. H. & Tompkins, R. L. (1995). Toward a life history of the hominidae. *Annual Review of Anthropology*, **25**, 257–79.

Smith, B. H., Crummett, T. L. & Brandt, K. L. (1994). Ages of eruption of primate teeth: a compendium for aging individuals and comparing life histories. *Yearbook of Physical Anthropology*, **37**, 177–231.

Smith, E. P. & Korach, K. S. (1996). Oestrogen receptor deficiency: consequences for growth. *Acta Paediatrica*, Supplement 417, 39–43.

Smith, M. T. (1984). The effects of migration on sampling in genetical surveys. In *Migration and Mobility*, ed. A. J. Boyce, pp. 97–110. London: Taylor & Francis

Snow, M. H. L. (1986). Control of embryonic growth rate and fetal size in mammals. In *Human Growth*, Vol. 1, 2nd edn, ed. F. Falkner & J. M. Tanner, pp. 67–82. New York: Plenum.

Sommerville, J. (1982). *The Rise and Fall of Childhood*. Beverly Hills, California: Sage.

Sontag, L. W. (1971). The history of longitudinal research: implications for the future. *Child Development*, **42**, 987–1002.

Spencer, F. (ed.) (1997). *History of Physical Anthropology: An Encyclopedia*. New York: Garland.

Spencer, H. (1886). *The Principles of Biology*, Vols. I and II. New York: D. Appleton.

Spies, H., Dreizen, S., Snodgrasse, R. M., Arnett, C. M. & Webb-Peploe, H. (1959). Effect of dietary supplement of non fat milk on human growth failure. *American Journal of the Diseases of Childhood*, **98**, 187–97.

Spitz, R. A. (1945). Hospitalism: an inquiry into the genesis of psychiatric conditions in early childhood. *Psychoanalytic Study of the Child*, **1**, 53–74.

Spurr, G. B. (1983). Nutritional status and physical work capacity. *Yearbook of Physical Anthropology*, **26**, 1–35.

Staines A., Bodansky, H. J., McKinney, P. A.,

Alexander, F. E., McNally, R. J., Law, G. R. Lilley, H. E., Stephenson, C. & Cartwright, R. A. (1997). Small area variation in the incidence of childhood insulin-dependent diabetes mellitus in Yorkshire, UK: links with overcrowding and population density. *International Journal of Epidemiology*, **26**, 1307–13.

Stallings, V. A., Oddleifson, N. W., Negrini, B. Y., Zemel, B. S. & Wellens, R. (1994). Bone mineral content and dietary calcium intake in children prescribed a low-lactose diet. *Journal of Pediatric Gastroenterology and Nutrition*, **18**, 440–5.

Stamp, T. C. B. & Round, J. M. (1974). Seasonal changes in human plasma levels of 25-hydroxy vitamin D. *Nature*, **247**, 563–5.

Starck, D. & Kummer, B. (1962). Zur Ontogenese des Schimpansenschadels. *Anthropologie Anzieger*, **25**, 204–15.

Stearns, S. C. (1992). *The Evolution of Life Histories*. Oxford: Oxford University Press.

Stebor, A. (1992). Infant Development among Guatemalan Refugee Families in South Florida. Ph.D. dissertation, University of Florida, Gainesville.

Steegmann, A. T., Jr. (1985). 18th century British military stature: growth cessation, selective recruiting, secular trends, nutrition at birth, cold and occupation. *Human Biology*, **57**, 77–95.

Stein, Z., Susser, M., Saenger, G. & Marolla, F. (1975). *Famine and Human Development. The Dutch Hunger Winter of 1944–1945*. London: Oxford University Press.

Stewart, T.D. (1949). Notas sobre esqueletos humanos prehistoricos hallados en Guatemala. *Antropologia y Historia de Guatemala*, **1**, 23–4.

Stini, W. A. (1975). Adaptive strategies of human populations under nutritional stress. In *Biosocial Interrelations in Population Adaptation*, ed. E. S. Watts, F. E. Johnston & G. W. Lasker, pp. 19–41. The Hague: Mouton.

Stinson, S. (1980). Child growth and the economic value of children in rural Bolivia. *Human Ecology*, **8**, 89–103.

Stinson, S. (1982). The effect of high altitude on the growth of children of high socioeconomic status in Bolivia. *American Journal of Physical Anthropology*, **59**, 61–71.

Stinson, S. (1985). Sex differences in environmental sensitivity during growth. *Yearbook of Physical Anthropology*, **28**, 123–47.

Stratakis, C. A., Gold, P. W. & Chrousos, G. P. (1995). Neuroendocrinology of stress: implications for growth and development. *Hormone Research*, **43**, 162–7.

Stratz, C. H. (1909). Wachstum und Proportionen desMenschen vor und nach der Geburt. *Archiv für Anthropologie*, **8**, 287–97.

Struhsaker, T. T. & Leyland, L. (1987). Colobines: infanticide by adult males. In *Primate Societies*, ed. Smuts, B. B., Cheney, D. L., Seyfarth, R. M., Wrangham, R. W. & Struhsaker, T.T., pp. 83–97. Chicago: University of Chicago Press.

Stutzle, W., Gasser, Th., Molinari, L., Largo, R. H., Prader, A. & Huber, P. S. (1980). Shape-invariant modelling of human growth. *Annals of Human Biology*, **7**, 507–28.

Styne, D. M. & McHenry, H. (1993). The evolution of stature in humans. *Hormone Research*, **39**, supplement 3, 3–6.

Susanne, C. (1980). Interrelations between some social and familial factors and stature and weight of young Belgian male adults. *Human Biology*, **52**, 701–9.

Szathmary, E. J. E. & Holt, N. (1983). Hypoglycemia in Dogrib Indians of the Northwest Territories, Canada: association with age and a centripetal distribution of body fat. *Human Biology*, **55**, 493–515.

Taffel, S. (1980). *Factors associated with low birth weight. United States, 1976*. DHEW Publication No. (PHS) 80-1915, Washington, DC: US Government Printing Office.

Tague, R. G. (1994). Maternal mortality or prolonged growth: age at death and pelvic size in three prehistoric Amerindian populations. *American Journal of Physical Anthropology*, **95**, 27–40.

Takahashi, E. (1984). Secular trend in mild consumption and growth in Japan. *Human Biology*, **56**, 427–37.

Takaishi, M. (1995). Growth standards for Japanese children – an overview with special reference to secular change in growth. In *Essays on Auxology*, ed. R. Hauspie, G. Lindgren & F. Falkner, pp. 302–11. Welwyn Garden City, UK: Castlemead.

Tamarkin, L., Baird, C. J. & Almeida, O. F. X. (1985). Melatonin: a coordinating signal for mammalian reproduction? *Science*, **227**, 714–20.

Tanner, J. M. (1947). The morphological level of personality. *Proceedings of the Royal Society of Medicine*, **40**, 301–3.

Tanner, J. M. (1955). *Growth and Adolescence*. Oxford: Blackwell Scientific Publications.

Tanner, J. M. (1962). *Growth and Adolescence*,

2nd edn. Oxford: Blackwell Scientific Publications.

Tanner, J. M. (1963). The regulation of human growth. *Child Development*, **34**, 817–47.

Tanner, J. M. (1965). Radiographic studies of body composition in children and adults. In *Human Body Composition*, ed. J. Brozek, pp. 211–36. Oxford: Pergamon Press.

Tanner, J. M. (1969). Relation of body size, intelligence, test scores, and social circumstances. In *Trends and Issues in Developmental Psychology*, ed. P. Mussen, J. Langer & M. Covington, pp. 182–201. New York: Holt, Rinehart & Winston.

Tanner, J. M. (1978). *Fetus Into Man*. Cambridge: Harvard University Press.

Tanner, J. M. (1981). *A History of the Study of Human Growth*. Cambridge, UK: Cambridge University Press.

Tanner, J. M. (1986). Growth as a mirror for the conditions of society: secular trends and class distinctions. In *Human Growth: A Multidisciplinary Review*, ed. A. Demirjian, pp. 3–34. London: Taylor and Francis.

Tanner, J. M. (1990). *Fetus Into Man*, 2nd edn. Cambridge: Harvard University Press.

Tanner, J. M. (1994). A historical perspective in human auxology. *Humanbiologia Budapestinensis*, **25**, 9–22.

Tanner, J. M. & Cameron, N. (1980). Investigation of the mid-growth spurt in height, weight and limb circumferences in single-year velocity data from the London 1966–67 growth survey. *Annals of Human Biology*, **7**, 565–77.

Tanner, J. M. & Whitehouse, R. H. (1975). Clinical longitudinal standards for height, weight, height velocity, weight velocity, and the stages of puberty. *Archives of Disease in Childhood*, **51**, 170–9.

Tanner, J. M., Healy, M. J. R., Lockhart, R. D., MacKenzie, J. D. & Whitehouse, R. H. (1956). Aberdeen Growth Study. I. The prediction of adult body measurements from measurements taken from birth to five years. *Archives of Disease in Childhood*, **31**, 382–481.

Tanner, J. M., Prader, A., Habich, H. & Ferguson-Smith, M. A. (1959). Genes on the Y chromosome influencing rate of maturation in man: skeletal age studies in children with Klinefelter's (XXY) and Turner's (XO) syndromes. *Lancet*, **2**, 141–4.

Tanner, J. M., Whitehouse, R. H., Hughes, P. C. R. & Carter, B. S. (1976). Relative importance of growth hormone and sex steroids for the growth at puberty of trunk length, limb length, and muscle width in growth hormone deficient children. *Journal of Pediatrics*, **89**, 1000–8.

Tanner, J. M., Hayashi, T., Preece, M. A. & Cameron, N. (1982). Increase in length of leg relative to trunk in Japanese children and adults from 1957 to 1977: comparison with British and with Japanese Americans. *Annals of Human Biology*, **9**, 411–23.

Tanner, J. M., Landt, K. W., Cameron, N., Carter, B. S. & Patel, J. (1983a). Predicting adult height from height and bone age in childhood. *Archives of Disease in Childhood*, **58**, 767.

Tanner, J. M., Whitehouse, R. H., Cameron, N., Marshall, W. A., Healy, M. J. R. & Goldstein, H. (1983b). *Assessment of Skeletal Maturity and Prediction of Adult Height*, 2nd edn. London: Academic Press.

Tanner, J. M., Wilson, M. E. & Rudman, C. G. (1990). Pubertal growth spurt in the female rhesus monkey: relation to menarche and skeletal maturation. *American Journal of Human Biology*, **2**, 101–6.

Taranger, J., Engstrom, I., Lichenstein, H. & Svennberg-Redegren, I. (1976). Somatic pubertal development. *Acta Paediatrica Scandinavica*, Suppl., 258, 121–35.

Tardieu, C. (1998). Short adolescence in early hominids: infantile and adolescent growth of the human femur. *American Journal of Physical Anthropology*, **107**, 163–78.

Teleki, G. E., Hunt, E. & Pfifferling, J. H. (1976). Demographic observations (1963–1973) on the chimpanzees of the Gombe National Park, Tanzania. *Journal of Human Evolution*, 5, 559–98.

Thom, R. (1983). *Mathematical models of morphogenesis*. Translated by W. M. Brookes & D. Rand. New York: Halsted Press/John Wiley.

Thompson, D'Arcy. W. (1917). *On Growth and Form*. Cambridge: Cambridge University Press.

Thompson, D'Arcy. W. (1942). *On Growth and Form*, revised edition. Cambridge: Cambridge University Press.

Thompson, D'Arcy. W. (1992). *On Growth and Form*, reissue of the 1942 revised edition. Cambridge: Cambridge University Press.

Thompson, J. L. (1995). Terrible teens: the use of adolescent morphology in the interpretation of Upper Pleistocene human evolution. *American Journal of Physical Anthropology*, Supplement 20, 210.

Thompson, J. L. & Bilsborough, A. (1997). The current state of the Le Moustier 1 skull. *Acta Perhistorica et Archaeologica*, **29**, 17–38.

Thompson, J. L. & Nelson, A. J. (1997). Relative postcranial development of Neandertals. *Journal of Human Evolution*, **32**, A23–A24.

Tiainen, J. M., Nuutinen, O. M. & Kalavainen, M. P. (1995). Diet and nutritional status in children with cow's milk allergy. *European Journal of Clinical Nutrition*, **49**, 605–12.

Timiras, P. S. (1972). *Developmental Physiology and Aging*. New York: MacMillan Publishing Co.

Timiras, P. S. & Valcana, T. (1972). Body growth. In *Developmental Physiology and Aging*, ed. P. S. Timiras, pp. 273–302. New York: MacMillan Publishing Co.

Tisserand-Perier, M. (1953). Etudes de certains processus de croissance chex les jum eaux. *Journal de Genetic Humaine*, **2**, 87–102.

Tobe, H., Arai, K. & Togo, M. (1994). Seasonal growth in body weight of Japanese children and its relation to physique. *American Journal of Human Biology*, **6**, 227–35.

Tobias, P. V. (1970). Puberty, growth, malnutrition and the weaker sex – and two new measures of environmental betterment. *The Leech*, **40**, 101–7.

Tobias, P. V. (1975). Stature and secular trend among Southern African Negroes and San (Bushmen). *The South African Journal of Medical Sciences*, **40**, 145–64.

Tobias, P. V. (1985). The negative secular trend. *Journal of Human Evolution*, **14**, 347–56.

Todd, J. T., Mark, L. S., Shaw, R. E. & Pittenger, J. B. (1980). The perception of human growth. *Scientific American*, **242**(2), 132–44.

Todd, T. W. (1937). *Atlas of Skeletal Maturation*. St Louis: C. V. Mosby.

Togo, M. (1995). Time-series analysis in human growth studies. In *Essays on Auxology*, ed. R. Hauspie, G. Lindgren, F. Falkner, pp. 100–105. Welwyn Garden City, UK: Castlemead.

Togo, M. & Togo, T. (1982). Time-series analysis of statue and body weight in five siblings. *Annals of Human Biology*, **9**, 425–40.

Trevathan, W. R. (1987). *Human Birth: An Evolutionary Perspective*. New York: Aldine de Gruyter.

Trevathan, W. R. (1996). The evolution of bipedalism and assisted birth. *Medical Anthropology Quarterly*, **10**, 287–98.

Trinkaus, E., Churchill, S. E. & Ruff, C. B.

(1994). Postcranial robusticity in *Homo*. II: Humeral bilateral asymmetry and bone plasticity. *American Journal of Physical Anthropology*, **93**, 1–34.

Turnbull, C. M (1983a). *The Human Cycle*. New York: Simon & Schuster.

Turnbull, C. M (1983b). *The Mbuti Pygmies*. New York: Holy, Rinehart & Winston.

Ubelaker, D. H. (1994). The biological impact of European contact in Ecuador. In *In the Wake of Conquest*, ed. C. S. Larsen & G. R. Milner, pp. 147–60. New York: Wiley-Liss.

Udry, J. R. (1994). The nature of gender. *Demography*, **31**, 561–73.

Udry, J. R. & Talbert, L. M. (1988). Sex hormone effects on personality at puberty. *Journal of Personality and Social Psychology*, **54**, 291–5.

Udry, J. R., Billy, J. D., Morris, N. M., Groff, T. R. & Raj, M. H. (1985). Serum androgenic hormones motivate sexual behavior in adolescent boys. *Fertility and Sterility*, **43**, 90–4.

Ulijaszek, S. J. & Strickland, S. S. (1993). *Nutritional Anthropology: Prospects and Perspectives*. London: Smith Gordon.

Ulijaszek, S., Johnston, F. E. & Preece, M., eds. (1998). *Cambridge Encyclopedia of Human Growth and Development*. Cambridge: Cambridge University Press.

United Nations (1980). *Patterns of Urban and Rural Growth*. Population Studies, No. 68. New York: United Nations.

Van de Hulst, H. C. (1957). *Light Scattering by Small Particles*. New York: Wiley.

Van de Koppel, J. C. H. & Hewlett, B. S. (1986). Growth of Aka pygmies and Bagandus of the Central African Republic. In *African Pygmies*, ed. L. L. Cavalli-Sforza, pp. 95–102. Orlando, Florida: Academic Press.

Van Loon, H., Saverys, V., Vuylsteke, J. P., Vlietinck, R. F. & Eeckels, R. (1986). Local versus universal growth standards: the effect of using NCHS as a universal reference. *Annals of Human Biology*, **13**, 347–57.

Varrela, J. (1984). Effects of X chromosome on size and shape of body: an anthropometric investigation in 47,XXY males. *American Journal of Physical Anthropology*, **64**, 233–42.

Varrela, J. & Alvesalo, L. (1985). Effects of the Y chromosome on quantitative growth: an anthropometric study of 47,XYY males. *American Journal of Physical Anthropology*, **68**, 239–45.

Varrela, J., Alvesalo, L. & Vinkka, H. (1984a). The phenotype of 45,X females: an anthropometric quantification. *Annals of Human Biology*, **11**, 53–66.

Varrela, J., Alvesalo, L. & Vinkka, H. (1984b). Body size and shape in 46,XY females with complete testicular feminization. *Annals of Human Biology*, **4**, 291–301.

Vaughn, J. M. (1975). *The Physiology of Bone*, 2nd edn. Oxford: Clarendon Press.

Vavra, H. M. & Querec, L. J. (1973). *A study of infant mortality from linked records by age of mother, total-birth order, and other variables*. DHEW Publication No. (HRA). 74-1851. Washington DC: US Government Printing Office.

Villar, J. & Belizan, J. M. (1982). The relative contribution of prematurity and fetal growth retardation to low birth weight in developing and developed countries. *American Journal of Obstetrics and Gynecology*, **143**, 793–98A.

Vincent, M. & Dierickx, J. (1960). Etude sur la croissance saisonnaire des escoliers de Leopoldville. *Annales de la Société Belge de Médecine Tropicale*, **40**, 837–44.

Vrba, E. S. (1996) Climate, heterochrony, and human evolution. *Journal of Anthropological Research*, **52**, 1–28.

Walcher, G. (1905). Ueber die Entstehung von Brachy- und Dolichocephalie durch willkürliche Beinflussung des kindlichen Schadels. *Zentralblatt für Gynakologie*, **29**, 193–6.

Walker, A. C. & Leakey, R. F. (eds.) (1993). *The Nariokotome Homo erectus Skeleton*. Cambridge, Massachusetts: Belknap Press.

Wan Nazaimoon, W. M., Osman, A., Wu, L. L. & Khalid, B. A. (1996). Effects of iodine deficiency on insulin-like growth factor-1, insulin-like growth factor-binding protein-3 levels and height attainment in malnourished children. *Clinical Endocrinology*, **45**, 79–83.

Warren, M. P. (1980). The effects of exercise on pubertal progression and reproductive function in girls. *Journal of Clinical Endocrinology and Metabolism*, **51**, 1150–7.

Waterlow, J. C. & Payne, P. R. (1975). The protein gap. *Nature*, **258**, 113–17.

Watts D. P. & Pusey A. E. (1993). Behavior of juvenile and adolescent great apes. In *Juvenile Primates*, ed. M. E. Pereira & L. A. Fairbanks, pp. 148–70. Oxford: Oxford University Press.

Watts, E. S. (1985). Adolescent growth and development of monkeys, apes and humans. In *Nonhuman Primate Models for Human Growth and Development*, ed. E. S. Watts, pp.

41–65. New York: Alan R. Liss.

Watts, E. S. (1990). Evolutionary trends in primate growth and development. In *Primate Life History and Evolution*, ed. C. J. DeRousseau, pp. 89–104. New York: Wiley-Liss.

Watts, E. S. & Gavan J. A. (1982). Postnatal growth of nonhuman primates: the problem of adolescent spurt. *Human Biology*, **54**, 53–70.

Weber, G. W., Prossinger, H. & Seidler, H. (1998). Height depends on month of birth. *Nature*, **391**, 754–5.

Webster, D. L., Evans, S. T. & Sanders, W. T. (1993). *Out of the Past: An Introduction to Archaeology*. Mountain View, California: Mayfield.

Weirman, M. E. & Crowley, W. F. Jr (1986). Neuroendocrine control of the onset of puberty. In *Human Growth*, 2nd edn, Vol. 2, ed. F. Falkner & J. M. Tanner, pp. 225–41. New York: Plenum.

Weisner, T. S. (1987). Socialization for parenthood in sibling caretaking societies. In *Parenting Across the Life Span: Biosocial Dimensions*, ed. J. B. Lancaster, J. Altmann, A. S. Rossi & L. R. Sherrod, pp. 237–70. New York: Aldine de Gruyter.

Weisner, T. S. (1996). The 5–7 transition as an ecocultural project. In *Reason and Responsibility: The Passage Through Childhood*, ed. A. Samaroff & M. Haith, pp. 295–326. Chicago: University of Chicago Press.

Weisner, T. S. & Gallimore, R. (1977). My brother's keeper: child and sibling caretaking. *Current Anthropology*, **18**, 169–90.

Weiss, P. & Kavanau, J. L. (1957). A model of growth and growth control in mathematical terms. *Journal of General Physiology*, **41**, 1–47.

Weiss, R. E. & Refetoff, S. (1996). Effect of thyroid hormone on growth. Lessons from the syndrome of resistance to thyroid hormone. *Endocrinology and Metabolism Clinics of North America*, **25**, 719–30.

Weitzman, E. B., Boyar, R. M., Kapen, S. & Hellman, L. (1975). The relationship of sleep and sleep stages to neuroendocrine secretion and biological rhythms in man. *Recent Progress in Hormone Research*, **31**, 399–446.

Werner, E. E., Bierman, J. M. & French, F. E. (1971). *The Children of Kauai*. Honolulu: University of Hawaii Press.

White, S. H. (1965). Evidence for a hierarchical arrangement of learning processes. In *Advances in Child Development and Behavior*,

Vol. 2, ed. L. P. Lipsitt & C. C. Spiker, pp. 187–220. New York: Academic Press.

Whiting, B. B. & Edwards, C. P. (1988). *Children of Different Worlds*. Cambridge: Harvard University Press.

Whiting, B. B. & Whiting, J. W. M. (1975). *Children of Six Cultures: A Psycho-cultural Analysis*. Cambridge, MA: Harvard University Press.

Widdowson, E. M. (1951). Mental contentment and physical growth. *Lancet*, **1**, 1316–18.

Widdowson, E. M. (1970). The harmony of growth. *Lancet*, **1**, 901–5.

Widdowson, E. M. (1976). Pregnancy and lactation: the comparative point of view. In *Early Nutrition and Later Development*, ed. A. W. Wilkinson, pp. 1–10. Chicago: Year Book Medical Publishers.

Widdowson, E. M. & Dickerson, J. W. T. (1964). The chemical composition of the body. In *Mineral Metabolism*, Vol. 2A, ed. C. L. Comar & F. Bronner, pp. 1–247. New York: Plenum.

Wierman, M. E. & Crowley, W. F., Jr (1986). Neuroendocrine control of the onset of puberty. In *Human Growth, Vol. 2, Postnatal Growth*, ed. F. Falkner & J. M. Tanner, pp. 225–41. New York: Plenum.

Williams, G. C. (1957). Pleiotropy, natural selection and the evolution of senescence. *Evolution*, **11**, 398–411.

Wilson, M. E., Gordon, T. F., Rudman, C. G. & Tanner, J. M. (1988). Effects of natural *versus* artificial environment on the tempo of maturation in female rhesus monkeys. *Endocrinology*, **123**, 2653–61.

Wilson, R. S. (1979). Twin growth: initial deficit, recovery, and trends in concordance from birth to nine years. *Annals of Human Biology*, **6**, 205–20.

Winkler, L. A. & Anemone, R. L. (1996). Recent development in hominoid ontogeny: an overview and summation. *American Journal of Physical Anthropology*, **99**, 1–8.

Winter, J. S. D. (1978). Prepubertal and pubertal endocrinology. In *Human Growth, Vol. 2, Postnatal Growth*, ed. F. Falkner & J. M. Tanner, pp. 183–213. New York: Plenum.

Wolanski, N. (1967). Basic problems in physical development in man in relation to the evaluation of development of children and youth. *Current Anthropology*, **8**, 355–60.

Wolanski, N. (1979). Parent–offspring similarity in body size and proportions. *Studies in Human Ecology*, **3**, 7–26.

Wolanski, N. (1980). Secular changes in contemporary man. *Anthropologica Contemporanea*, **3**, 427–50.

Wolanski, N. (1985). Secular trend, secular changes, or long-term adaptational fluctuations? *Acta Medica Auxologica*, **17**, 7–19.

Wolanski N. (1990). Human population as bioindicator of environmental conditions (environmental factors in biological status of population of Poland). *Studies in Human Ecology*, **9**, 295–322.

Wolanski N. (1995). Household and family as environment for child growth (Cross cultural studies in Poland, Japan, South Korea and Mexico). In *Human Ecology: Progress Through Integrative Perspectives*, ed. S.D. Wright, D.E. Meeker & R. Griffore, pp. 140–52. Bar Harbor, Maine: The Society for Human Ecology.

Wolanski, N. & Bogin, B. (eds.) (1996). *Family As An Environment For Human Development*. New Delhi: Kamal Raj.

Wolanski, N. & Kasprzak E. (1976). Stature as a measure of effects of environmental change. *Current Anthropology*, **17**(3), 548–52.

Wolanski, N., Dickinson, F. & Siniarska, A. (1993). Biological traits and living conditions of Maya Indian and non-Maya girls from Merida, Mexico. *International Journal of Anthropology*, **8**, 233–46.

Wolff, G. (1935). Increased bodily growth of school-children since the war. *Lancet*, **1**, 1006–11.

World Resources Institute (1994). *World Resources 1994–96*. Oxford: Oxford University Press.

World Resources Institute (1996). *World Resources 1996–97*. Oxford: Oxford University Press.

Worthman, C. M. (1986). Later-maturing populations and control of the onset of puberty. *American Journal of Physical Anthropology*, **69**, 282 (abstract).

Worthman, C. M. (1993). Biocultural interactions in human development. In *Juvenile Primates: Life History, Development, and Behavior*, ed. M. E. Perieira & L. A. Fairbanks, pp. 339–57. New York: Oxford University Press.

Wurtman, R. J. (1975). The effects of light on the human body. *Scientific American*, **233**(1), 68–77.

Wurtman, R. J. & Axelrod, J. (1965). The pineal gland. *Scientific American*, **233**, 50–60.

Wyatt, D. T., Simms, M. D. & Horwitz, S. M. (1997). Widespread growth retardation and variable growth recovery in foster children in the first year after intial placement. *Archives of Pediatric and Adolescent Medicine,* **15,** 813–16.

Young, V. R., Steffee, W. P., Pencharz, P. B., Winterer, J. C. & Scrimshaw, N.S. (1975). Total human body protein synthesis in relation to protein requirements at various ages. *Nature,* **253,** 192–4.

Zemel, B.S. & Katz, S.H. (1986). The contribution of adrenal and gonadal androgens to the growth in height of adolescent males. *American Journal of Physical Anthropology,* **71,** 459–66.

Zezulak, K. M. & Green, H. (1986). The generation of insulin-like growth factor-1-sensitive cells by growth hormone action. *Science,* **233,** 551–3.

Zihlman, A. L. (1982). *Human Evolution Coloring Book.* New York: Barnes & Noble Books.

Zihlman, A. L. (1997). Women's bodies, women's lives: an evolutionary perspective. In *The Evolving Female: A Life History Perspective,* ed. M. E. Morbeck, A. Galloway & A. L. Zihlman, pp. 185–97. Princeton: Princeton University Press.

Index

442

Printed in the United Kingdom by
Lightning Source UK Ltd., Milton Keynes
136605UK00001B/325-330/A